ESSENTIAL
MATH
FOR COLLEGE-BOUND STUDENTS

ESSENTIAL MATH

FOR COLLEGE-BOUND STUDENTS

NORMAN LEVY
JOAN U. LEVY

College Preparation

MACMILLAN • USA

For Joshua Seth and Jessica Dawn
two of our best reasons for writing this book

First Edition

Macmillan General Reference
A Simon & Schuster Macmillan Company
1633 Broadway
New York, NY 10019-6785

An Arco Book

MACMILLAN is a registered trademark of Macmillan, Inc.
ARCO is a registered trademark of Prentice-Hall, Inc.

Library of Congress Cataloging-in-Publication Data

Levy, Norman.
    Essential math for college-bound students / by Norman Levy, Joan U. Levy.
    p.   cm.
    ISBN 0-671-79971-1
    1. Mathematics. I. Levy, Joan U. II. Title.
QA39.2.L47   1989   89-27643
610—dc20   CIP

Manufactured in the United States of America

10 9 8 7 6

CONTENTS

Introduction

This book, as the title suggests, is designed to review all the important components of high school mathematics oriented toward college-bound students. High school math curricula are in the process of change. Some school districts adhere to the traditional sequence of algebra, geometry, intermediate algebra and trigonometry. Other school districts have changed to the new integrated mathematics sequences I, II and III. Some topics have been deleted, some transposed and some added. Topics such as logic, probability and statistics are now included in these math sequences. Twelfth grade math, once known as advanced algebra or college algebra, is now called precalculus. This book encompasses both the old and the new spectrums and integrates them into unified topics.

This book is not a textbook. It makes no attempt to provide detailed analyses of the whys and wherefores of each process. Rather, it offers a comprehensive review of high school topics. It is based on the premise that mathematics is learned best by solving problems, not merely by reading about them. Thus, hundreds of author-directed examples and student problems are given, so that the college-bound student can review a subject thoroughly.

Each chapter is succinct. Important material is boxed. Every example is worked out in detail. For each example, the problem is given in the left column with sufficient space for the student to work out his own solution. In the right column is the author-provided solution as shown below.

Example:

Solve for x: $3x - 7 = 2x + 4$

Solution:

$$
\begin{array}{rl}
3x - 7 = & 2x + 4 \\
-2x & -2x \\
\hline
x - 7 = & 4 \\
+7 & +7 \\
\hline
x = & 11
\end{array}
$$

The student should attempt to work out each example on his own before looking at the solution provided. At the end of each review section are multiple-choice problems similar to the problems you will face on college-entrance examinations. Detailed solutions to the practice problems are provided every four chapters.

Essential Math for College-Bound Students begins with a 50-question Pretest and a **Diagnostic Chart** indicating the subject area of each test question. Cumulative Review Tests of 20 questions each are provided every four chapters. These tests, too, are accompanied by a diagnostic analysis to help evaluate your progress. A 50-question Post-test at the end of the book serves as a final review of high school math. Throughout the text, over 1000 examples provide a "hands on" approach to mastering essential mathematical concepts.

This self-teaching text is designed to answer the common student lament: "If only I had a review sheet with all the important information." Here you have it—a capsule review of each topic covered in high school mathematics complete with practice questions to prepare you to score high on college entrance and college placement tests in math.

Special Features of This Book

- Capsule reviews summarize the essential concepts of each area of high school math
- Key concepts are boxed for added emphasis
- Examples allow sufficient work space for students to try each problem before looking at the detailed solutions provided
- Cumulative review tests are offered every four chapters with extensive Diagnostic Charts for self-evaluation
- Diagnostic charts indicate subject area of each question, thus providing an instant analysis of individual trouble spots
- Author's Notes, indicated by the symbol on the right, provide additional clarification of important points
- Author's Warnings, indicated by the symbol on the right, point out common errors to be avoided
- Carefully worked through solutions are provided for every problem
- All levels of high school math are integrated by topic

This book is designed to serve as:
1. A review of high school math.
2. Preparation for standardized college entrance examinations.
3. Preparation for Math Level I and II Achievement tests.
4. Preparation for college placement tests in mathematics.

Standardized Tests in Mathematics

Aside from the mathematics curriculum covered in the high school, college-bound students face a number of very important standardized tests during their high school years. These examinations are used by the colleges as part of their admissions criteria. Thus the standardized tests help to determine which students will gain admission to a given college. The importance of these tests should be stressed to every high school student. The standardized tests assist the college officials with a non-demographic, non-geographic comparison of skills.

The chart on page 9 summarizes the various tests.

Preliminary Scholastic Aptitude Test/National Merit Scholarship Qualifying Test (PSAT/NMSQT)

The Preliminary Scholastic Aptitude Test/National Merit Scholarship Qualifying Test (PSAT/NMSQT) is taken by high school students in October of their junior year. The test is an hour and forty minutes long and consists of two 50 minute sections:

Verbal—65 questions
Mathematics—50 questions

The **Mathematics Section** of the PSAT/NMSQT has two types of questions:

Regular Multiple Choice:
A mathematical situation is presented involving algebra, arithmetic, geometry, spatial relationships or analytic reasoning. The student must select the correct solution from five choices.

Quantitative Comparison:
The student is presented with two relationships, one in column A and the other in column B. He is asked to evaluate the quantities under all allowed conditions of the problem and to determine whether:

- Column A is always greater than column B (answer choice A)
- Column B is always greater than column A (answer choice B)
- The two columns are always equal (answer choice C)
- An "always" relationship is not defined (answer choice D)

Scoring for both Verbal and Mathematics sections is from 20 to 80. Each correct answer contributes upward to the score; each incorrect answer contributes fractionally downward from the score, and omitted questions neither raise nor lower the score. The scoring is designed so that random guessing cannot significantly improve your score. It is generally better to review your completed work than to waste time guessing. However, if you are able to eliminate one or more choices, it may be advantageous to make an educated guess.

The **purpose** of the PSAT/NMSQT is multifold:

- First, it is, as its name implies, a preliminary measure of scholastic aptitude. It is meant to give the student some guidelines about his verbal and mathematical abilities prior to taking the Scholastic Aptitude Test (SAT).* The PSAT score is *not* sent to colleges and does not affect college admission.
- It is an opportunity for the student to compare his abilities to those of other students nationwide who are competing for similar goals. The reported percentile scores are useful toward this end.
- Entrance to competition for National Merit scholarships and recognition is based on a selection index formulated by combining twice the verbal score plus the mathematics score. Thus the PSAT/NMSQT can have financial and/or prestige value for the college-bound student.

The PSAT/NMSQT is administered by the Educational Testing Service (ETS) on behalf of the College Board and National Merit Scholarship Corporation (NMSC). Further information and details are available from:

PSAT/NMSQT	NMSC	College Board
P.O. Box 24700	1 Rotary Center	PSAT/NMSQT
Oakland, CA 94623-1700	1560 Sherman Avenue	45 Columbus Ave.
	Evanston, IL 60201	New York, NY 10023

Scholastic Aptitude Test (SAT)

The Scholastic Aptitude Test (SAT) is a three hour, multiple choice test offered by the Admissions Testing Program (ATP) of the College Board to help colleges compare the potential of students from different geographic and demographic circumstances. Since the SAT is one of the vehicles used by the colleges and universities to help decide on admissions, the SAT assumes a great importance to the student.

The SAT is a longer and more difficult version of the PSAT. It consists of six sections, each 30 minutes in length. They are:

Verbal—40 questions ⎫
Verbal—45 questions ⎬ 85 questions

Mathematics—25 questions ⎫
Mathematics—35 questions ⎬ 60 questions

Test of Standard Written English (TSWE)—50 questions

Equating or Experimental Sections—varies in length

 Only the Verbal 40 and 45 and Mathematics 25 and 35 question sections are included in the SAT score.

The two verbal sections are combined to yield a Verbal Score ranging from 200 to 800. Similarly, the two mathematics sections are combined to yield a Mathematics Score ranging from 200 to 800. Each correct answer contributes toward raising the score; each incorrect answer fractionally lowers the score, and omitted questions neither raise nor lower the

*SAT and Scholastic Aptitude Test are registered trademarks that are the exclusive property of The College Entrance Examination Board.

score. SAT scoring is designed so that random guessing will not significantly improve your score. It is generally better to review completed work for potential errors than to waste time on random guesses. However, if you are able to eliminate one or more choices it may be advantageous to make an educated guess.

The **Mathematics Section** of the SAT has two types of questions:

Regular Multiple Choice:

A mathematical situation is presented involving analytic reasoning, algebra, geometry, spatial relationships or arithmetic processes. The student must select the correct solution to the given question from five choices. There are 40 such questions on the SAT. For example:

If a apples cost b cents, then how many apples can be bought for b^2 dollars?

A) a^2 B) ab C) $100a^2$ D) $100ab$ E) $\dfrac{a^2}{100}$

The correct solution is D. Establish a proportion of apples to dollars:

$$\frac{\text{apples}}{\text{dollars}} : \quad \frac{a}{\dfrac{b}{100}} = \frac{x}{b^2}$$

$$x = 100ab$$

Quantitative Comparison:

Two relationships are presented, one in column A and the other in column B. The student is asked to evaluate the two quantities using any provided constraints and to determine whether:

- Column A is always greater than column B (answer choice A)
- Column B is always greater than column A (answer choice B)
- The two columns are always equal (answer choice C)
- An "always" relationship does not exist (answer choice D)

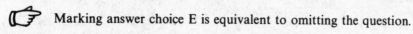 Marking answer choice E is equivalent to omitting the question.

There are 20 such questions on the SAT. For example:

Column A	Column B
	$x \neq -4$
$x + 4$	$\dfrac{x^2 - 16}{x - 4}$

The correct solution is C. The two quantities are always equal (when $x \neq -4$).

$$\frac{x^2 - 16}{x - 4} = \frac{(x + 4)(x - 4)}{(x - 4)} = x + 4$$

The major **purpose** of the SAT is to provide college officials with a standardized basis of comparison of an individual's college potential. Colleges and universities use the results, together with grade point average, class rank, extracurricular activities, personal interview and application data, to evaluate admission to their respective schools.

Further information and details are available from:

College Board ATP
CN 6200
Princeton, NJ 08541-6200

American College Testing (ACT) Program

The American College Testing (ACT) Program is a standardized test that is used by colleges to help determine which students will be admitted to a particular college or university. The ACT test battery can be taken, in many cases, in lieu of the SAT.

The ACT consists of four discrete sections:

English Test—75 questions—45 minutes
Mathematics Test—60 questions—60 minutes
Reading Test—40 questions—35 minutes
Science Reasoning Test—40 questions—35 minutes

The **Mathematics** test presents mathematics problems encompassing pre-algebra, intermediate algebra, plane geometry, coordinate geometry and trigonometry. The student must select the solution from five possible choices. For example:

If $a = b$ and $\dfrac{1}{c} = b$, then $c =$

A) a B) $-a$ C) b D) $\dfrac{1}{a}$ E) $-b$

The correct answer is D.

$$a = b = \frac{1}{c}$$

$$a = \frac{1}{c} \rightarrow ac = 1$$

$$c = \frac{1}{a}$$

The **scoring** on the ACT ranges from 1 to 36 for the individual test scores and the composite score and from 1 to 18 for the subscores. Unlike the SAT, no penalty is applied for incorrect answers. The student taking the ACT should always guess. No response should go unanswered.

The **purpose** of the ACT is to provide the college and university admissions officials with a standardized basis of comparison of an individual's college potential. These results, together with grade point average, class rank, extracurricular activities, personal interview and application data, contribute to the admissions evaluation process.

Additional information and details are available from:

American College Testing Registration
P.O. Box 168
Iowa City, IA 52243

Achievement Tests

Some colleges and universities require further standardized testing. They ask the student to take from one to three Achievement Tests in different subject areas. There are a total of sixteen achievement tests given in fourteen subject areas. Each test is 60 minutes in duration.

Scoring for each test ranges from 200 to 800. As with the SAT, each correct answer contributes toward raising the score; each incorrect answer fractionally lowers the score, and omitted questions neither raise nor lower the score. The scoring is designed so that random guessing will not significantly improve your score. However, if you are able to eliminate one or more choices, it may be advantageous to make an educated guess.

The Achievement Tests include:

Languages	**Sciences**
French	Biology
German	Chemistry
Hebrew	Physics
Latin	
Spanish	
Italian	

Mathematics

Mathematics—Level I
This test includes second year algebra, geometry, trigonometry, elementary functions, reasoning, logic and elementary number theory.

Mathematics—Level II
This test includes third year algebra, geometry, trigonometry, elementary functions, sequences, series, logic, probability, statistics and elementary number theory.

Mathematics—Level I and Mathematics —Level II count as one Achievement Test. Thus the student should choose between the two tests.

Social Studies	**English**
American History and Social Studies	English Composition
European History and World Cultures	English Composition with Essay
	Literature

The two English Composition Achievement Tests also count as one. Therefore, the student should choose between the two tests.

The **purpose** of the achievement tests is defined by the college or university requesting that the tests be taken. They can serve either to aid the college officials in the admissions process or they can be used for placement in their respective subject areas for an entering freshman.

Further information can be obtained from:

College Board ATP
CN 6200
Princeton, NJ 08541-6200

Mathematics Placement Test

Many colleges and universities require their students to take a placement test in mathematics prior to beginning their freshman year. The examination usually tests mathematical concepts and applications through precalculus. It is used for placement in beginning college mathematics courses. Further information can be obtained from your admitting college.

Summary of Standardized Tests

Test	Sections	Number of Questions	Minutes	Mathematics Level	For Further Information
Preliminary Scholastic Aptitude Test/ National Merit Scholarship Qualifying Test (PSAT/NMSQT)	Verbal Math	65 50	50 50	Through: Arithmetic Reasoning, Elementary Algebra and Geometry	PSAT/NMSQT P.O. Box 24700 Oakland, CA 94623-1700 *N.J.:* (609)683-0449 *Calif.:* (415)653-5595
Scholastic Aptitude Test (SAT)	Verbal Verbal Math Math	40 45 25 35	30 30 30 30	Through: Arithmetic Reasoning, Elementary Algebra and Geometry	College Board ATP CN 6200 Princeton, NJ 08541-6200 *N.J.:* (609)771-7588 *Calif.:* (415)654-1200
Test of Standard Written English (TSWE)	Grammar	50	30		
Equating	English or Math or TSWE	Varies	30		
American College Testing Program (ACT)	English Mathematics Reading Science Reasoning	75 60 40 40	45 60 35 35	Through: Second Yr. Algebra, Geometry, Number Concepts, Arithmetic and Algebraic Reasoning, Trigonometry	American College Testing Registration P.O. Box 168 Iowa City, IA 52243 (319)337-1270
Achievement Tests (ACH)	French German Hebrew (Modern) Italian Latin Spanish Biology Chemistry Physics Mathematics—Level I Mathematics—Level II American History and Social Studies European History and World Cultures English Composition All multiple-choice English Composition with Essay Literature	85 80 85 85 70–75 85 95 85 75 50 50 95 95 85–90 70 + Essay 55–65	60 60 60 60 60 60 60 60 60 60 60 60 60 60 60 60	Level I Through: Second Yr. Algebra, Geometry, Trigonometry, Elementary Functions, Reasoning, Logic and Elementary Number Theory Level II Through: Third Yr. Algebra, Geometry, Trigonometry, Elementary Functions, Sequences, Series, Logic, Probability, Statistics and Elementary Number Theory	College Board ATP CN 6200 Princeton, NJ 08541-6200 *N.J.:* (609)771-7558 *Calif.:* (415)654-1200
Mathematics Placement Test	Math			Usually through Precalculus	Your Admitting College

Pretest

The questions that follow represent each of the topics covered in this book. Solve each problem and select the correct answer from the five choices given. When you have completed the test, compare your answers with the solutions at the end of the test. Use the Diagnostic Chart on page 20 to evaluate your strengths and weaknesses.

1. The $\sqrt{17}$ is a member of which set of numbers?

   A) rational   B) integers   C) irrational
   D) counting   E) imaginary

2. The inverse of $\sim p \to q$ is

   A) $p \wedge q$ B) $p \to \sim q$ C) $q \to \sim p$
   D) $q \to p$ E) $p \vee \sim q$

3. Factor $x^2 - 7x + 10$.

   A) $(x - 10)(x + 1)$ B) $(x + 5)(x + 2)$
   C) $(x - 5)(x - 2)$
   D) $(x + 5)(x - 2)$ E) $(x + 10)(x + 1)$

4. How many degrees are $\dfrac{2\pi}{15}$ radians?

   A) 12   B) 24   C) 36   D) 156   E) 180

5. If $\log 2 = A$ and $\log 3 = B$, then $\log 6$ is equal to

   A) $A + B$ B) $A - B$ C) AB D) $\dfrac{A}{B}$ E) A^B

6. For which value of c will the roots of the equation $f(x) = x^2 + 4x + c$ be real and equal?

   A) 1   B) 2   C) 3   D) 4   E) 5

7. Which is the fourth term in the expansion of $(x + 3)^5$?

   A) $270x^2$ B) $135x^2$ C) $405x$
   D) $135x$ E) $-135x$

8. The circumference of a circle is 20π. What is the area of the circle?

   A) 10π B) 20π C) 30π D) 100π E) 400π

9. Perform the indicated operation and express in simplest form:

$$\frac{x^2 - 3x}{2x^2 + x - 6} \div \frac{x^2 - 5x + 6}{x^2 - 4}$$

   A) $\dfrac{x}{x + 6}$ B) $\dfrac{x - 3}{x + 1}$ C) $\dfrac{x(x - 3)}{(x - 2)}$

   D) $\dfrac{x}{2x - 3}$ E) $\dfrac{1}{2x - 3}$

10. What are the coordinates of the center of a circle whose equation is $(x - 1)^2 + (y + 5)^2 = 7$?

   A) $(-1, 5)$ B) $(\sqrt{7}, 0)$ C) $(1, -5)$
   D) $(-5, 1)$ E) $(5, -1)$

11. The roots of the equation $x^2 - 6x - 2 = 0$ are

   A) $3 \pm \sqrt{7}$ B) $-3 \pm \sqrt{7}$ C) $3 \pm \sqrt{2}$
   D) $3 \pm \sqrt{11}$ E) $-3 \pm \sqrt{11}$

12. The expression $\dfrac{5}{2 - \sqrt{3}}$ is equivalent to

   A) $10 + 5\sqrt{3}$ B) $-2 - \sqrt{3}$ C) $-10 - 5\sqrt{3}$
   D) $2 + \sqrt{3}$ E) $2 - \sqrt{3}$

13. Find the mode of the following group of numbers:
    8, 8, 9, 10, 11

   A) 8   B) 9   C) $8\frac{1}{2}$ D) $9\frac{1}{5}$ E) 11

14. What is the solution set of the equation

$$|3x + 2| = 5?$$

   A) $\{1\}$ B) $\{\frac{7}{3}\}$ C) $\{-\frac{7}{3}\}$
   D) $\{1, -\frac{7}{3}\}$ E) $\{-1, \frac{7}{3}\}$

15. In triangle ABC, if $a=2$, $b=4$ and $m\angle C=60$, what is the length of side c?

    A) $2\sqrt{7}$ B) 2 C) $2\sqrt{3}$ D) $\sqrt{21}$ E) $4\sqrt{7}$

16. If $m=8$, find the value of $m^{-2/3}$.

    A) $-\frac{16}{3}$ B) -4 C) 4 d) $\frac{1}{4}$ E) 64

17. The square of $(2-2i)$ is

    A) 0 B) $-8i$ C) $4-4i$ D) 4 E) -4

18. The amplitude of the graph of $y=3\cos 2x$ is

    A) π B) 2 C) 3 D) 4π E) $\dfrac{2\pi}{3}$

19. What is the solution set for $(x+3)(x-2)>0$?

    A) ![number line open circle at -3, arrow left]
       -3
    B) ![number line open circles at -3 and 2, arrow right]
       -3 2
    C) ![number line open circle at 2, arrow right]
       2
    D) ![number line open circles at -3 and 2, arrows both]
       -3 2
    E) ![number line closed circles at -3 and 2]
       -3 2

20. How many different 6-player combinations are possible from a group of 10 students trying out for a team?

    A) 60 B) 105 C) 210 D) 360 E) 720

21. The locus of points in a plane that are a given distance d from a point P is

    A) one circle   B) two circles
    C) one circle and one point
    D) two parallel lines   E) a triangle

22. If 35% of number is 70, what is the number?

    A) 200 B) 20 C) 2000 D) 245
    E) 24.5

23. If $f(x)=2x-1$ and $g(x)=3x+1$, then $g(f(x))=$

    A) $2x+1$ B) $3x-1$ C) $6x-2$ D) x
    E) $6x^2-x-1$

24. Find the lateral area of a right circular cone whose height is 10 and base radius is 2.

    A) 400π B) 20π C) 40π D) $2\sqrt{26}$
    E) $4\pi\sqrt{26}$

25. Eliminate the parameter n and identify the curve.

    $$x = 3 - \tfrac{1}{2}n^2, \; y = 1 + n$$

    A) line   B) circle   C) ellipse
    D) parabola   E) hyperbola

26. What statement will this mini-program print?
    ```
    1  LET A=3
    2  LET B=4
    3  LET C=(B-A)²
    4  IF C>0 GO TO INSTRUCTION 7
    5  IF C=0 GO TO INSTRUCTION 9
    6  IF C<0 GO TO INSTRUCTION 11
    7  PRINT"A=3"
    8  STOP
    9  PRINT"B=4"
    10 STOP
    11 PRINT"B-A=1"
    12 STOP
    ```

    A) STOP   B) A=3   C) B=4
    D) B-A=1   E) B-A=-1

27. $\displaystyle\prod_{k=1}^{3}(2k)=$

    A) 8 B) 12 C) $2k+1$ D) $6k-1$
    E) 48

28. $1+i$ is equivalent to

    A) $\sqrt{2}$ B) $\sqrt{2}(\cos 0° + i \sin 0°)$
    C) $\sqrt{2}(\cos 30° + i \sin 30°)$
    D) $\sqrt{2}(\cos 45° + i \sin 45°)$
    E) $\sqrt{2}(\cos 60° + i \sin 60°)$

29. Evaluate $e^{-i\pi}$.

    A) 0 B) 1 C) -1 D) π E) $-\pi$

30. If $\log 3.9=.5911$ and $\log 1.97=.2955$, find $\sqrt{390}$.

    A) 18.6 B) 19.3 C) 19.7 D) 19.9 E) 20.1

31. Sinh $x =$

    A) $\dfrac{1}{\sec x}$ B) $\dfrac{e^x - e^{-x}}{e^x + e^{-x}}$ C) $\dfrac{e^x + e^{-x}}{2}$

    D) $\dfrac{e^x - e^{-x}}{2}$ E) $1 - \cosh x$

32. x varies jointly as j and the $\sqrt{l}$. If x is 3 when j is 4 and l is 4, find x when l is 9 and j is 9.

    A) $\frac{81}{8}$ B) 72   C) $\frac{9}{8}$ D) $\frac{8}{3}$ E) 6

33. $y = \dfrac{(x+2)(x-3)(x-1)}{(x-7)(x+5)}$ has a vertical asymptote at

    A) $x = -2$ B) $x = 3$ C) $x = 1$
    D) $x = 7$ E) $x = 5$

34. Of 18 dogs in a show, 6 dogs have black spots, 7 dogs have brown spots and 3 dogs have both black and brown spots. How many dogs have neither black nor brown spots?

    A) 2   B) 4   C) 7   D) 8   E) 9

35. If $f(x) = x^2 - 2x + 1$, $f(x + 1) =$

    A) $x^3 - 2x^2 + 2$ B) $x^2 - 2x + 2$
    C) $x^2 - x + 2$ D) $x^2 - 2x + 3$ E) x^2

36. The expression $\log \sqrt{xy}$ is equivalent to

    A) $2 \log x + \log y$ B) $2(\log x + \log y)$
    C) $\frac{1}{2} \log x + \log y$ D) $\frac{1}{2}(\log x + \log y)$
    E) $2 \log x + \frac{1}{2} \log y$

37. Solve for x: $27^{2x} = 3^{3x-5}$.

    A) $-\frac{3}{5}$ B) $-\frac{5}{3}$ C) 5   D) -5
    E) 15

38. Find the value of $\sin 210°$.

    A) $-\dfrac{\sqrt{3}}{2}$ B) $\dfrac{\sqrt{3}}{2}$ C) $-\dfrac{1}{2}$ D) $\dfrac{1}{2}$

    E) $-\dfrac{7}{2}$

39. Express the sum of $(3 + 2\sqrt{-16})$ and $(7 - 3\sqrt{-81})$ in $a + bi$ form.

    A) $10 + \sqrt{-97}$ B) $10 + 5i$ C) $10 - 2i$
    D) $10 + 10i$ E) $10 - 19i$

40. Chords AB and CD of circle O intersect at E. If $AE = 4$, $AB = 5$, $CE = 2$, find ED.

    A) 2   B) 3   C) 7   D) 8   E) 10

41. The greatest common monomial factor of $16x^5 + 6x^4 + 10x^3$ is

    A) 2   B) $4x$ C) $4x^2$ D) $2x^3$
    E) x^4

42. Express $.22222\bar{2}$ as a rational number.

    A) $.22$ B) $\frac{22}{100}$ C) $\frac{222}{1000}$ D) $\frac{2}{90}$ E) $\frac{2}{9}$

43. What is the length of the line segment joining the points $N(1, -1)$ and $L(-3, 3)$?

    A) $2\sqrt{5}$ B) $4\sqrt{2}$ C) $5\sqrt{2}$ D) 4
    E) $2\sqrt{2}$

44. What is the solution set for the following system of equations?

    $$2x - 3y = 4$$
    $$x - 2y = 3$$

    A) $\{2, -1\}$ B) $\{-1, -2\}$ C) $\{-1, 3\}$
    D) $\{3, -2\}$ E) $\{-2, -3\}$

45. What is the mean of the following data?

Score	Frequency
7	1
11	2
12	3

    A) 7   B) $10\frac{2}{3}$ C) 11   D) $10\frac{5}{6}$ E) 12

46. In triangle JSL, if $s = 10$, $l = 16$ and $m \angle J = 30$, the area of the triangle is

    A) 160   B) $80\sqrt{3}$ C) $40\sqrt{3}$ D) 40
    E) $40\sqrt{2}$

47. The product of the digits of a positive two digit number is 8. If the number is four times the sum of the digits, find the number.

    A) 42   B) 24   C) 81   D) 18   E) 16

48. What is the probability of exactly 2 heads out of 5 tosses of a fair coin?

   A) $_5C_2(\frac{1}{2})^2(\frac{1}{2})^3$ B) $_3C_2(\frac{1}{2})^2(\frac{1}{2})^3$ C) $_5C_5(\frac{1}{2})^2(\frac{1}{2})^3$
   D) $\frac{2}{5}$ E) $_5C_2(\frac{1}{2})$

49. $xy^2 = k$, where k is a constant, is an example of what type of variation?

   A) direct   B) inverse   C) joint
   D) combined   E) sine wave

50. What is the sum of an arithmetic progression of 10 terms whose first term is 8 and whose common difference is -3?

   A) -12 B) -47 C) -11
   D) -95 E) -55

Solutions to Pretest

1. **(C)** The $\sqrt{17}$ is an infinite non-repeating decimal. It is a member of the set of irrational numbers.

2. **(B)** The inverse is the statement with both the hypothesis and conclusion negated.

3. **(C)** $x^2 - 7x + 10 = (x-2)(x-5)$

4. **(B)** $\dfrac{2\pi}{15} \times \dfrac{180}{\pi} = \dfrac{2(\overset{12}{\cancel{180}})}{\cancel{15}} = 24°$

5. **(A)** $\log 6 = \log (2 \cdot 3) = \log 2 + \log 3$
$\qquad\qquad\qquad\qquad = A + B$

6. **(D)** If the discriminant $(b^2 - 4ac) = 0$, then the roots are real and equal.
$a = 1, b = 4, c = c$
$b^2 - 4ac = (4)^2 - 4(1)(c) = 0 \quad 16 = 4c$
$\qquad\qquad\qquad\qquad\qquad\qquad c = 4$

7. **(A)** kth term $= {}_nC_{k-1}(a)^{n-(k-1)}(b)^{k-1}$
$k = 4, n = 5, a = x, b = +3, k - 1 = 3$
4th term $= {}_5C_3(x)^{5-3}(3)^3$
$\qquad = \dfrac{5 \cdot 4 \cdot \cancel{3}}{\cancel{3} \cdot 2 \cdot 1} x^2(27)$
$\qquad = 270x^2$

8. **(D)** $C = 2\pi r = 20\pi$
$\qquad r = 10$
$\qquad A = \pi r^2 = \pi (10)^2 = 100\pi$

9. **(D)** $\dfrac{x^2 - 3x}{2x^2 + x - 6} \div \dfrac{x^2 - 5x + 6}{x^2 - 4}$

$= \dfrac{x^2 - 3x}{2x^2 + x - 6} \cdot \dfrac{x^2 - 4}{x^2 - 5x + 6}$

$= \dfrac{x(x-3)^{(1)}}{(2x - 3)(x+2)} \cdot \dfrac{(x+2)^{(1)}(x-2)^{(1)}}{(x-3)^{(1)}(x-2)^{(1)}}$

$= \dfrac{x}{2x - 3}$

10. **(C)** $(x - h)^2 + (y - k)^2 = r^2$
Center: (h, k), radius: r
$(x - 1)^2 + (y + 5)^2 = r^2$
Center: $(1, -5)$, radius: $\sqrt{7}$

11. **(D)** $x = \dfrac{-b \pm \sqrt{b^2 - 4ac}}{2a}$

$a = 1, b = -6, c = -2$

$x = \dfrac{-(-6) \pm \sqrt{(-6)^2 - 4(1)(-2)}}{2(1)}$

$= \dfrac{6 \pm \sqrt{36 + 8}}{2}$

$= \dfrac{6 \pm \sqrt{44}}{2}$

$= \dfrac{6 \pm 2\sqrt{11}}{2} = 3 \pm \sqrt{11}$

12. **(A)** $\dfrac{5}{2 - \sqrt{3}} \cdot \dfrac{2 + \sqrt{3}}{2 + \sqrt{3}} = \dfrac{5(2 + \sqrt{3})}{4 - 3} = 10 + 5\sqrt{3}$

13. **(A)** The mode is the number which occurs the most frequently. The mode = 8.

14. **(D)** $\qquad |3x + 2| = 5$

$\qquad\qquad \downarrow \qquad\qquad \downarrow$

$\quad 3x + 2 = -5 \quad 3x + 2 = 5$

$\qquad x = -\tfrac{7}{3} \qquad\quad x = 1$

$\qquad\qquad\text{check}$

$|3x + 2| = 5 \qquad\quad |(3x + 2)| = 5$
$|3(-\tfrac{7}{3}) + 2| \overset{?}{=} 5 \qquad |3(1) + 2| = 5$
$|-7 + 2| \overset{?}{=} 5 \qquad\qquad |5| = 5$
$|-5| \overset{?}{=} 5$
$5 = 5$

2 solutions $\{1, -\tfrac{7}{3}\}$

15. **(C)** $c^2 = a^2 + b^2 - 2ab \cos C$
$c^2 = (2)^2 + (4)^2 - 2(2)(4) \cos 60°$
$c^2 = 4 + 16 - 16(\frac{1}{2})$
$c^2 = 12$
$c = 2\sqrt{3}$

16. **(D)** $8^{-2/3}$. The 3 indicates cube root
The 2 represents squaring
The "—" indicates reciprocal

$8^{-2/3} = \dfrac{1}{(\sqrt[3]{8})^2} = \dfrac{1}{4}$

17. **(B)** $(2-2i)^2 = (2-2i)(2-2i)$
$= 2(2-2i) - 2i(2-2i)$
$= 4 - 4i - 4i + 4i^2$
$= 4 - 8i + 4(-1)$
$= -8i$

18. **(C)** The amplitude is 3.

19. **(D)** $(x+3)(x-2) = 0$
$x = -3 | x = 2$
The critical values are -3 and 2.
They are excluded due to the $>$ in the original problem (the values -3 and 2 are represented by hollow circles).

Test the interval $-\infty < x < -3$, e.g., $x = -10$
$(-10+3)(-10-2) = +84 > 0$ True

Test the interval $-3 < x < 2$, e.g., $x = 0$
$(0+3)(0-2) = -6 > 0$ False

Test the interval $x > 2$, e.g., $x = 3$
$(3+3)(3-2) = 6 > 0$ True

20. **(C)** $_{10}C_6 = {_{10}C_4} = \dfrac{\overset{(3)}{10} \cdot \overset{(1)}{9} \cdot 8 \cdot 7}{4 \cdot 3 \cdot 2 \cdot 1} = 210$

21. **(A)**

A circle with center P and radius d

22. **(A)** $\dfrac{35}{100} \cdot x = 70$
$\dfrac{35x}{100} = 70$
$x = 200$

23. **(C)** $g(f(x)) = g(2x-1)$
$= 3(2x-1) + 1$
$= 6x - 3 + 1$
$= 6x - 2$

24. **(E)** L.A. $= \pi r l$
$= \pi(2)(2\sqrt{26})$
$= 4\pi\sqrt{26}$

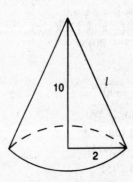

$l^2 = 10^2 + 2^2 = 104$
$l = \sqrt{104} = 2\sqrt{26}$

25. **(D)** $x = 3 - \frac{1}{2}n^2$
$y = 1 + n \Rightarrow n = y - 1$
$\therefore x = 3 - \frac{1}{2}(y-1)^2$ Substitute for n
$x - 3 = -\frac{1}{2}(y-1)^2$
This is the equation of a parabola.

26. **(B)**

Value of A	Value of B	Value of $(B-A)^2 = C$
3	4	$(4-3)^2 = 1$

Go to instruction 7
$A = 3$

27. **(E)** $\displaystyle\prod_{k=1}^{3}(2k) = [2(1)] \cdot [2(2)] \cdot [2(3)]$
$2 \cdot 4 \cdot 6 = 48$

28. (D) $x + yi = 1 + i$ $x = 1, y = 1$

$$r = \sqrt{x^2 + y^2} = \sqrt{2}$$

$$\text{Tan } \theta = \frac{y}{x} = \frac{1}{1} = 1, \theta = 45°$$

$$1 + i = \sqrt{2}(\cos 45° + i \sin 45°)$$

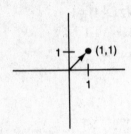

29. (C) $e^{i\theta} = \cos \theta + i \sin \theta$

$$e^{i(-\pi)} = \cos(-\pi) + i \sin(-\pi)$$
$$= -1 + i(0) = -1$$

30. (C) $x = \sqrt{390}$

$$\log x = \log\sqrt{390}$$
$$= \tfrac{1}{2} \log 390$$
$$= \tfrac{1}{2}(2.5911)$$
$$= 1.2955$$
$$x = \text{antilog}(1.2955)$$
$$x = 1.97 \times 10^1 = 19.7$$

31. (D) $\text{Sinh } x = \dfrac{e^x - e^{-x}}{2}$

32. (A) $\dfrac{x_1}{j_1\sqrt{l_1}} = \dfrac{x_2}{j_2\sqrt{l_2}}$

$$\frac{3}{4\sqrt{4}} = \frac{x}{9\sqrt{9}}$$

$$\frac{3}{8} = \frac{x}{27}$$

$$x = \frac{81}{8}$$

33. (D) Vertical asymptotes of $f(x) = \dfrac{g(x)}{h(x)}$ occur

when $h(x) = 0$
$$(x - 7)(x + 5) = 0$$
$$x = 7 | x = -5$$

34. (D)

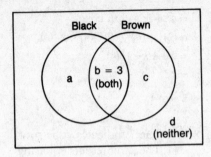

$$b = 3$$
$$a + b = 6 \quad \therefore a = 3$$
$$b + c = 7 \quad \therefore c = 4$$
$$\text{Total} = a + b + c + d = 18$$
$$3 + 3 + 4 + d = 18$$
$$d = 8$$

35. (E) $f(x) = x^2 - 2x + 1$

$$f(x + 1) = (x + 1)^2 - 2(x + 1) + 1$$
$$= x^2 + 2x + 1 - 2x - 2 + 1$$
$$= x^2$$

36. (D) $\log \sqrt{xy} = \log(xy)^{1/2}$
$$= \tfrac{1}{2} \log xy = \tfrac{1}{2}(\log x + \log y)$$

37. (B) $27^{2x} = 3^{3x-5}$
$$(3^3)^{2x} = 3^{3x-5}$$
$$3^{6x} = 3^{3x-5}$$
$$6x = 3x - 5$$
$$x = \frac{-5}{3}$$

38. (C)

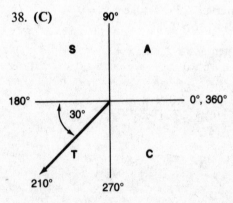

$$\sin 210° = -\sin 30°$$

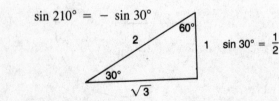

$$\therefore \sin 210° = -\sin 30° = -(\tfrac{1}{2}) = -\tfrac{1}{2}$$

39. **(E)** $3 + 2\sqrt{-16} = 3 + 2(4i) = 3 + 8i$
$7 - 3\sqrt{-81} = 7 - 3(9i) = 7 - 27i$
$(3 + 8i) + (7 - 27i) = 10 - 19i$

40. **(A)**

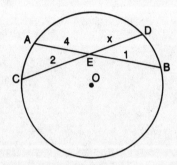

$(AE)(EB) = (CE)(ED)$
$(4)(1) = (2)x$
$x = 2$

41. **(D)**
1. The GCF $= 2x^3$
2. $\dfrac{16x^5 + 6x^4 + 10x^3}{2x^3} = 8x^2 + 3x + 5$
3. $16x^5 + 6x^4 + 10x^3 = 2x^3[8x^2 + 3x + 5]$

42. **(E)** $10x = 2.2222\bar{2}$

$$\dfrac{x = \quad .2222\bar{2}}{9x = 2}$$

$x = \tfrac{2}{9}$

43. **(B)** $d = \sqrt{(x_2 - x_1)^2 + (y_2 - y_1)^2}$
$= \sqrt{(-3 - 1)^2 + (3 - (-1))^2}$
$= \sqrt{(-4)^2 + (4)^2}$
$= \sqrt{16 + 16} = \sqrt{32} = \sqrt{16 \cdot 2}$
$= 4\sqrt{2}$

44. **(B)**

$$\begin{array}{c} 2x - 3y = 4 \\ (x - 2y = 3)(-2) \\ \hline 2x - 3y = 4 \\ -2x + 4y = -6 \\ \hline y = -2 \end{array}$$

Substitute $y = -2$ into one of the original equations

$x - 2y = 3$
$x - 2(-2) = 3$
$x = -1$
$\{-1, -2\}$

45. **(D)** mean = average $= \dfrac{\Sigma f_i x_i}{\Sigma f_i}$

$= \dfrac{1(7) + 2(11) + 3(12)}{1 + 2 + 3}$

$= \dfrac{7 + 22 + 36}{6} = \dfrac{65}{6} = 10\tfrac{5}{6}$

46. **(D)**

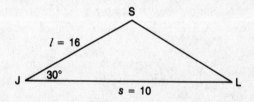

$A = \frac{1}{2} sl \sin J$
$= \frac{1}{2}(10)(16)\sin 30°$
$= 80 \sin 30° = 80(\tfrac{1}{2}) = 40$

47. **(B)** Let $u =$ unit's digit
$t =$ ten's digit
$(u)(t) = 8$
$10t + u = 4(t + u) \Rightarrow 10t + u = 4t + 4u$
$6t = 3u$
$2t = u$

Substitute for u

$u \cdot t = 8$
$(2t)t = 8$
$t^2 = 4$
$t = \pm 2$ reject $t = -2$
$t = 2$
$u = 4$
The number is $10t + u = 24$

48. **(A)** $p = p(\text{success}) = \frac{1}{2}$, $n = 5$ attempts
$q = p(\text{failure}) = \frac{1}{2}$, $r = 2$ successes
$$_5C_2(\tfrac{1}{2})^2(\tfrac{1}{2})^3$$

49. **(B)** $x = \dfrac{k}{y^2}$ or $xy^2 = k$ is an example of inverse square variation.

50. **(E)** $S = \dfrac{n}{2}(2a + (n - 1)d)$
$$= \frac{10}{2}(2(8) + (10 - 1)(-3))$$
$$= 5(16 - 27) = 5(-11) = -55$$

How to Use the Diagnostic Chart (page 20)

The numbers at the left correspond to the question numbers in the test. The words and numbers at the top of the chart correspond to the chapters and chapter names. The darkened square in the row for each question indicates the subject area from which each question is drawn. Mark the chart with a $\sqrt{}$ for each question answered correctly and an $\times$ for each question answered incorrectly. You'll be able to see at a glance which topics need further study.

Diagnostic Chart Pretest

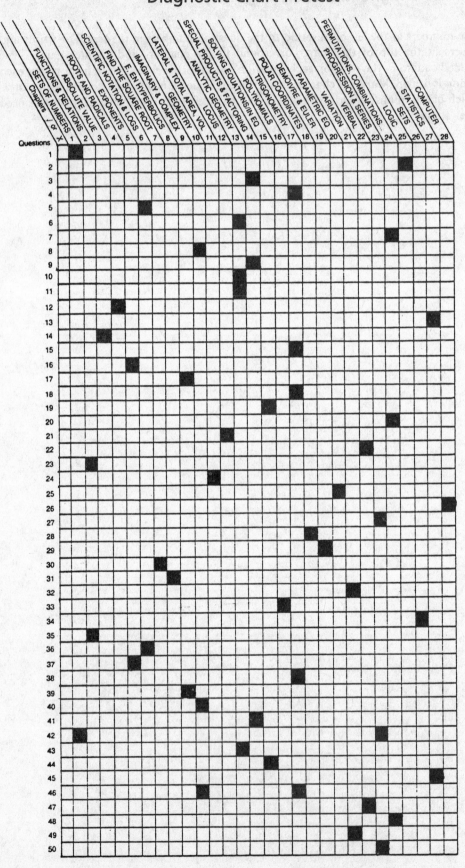

UNIT ONE

1. Sets of Numbers

1.1 Fundamental Facts

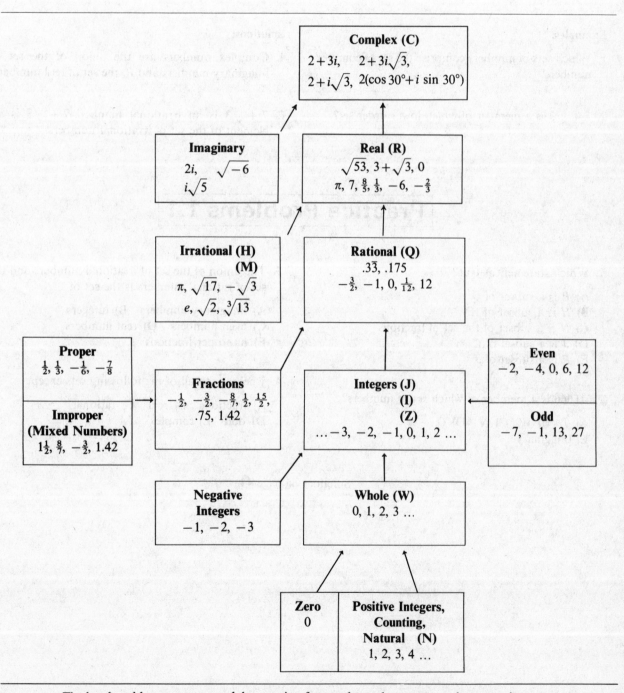

Complex (C)
$2+3i$, $2+3i\sqrt{3}$, $2+i\sqrt{3}$, $2(\cos 30° + i\sin 30°)$

Imaginary
$2i$, $\sqrt{-6}$
$i\sqrt{5}$

Real (R)
$\sqrt{53}$, $3+\sqrt{3}$, 0
π, 7, $\frac{8}{5}$, $\frac{1}{3}$, -6, $-\frac{2}{3}$

Irrational (H) (M)
π, $\sqrt{17}$, $-\sqrt{3}$
e, $\sqrt{2}$, $\sqrt[3]{13}$

Rational (Q)
$.3\overline{3}$, $.175$
$-\frac{3}{2}$, -1, 0, $\frac{1}{12}$, 12

Proper
$\frac{1}{2}$, $\frac{1}{3}$, $-\frac{1}{6}$, $-\frac{7}{8}$

Improper (Mixed Numbers)
$1\frac{1}{2}$, $\frac{8}{7}$, $-\frac{3}{2}$, 1.42

Fractions
$-\frac{1}{2}$, $-\frac{3}{2}$, $-\frac{8}{7}$, $\frac{1}{2}$, $\frac{15}{2}$, $.75$, 1.42

Integers (J) (I) (Z)
$\ldots -3, -2, -1, 0, 1, 2 \ldots$

Even
-2, -4, 0, 6, 12

Odd
-7, -1, 13, 27

Negative Integers
-1, -2, -3

Whole (W)
$0, 1, 2, 3 \ldots$

Zero
0

Positive Integers, Counting, Natural (N)
$1, 2, 3, 4 \ldots$

The bracketed letters represent alpha notation frequently used to represent the respective set.

23

Irrational numbers are real numbers which cannot be expressed in $\frac{a}{b}$ form, where a and b are integers. These include non-terminating, non-repeating decimals and square roots of non-perfect squares.

 A bar over the last digit or last group of digits in a decimal number indicates infinite repetition of the sequence under the bar. For example $.136\overline{36} = .13636363636\ldots$

Examples:

1. Which sets of numbers comprise the set of complex numbers?

2. $7 + \sqrt{3}$ is a member of which lower order set?

Solutions:

1. Complex numbers are the union of the set of imaginary numbers and R, the set of real numbers.

2. $7 + \sqrt{3}$ is an irrational number. $7 + \sqrt{3}$ is an element of the set of irrational numbers.

Practice Problems 1.1

1. Which statement is true?

   A) R is a subset of Q
   B) H is a subset of W
   C) N is a subset of the set of fractions
   D) J is a subset of C
   E) R is a subset of J

2. $.1666\overline{6}$ is a member of which set of numbers?

   A) J B) W C) N D) Q
   E) Imaginary

3. The union of the set of irrational numbers and the set of rational numbers is the set of

   A) imaginary numbers   B) integers
   C) even numbers   D) real numbers
   E) improper fractions

4. $\frac{5}{7}$ belongs to all of the following sets except

   A) fractions   B) real   C) rational
   D) odd   E) complex

Solutions on page 44

2. Functions and Relations

2.1 Essentials

Relation

A **relation** is a set of ordered pairs (x, y). The **domain** of the relation is the set of all first numbers of the relation; the **range** is the set of all resulting second numbers in the relation.

Function

A relation in which each value of the first number has a unique value for the second number.
(A given x value cannot have more than one y value.)

Domain

All acceptable values of the first number in a relation or function.

 x values which create division by 0 are not acceptable values.
x values which create imaginary values are not acceptable values.

Range

All values of the second number in a relation or function which can result from admissible values of the first number.

$y = f(x)$

Read as "y is a function of x" or "y equals f of x".
Values of x (the independent variable) represent numbers in the domain of the function. The resulting values of y or $f(x)$ (the dependent variable) represent numbers in the range of the function.

 $f(x)$ is not read as f times x.

$y = f(a)$

When a function $f(x)$ is defined by an equation, the value $f(a)$ is found by substituting the value 'a' for each appearance of the variable 'x'.

Vertical Line Test

The graphical technique to determine if a relation is a function. If a vertical line drawn anywhere on the coordinates can only cross the graph of the relation at one and only one intersection, then the relation is a function.

Examples:

1. If $f(x) = x^2 + 3x$, find $f(x + 2)$.

Solutions:

1.
$$f(x) = x^2 + 3x$$
$$f(x + 2) = (x + 2)^2 + 3(x + 2)$$
$$= x^2 + 4x + 4 + 3x + 6$$
$$= x^2 + 7x + 10$$

25

2. Does the graph of the relation shown below represent a function?

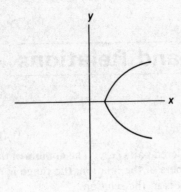

2. Solution: Use the Vertical Line Test. Since a vertical line *can* cross the graph at more than one point of intersection the relation is NOT a function.

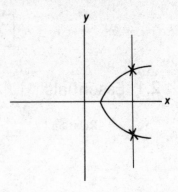

3. If $f(x) = 3x - 2$, find $\dfrac{f(x+h) - f(x)}{h}$, $h \neq 0$.

3.
$$f(x) = 3x - 2$$
$$f(x+h) = 3(x+h) - 2 = 3x + 3h - 2$$
$$\frac{f(x+h) - f(x)}{h} = \frac{(3x + 3h - 2) - (3x - 2)}{h} = \frac{3h}{h} = 3$$

4. If $f(x) = x^2 + 4x - 2$, find the domain and range of the function.

4. Sometimes a simple sketch is very useful as an aid for determining domain and range.

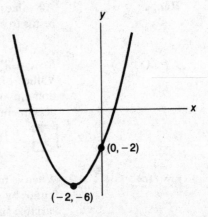

The domain: all real x (all values of x are good)
The range: $y \geq -6$ (there are no y values below $y = -6$)

Practice Problems 2.1

1. If $f(x) = 2x^2 - 2x + 1$, find $f(-3)$.
   A) -23 B) -11 C) -3 D) 13   E) 25

2. Which of the graphs shown on the next page does not represent a function?

A)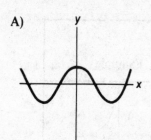

B)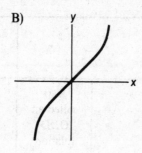

C)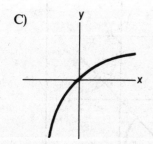

D)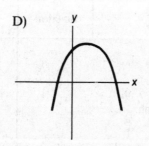

E)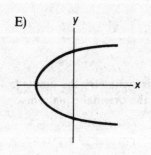

3. If $f(x) = 2x^2 - 2$, find $\dfrac{f(x+h) - f(x)}{h}$, $h \neq 0$.

A) $4x + 2h$ B) $4x + 2h - 4$
C) $4x^2 + 4x + 2h$ D) $4xh - 2h^2$ E) $2x + h$

4. Find the range of $f(x) = |x|$.

A) All real numbers B) $y \geq 0$ C) $y > 0$
D) $y \geq x$ E) $y < 0$

5. If $f(x) = -x^2 - 3x - 2$, find $f(-3)$.

A) -2 B) -20 C) 16
D) -17 E) -14

6. Does the relation $x^2 + y^2 = 9$ represent a function?

A) Never B) Always
C) Only when $x \geq 3$ D) Only when $y \geq 3$
E) When x and y are ≥ 3

Solutions on page 44

2.2 Symmetry

Type of Symmetry	Conditions	Example
y-axis	If (x,y) is on the graph of $f(x)$, then $(-x,y)$ is also on the graph. or If $f(-x) = f(x)$	also called an EVEN relation or function
x-axis	If (x,y) is on the graph of f, then $(x, -y)$ is also on the graph. Note: these are relations not functions	

Type of Symmetry	Conditions		Example
Origin (0,0)	If (x,y) is on the graph of $f(x)$, then $(-x,-y)$ is also on the graph. or If $f(-x) = -f(x)$	also called an ODD relation or function	
The line x = y	If (x,y) is on the graph of $f(x)$, then (y,x) is also on the graph.		

Examples:

1. Is $x^2 + y^2 = 9$ symmetric to the y-axis?

2. Does the equation $xy = 8$ have origin symmetry?

3. Does the equation $x = y^2$ have x-axis symmetry?

4. Does $xy = 8$ have y-axis symmetry?

Solutions:

1. For $y=$ axis symmetry the points (x,y) and $(-x,y)$ both yield the original equation.

   Test Points
   $$x^2 + y^2 = 9 \quad \begin{array}{l} (x,\ y) \Rightarrow x^2 + y^2 = 9 \\ (-x,\ y) \Rightarrow (-x)^2 + y^2 = 9 \Rightarrow x^2 + y^2 = 9 \end{array}$$

   There is y-axis symmetry.

2. For origin symmetry the points (x, y) and $(-x, -y)$ both yield the original equation.

   Test Points
   $$xy = 8 \quad \begin{array}{l} (x, y) \Rightarrow xy = 8 \\ (-x, -y) \Rightarrow (-x)(-y) = 8 \Rightarrow xy = 8 \end{array}$$

   There is origin symmetry.

3. For $x=$ axis symmetry the points (x,y) and $(x,-y)$ both yield the original equation.

   Test Points
   $$x = y^2 \quad \begin{array}{l} (x, y) \Rightarrow x = y^2 \\ (x, -y) \Rightarrow x = (-y)^2 = y^2 \end{array}$$

   There is x-axis symmetry.

4. For $y=$ axis symmetry the points (x,y) and $(-x,y)$ both yield the same equation.

   Test Points
   $$xy = 8 \quad \begin{array}{l} (x, y) \Rightarrow xy = 8 \\ (-x, y) \Rightarrow (-x)(y) = 8 \Rightarrow xy = -8 \end{array}$$

   There is *no* y-axis symmetry.

Practice Problems 2.2

1. The graph of $y = |x| + 3$ has what type of symmetry?

   A) x-axis only   B) y-axis only
   C) Origin only   D) $x = y$ only
   E) x- and y-axes only

2. The graph of the equation $y = 7$ has what type of symmetry?

   A) x-axis only   B) y-axis only
   C) Origin only   D) $x = y$ only
   E) x- and y-axes only

3. The graph of the relation $x^2 - y^2 = 4$ has what type of symmetry?

   A) x-axis, origin and $x = y$ only
   B) y-axis, origin and $x = y$ only
   C) Origin only
   D) Origin, x-axis and y-axis only
   E) Origin and x-axis only

4. Classify the equation $y = \sin x$.

   A) Even function   B) Odd function
   C) Even relation but not a function
   D) Odd relation but not a function
   E) Neither even nor odd

5. The graph of the equation $xy = 8$ has what type of symmetry?

   A) Origin, $x = y$ and x-axis only
   B) Origin, $x = y$ and y-axis only
   C) Origin and $x = y$ only
   D) $x = y$ and x-axis only
   E) $x = y$ and y-axis only

6. The graph of $y = 2x - 1$ has what type of symmetry?

   A) None   B) $x = y$ only
   C) x-axis only   D) y-axis only
   E) Origin only

Solutions on page 45

2.3 The Algebra of Functions

Given two functions $f(x)$ and $g(x)$	
Adding	$(f + g)(x) = f(x) + g(x)$
Subtracting	$(f - g)(x) = f(x) - g(x)$
Multiplying	$(f \cdot g)(x) = f(x) \cdot g(x)$
Dividing	$\left(\dfrac{f}{g}\right)(x) = \dfrac{f(x)}{g(x)}$
Composition	$(f \circ g)(x) = f(g(x))$
Inverse	$f^{-1}(x) =$ the inverse of $f(x)$

The **inverse** of a function denoted by $f^{-1}(x)$ represents the relation such that

$$f(f^{-1}(x)) = f^{-1}(f(x)) = x$$

Graphically, f^{-1} of x represents the reflection of $f(x)$ about the line $y = x$.

 $f^{-1}(x)$ does *not* mean $\dfrac{1}{f}$, nor f^{-1} times x.

Procedure for calculating the inverse of $f(x)$

Example: Find the inverse of $f(x) = 2x - 3$

Step	Description	Example
1	Replace $f(x)$ with y	$y = 2x - 3$
2	Swap every x and y	$x = 2y - 3$
3	Solve for y	$y = \dfrac{x + 3}{2}$
4	Replace y with $f^{-1}(x)$	$f^{-1}(x) = \dfrac{x + 3}{2}$

☞ $f(x)$, $f^{-1}(x)$ Notes:

If $f(x) = 2x - 3$ and $f^{-1}(x) = \dfrac{x + 3}{2}$

$$f(f^{-1}(x)) = 2\left(\dfrac{x + 3}{2}\right) - 3 = x$$

$$f^{-1}(f(x)) = \dfrac{(2x - 3) + 3}{2} = x$$

NOTE:

$f(f^{-1}(x)) = x$

$f^{-1}(f(x)) = x$

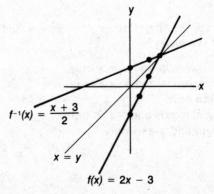

$f^{-1}(x) = \dfrac{x + 3}{2}$

$x = y$

$f(x) = 2x - 3$

NOTE: $f^{-1}(x)$ is the reflection of $f(x)$ about the line $x = y$.

Examples:

1. Find the inverse of $f(x) = \dfrac{x - 1}{x + 2}$.

Solutions:

1. Let $\quad y = \dfrac{x - 1}{x + 2}$

Swap x and y

$$x = \dfrac{y - 1}{y + 2}$$

Solve for y

$$xy + 2x = y - 1$$
$$xy - y = -2x - 1$$
$$y(x - 1) = -2x - 1$$
$$y = \dfrac{2x + 1}{1 - x}$$

replace y with $f^{-1}(x)$

$$f^{-1}(x) = \dfrac{2x + 1}{1 - x}$$

2. If $f(x)=x^2-1$ and $g(x)=x+1$, find

A) $(f+g)(x)$ B) $(f-g)(x)$ C) $(f\cdot g)(x)$

D) $\left(\dfrac{f}{g}\right)(x)$ E) $(f\circ g)(x)$ F) $(g\circ f)(x)$

G) $(f\circ f)(x)$ H) $g^{-1}(x)$

2. A) $(f+g)(x) = f(x) + g(x) = (x^2-1)+(x+1)$
$= x^2 + x$

B) $(f-g)(x) = f(x) - g(x) = (x^2-1)-(x+1)$
$= x^2 - x - 2$

C) $(f\cdot g)(x) = f(x)\cdot g(x) = (x^2-1)(x+1)$
$= x^3 + x^2 - x - 1$

D) $\left(\dfrac{f}{g}\right)(x) = \dfrac{f(x)}{g(x)} = \dfrac{x^2-1}{x+1} = \dfrac{(x+1)(x-1)}{(x+1)} = x-1$

E) $(f\circ g)(x) = f(g(x)) = f(x+1) = (x+1)^2 - 1$
$= x^2 + 2x$

F) $(g\circ f)(x) = g(f(x)) = g(x^2-1) = (x^2-1)+1$
$= x^2$

G) $(f\circ f)(x) = f(f(x)) = f(x^2-1) = (x^2-1)^2 - 1$
$= x^4 - 2x^2$

H) $g^{-1}(x)$ $g(x) = x+1$

STEP	
1	$y = x + 1$
2	$x = y + 1$
3	$y = x - 1$
4	$g^{-1}(x) = x - 1$

Practice Problems 2.3

1. If $f(x)=3x-3$ and $g(x)=2x+1$, find $(f+g)(x)$.

A) $x-2$ B) $5x-2$ C) $6x-5$
D) $6x^2-3x-3$ E) $6x$

2. If $f(x)=3x-3$ and $g(x)=2x+1$, find $(f\cdot g)(x)$.

A) $x-2$ B) $5x-2$ C) $6x-5$
D) $6x^2-3x-3$ E) $6x$

3. If $f(x)=3x-3$ and $g(x)=2x+1$, find $(f\circ g)(x)$.

A) $x-2$ B) $5x-2$ C) $6x-5$
D) $6x^2-3x-3$ E) $6x$

4. If $f(x)=3x-3$ and $g(x)=2x+1$, find $(g\circ f)(x)$.

A) $x-2$ B) $5x-2$ C) $6x-5$
D) $6x^2-3x-3$ E) $6x$

5. If $f(x)=\dfrac{x-1}{x+1}$, find $f^{-1}(x)$.

A) $\dfrac{x+1}{x-1}$ B) 1 C) $\dfrac{x-2}{x+2}$

D) $\dfrac{-(x+1)}{x-1}$ E) x

6. The graph which best represents the inverse of $f(x)$ is

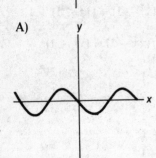

$y = f(x)$

C)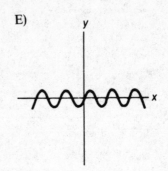

D)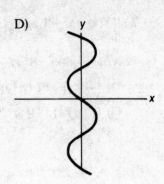

A)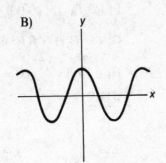

B)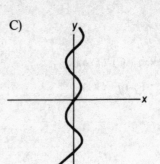

E)

Solutions on page 46

2.4 Summary of Basic Function Types

Type	Examples of Algebraic Representation	Examples of Graphic Representation
Identity function	$f(x) = x$	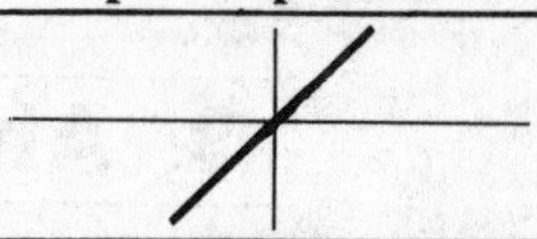
Constant function	$f(x) = a$	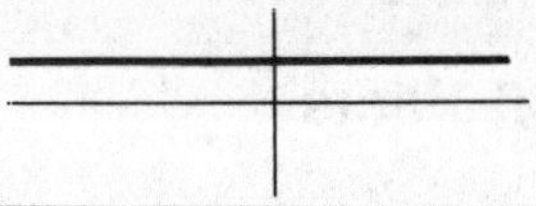
Absolute Value function	$f(x) = \|x\|$	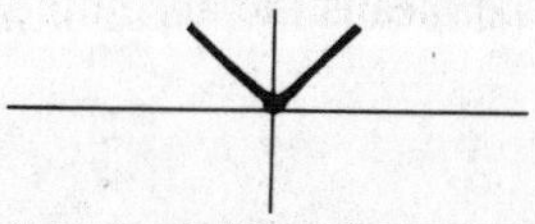
Linear function	$f(x) = mx + b$ m = slope b = y intercept	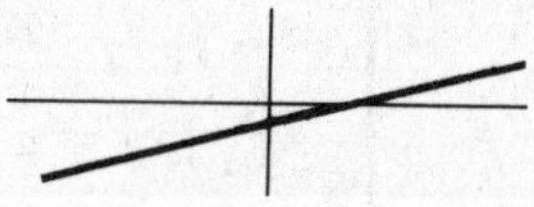
Quadratic function	$f(x) = ax^2 + bx + c$	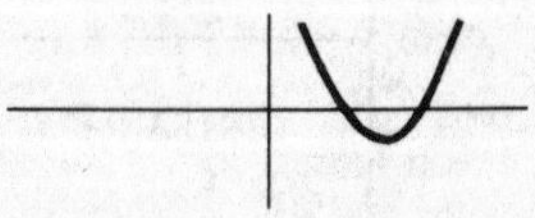
Polynomial function	$f(x) = a_n x^n + a_{n-1} x^{n-1} \ldots$	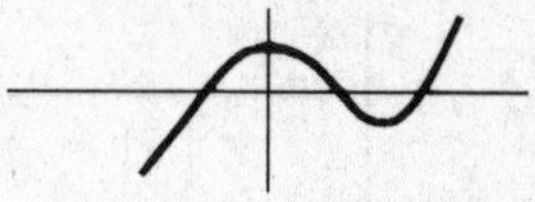
Greatest Integer function (Step function)	$f(x) = [x]$ { $[x]$ is the greatest integer less than or equal to x }	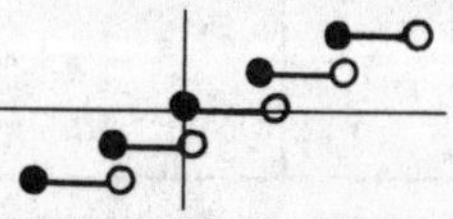
Rational function	$f(x) = \dfrac{g(x)}{h(x)}$	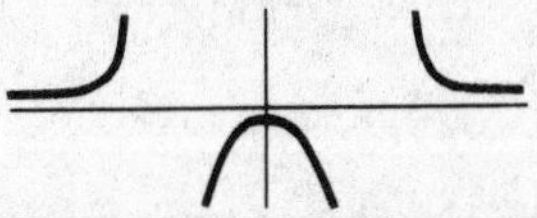
Inverse function	$f^{-1}(x)$	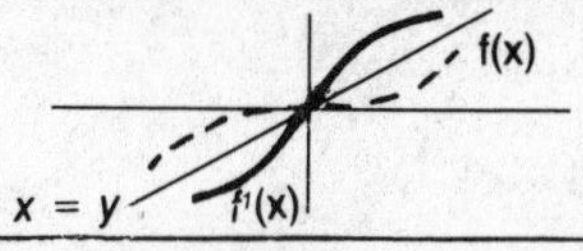
Trigonometric, Periodic, Circular or Wrapping functions	$f(x) = \sin x$	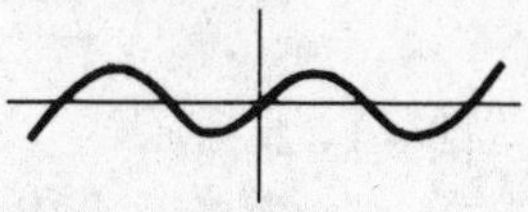
TRANSCENDENTOL FUNCTIONS: **Exponential functions**	$f(x) = b^x$	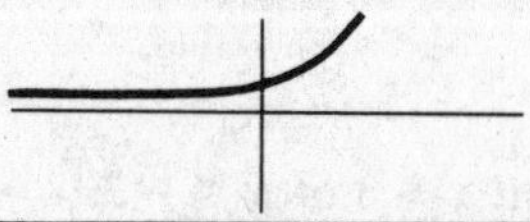
Logarithmic functions	$f(x) = \log_b x$	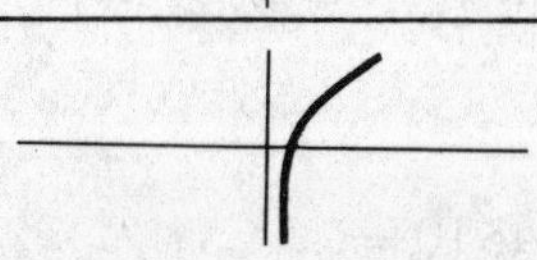
Hyperbolic functions	$f(x) = \sinh x = \dfrac{e^x - e^{-x}}{2}$	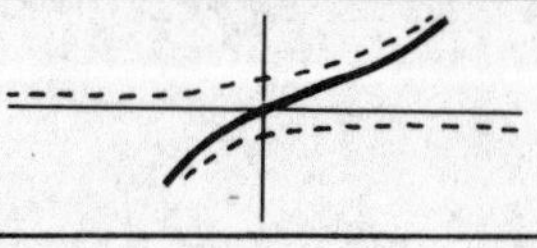

NOTE: Circles and ellipses are relations not functions.

3. Absolute Value

3.1 Fundamental Facts

The symbol $|x|$ means the absolute value of x.

	$\lvert x \rvert = a$	$x = a$ $x = -a$	 $-a \qquad a$
$\lvert x \rvert = x \quad x \geq 0$	$\lvert x \rvert > a$	$x > a$ or $x < -a$	$-a \qquad a$
	$\lvert x \rvert \geq a$	$x \geq a$ or $x \leq -a$ $(a \geq 0)$	$-a \qquad a$
$\lvert x \rvert = x \quad x < 0$	$\lvert x \rvert < a$	$x < a$ and $x > -a$ $(-a < x < a)$	$-a \qquad a$
	$\lvert x \rvert \leq a$ $(a > 0)$	$x \leq a$ and $x \geq -a$ $(-a \leq x \leq a)$	$-a \qquad a$

Examples:

1. Evaluate $\lvert 5 - 3 \rvert - \lvert 3 - 5 \rvert$.

2. Evaluate $\lvert 0 \rvert + \lvert -2 \rvert - \lvert 2 \rvert - \lvert -2 \rvert$.

3. Graphically represent $x > 3$ or $x \leq -2$.

4. Graph the solution to $\lvert x \rvert = 5$.

Solutions:

1. $\lvert 5 - 3 \rvert = \lvert 2 \rvert = 2 \qquad \lvert 3 - 5 \rvert = \lvert -2 \rvert = 2$
   $\lvert 5 - 3 \rvert - \lvert 3 - 5 \rvert = 2 - 2 = 0$

2. $\lvert 0 \rvert = 0, \lvert -2 \rvert = 2, \lvert 2 \rvert = 2$
   $\lvert 0 \rvert + \lvert -2 \rvert - \lvert 2 \rvert - \lvert -2 \rvert$
   $= 0 + 2 - 2 - 2 = -2$

3.
   $-2 \qquad 3$

4. $\lvert x \rvert = 5 \qquad x = 5$
   $x = -5$

$-5 \qquad 5$

Absolute Value Equations (Graphs)

The absolute value function is generally in the shape of a "$\vee$" with the vertex at the value of x which makes the sentence inside the absolute value sign equal to zero. This is the critical value of x for the absolute value graph. To the left of the critical value the $|f(x)| = g(x)$ represents the equation $f(x) = -g(x)$, to the right of the critical value $|f(x)| = g(x)$ represents the equation $f(x) = g(x)$.

$$|f(x)| = g(x)$$

The vertex occurs at the value(s) such that $f(x) = 0$.

This region represents the equation $f(x) = -g(x)$	This region represents the equation $f(x) = g(x)$

Critical
value
of x

A chart can be developed to graph an absolute value function. Be sure to choose values above, below and at the critical value.

Examples:

5. Graph $|x - 1| = y$.

6. Graph $y = |x - 2| + 3$.

Solutions:

5. The critical value is $x = 1$.
Using a chart

x	y
-2	3
-1	2
0	1
1	0
2	1
3	2
4	3

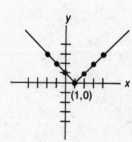

6. The critical value is $x = 2$.
Using a chart

x	y
-1	6
0	5
1	4
2	3
3	4
4	5
5	6

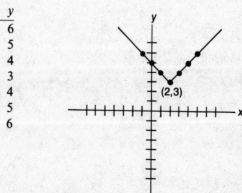

Practice Problems 3.1

1. Evaluate  $+ |6(-2)| + |6| \cdot |-2|.$

   A) 0   B) 6   C) 18   D) 30   E) 24

2. Graphically represent $|x| \leqslant 3$.

   A) 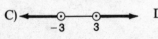   B)

   C) 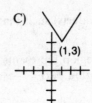   D)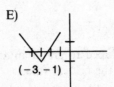

   E)

3. Given the equation $y = |2x - 6| + 5$, the equation which represents the left leg of the "**V**" is

   A) $y = -(2x - 6) + 5$ B) $y = (2x - 6) - 5$
   C) $y = (2x - 6) + 5$ D) $y = -(2x - 6) - 5$
   E) $y = (2x - 6)$

4. Graph $y = |x + 3| - 1.$

   A) 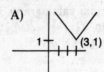   B)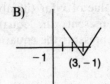

   C)   D)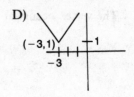

   E)

Solutions on page 46

4. Roots and Radicals

4.1 Basics and Beyond

To find the square root of a number means to find one of its two equal factors.

$\sqrt{}$ is the symbol for square root and is called a radical sign.

$\sqrt[3]{}$ is the notation for cube root and means to find one of its three equal factors.

 The $\sqrt{}$ symbol indicates the principal square root. That is, the positive root only: $\sqrt{4} = 2$ not -2.

Addition and Subtraction:	To add or subtract square roots, the value under the radical must be the same. Add or subtract the coefficients and retain the radical. $$m\sqrt{a} + n\sqrt{a} = (m + n)\sqrt{a}$$ $$m\sqrt{a} - n\sqrt{a} = (m - n)\sqrt{a}$$
Multiply and Divide:	$\sqrt{a} \cdot \sqrt{b} = \sqrt{ab}\ (a, b \geqslant 0), \quad m\sqrt{a} \cdot n\sqrt{b} = mn\sqrt{ab}$ $$\frac{\sqrt{a}}{\sqrt{b}} = \sqrt{\frac{a}{b}}\left(\begin{array}{l}a \geqslant 0\\ b > 0\end{array}\right), \quad \frac{m\sqrt{a}}{n\sqrt{b}} = \frac{m}{n}\sqrt{\frac{a}{b}}$$

 A square root should be expressed in simplest form: $2\sqrt{2}$, not $\sqrt{8}$.

 Common Error: $\sqrt{2} + \sqrt{3} = \sqrt{5}$. Should be $\sqrt{2} + \sqrt{3} = \sqrt{2} + \sqrt{3}$. These radicals cannot be added. To add radicals the values under the radical sign must be the same.

Examples:

1. Find $\sqrt{16}$.

2. Combine $3\sqrt{2} + 7\sqrt{2} - 5\sqrt{2}$.

Solutions:

1. $\sqrt{16} = 4$. Note: the $\sqrt{}$ indicates the positive root only.

2. $3\sqrt{2} + 7\sqrt{2} - 5\sqrt{2} = (3 + 7 - 5)\sqrt{2} = 5\sqrt{2}$

 To simplify a square root, factor the value under the radical sign into two numbers, one of which is a perfect square. Then separate the original square root into the product of the square roots of the factors, and take the square root of the perfect square.

37

Examples:

3. Simplify $\sqrt{18}$.

4. Combine $\sqrt{125} + \sqrt{20} - \sqrt{500}$.

5. Multiply $(4\sqrt{3})(2\sqrt{6})$.

6. Find $(2\sqrt{5})^2$.

7. Divide $\dfrac{10\sqrt{20}}{2\sqrt{4}}$.

8. $(3 + \sqrt{2})(2 + \sqrt{10}) =$

9. $(2 + \sqrt{7})^2 =$

10. Simplify $\sqrt{a^4 b^5 c^3}$.

Solutions:

3. $\sqrt{18} = \sqrt{9 \cdot 2} = \sqrt{9}\sqrt{2}$
$\quad = 3\sqrt{2}$

4. First simplify each term:
$\sqrt{25 \cdot 5} + \sqrt{4 \cdot 5} - \sqrt{100 \cdot 5}$
$= \sqrt{25}\sqrt{5} + \sqrt{4}\sqrt{5} - \sqrt{100}\sqrt{5}$
$= 5\sqrt{5} + 2\sqrt{5} - 10\sqrt{5}$
$= -3\sqrt{5}$

5. $(4\sqrt{3})(2\sqrt{6}) = 4 \cdot 2\sqrt{3 \cdot 6} = 8\sqrt{18}$
$= 8\sqrt{9 \cdot 2} = 8\sqrt{9}\sqrt{2}$
$= 8(3)\sqrt{2}$
$= 24\sqrt{2}$

6. $(2\sqrt{5})(2\sqrt{5}) = 2 \cdot 2\sqrt{5 \cdot 5}$
$= 4\sqrt{25}$
$= 4(5)$
$= 20$

☞ For non-negative value of x: $\sqrt{x} \cdot \sqrt{x} = x$. For example, $\sqrt{1011} \cdot \sqrt{1011} = 1011$.

7. $\dfrac{10}{2}\sqrt{\dfrac{20}{4}} = 5\sqrt{5}$

8. $(3 + \sqrt{2})(2 + \sqrt{10}) =$
$3(2) + 3\sqrt{10} + 2\sqrt{2} + \sqrt{2 \cdot 10}$
$6 + 3\sqrt{10} + 2\sqrt{2} + \sqrt{20}$
$6 + 3\sqrt{10} + 2\sqrt{2} + \sqrt{4 \cdot 5}$
$6 + 3\sqrt{10} + 2\sqrt{2} + \sqrt{4}\sqrt{5}$
$6 + 3\sqrt{10} + 2\sqrt{2} + 2\sqrt{5}$
$6 + 2\sqrt{2} + 2\sqrt{5} + 3\sqrt{10}$

9. $(2 + \sqrt{7})(2 + \sqrt{7})$
$= 2(2) + 2\sqrt{7} + 2\sqrt{7} + \sqrt{7}\sqrt{7}$
$= 4 \quad + 4\sqrt{7} \qquad + 7$
$= 11 + 4\sqrt{7}$

10. Note: $x^{\text{Positive and Even}}$ is a perfect square.
$\sqrt{a^4 b^4 b^1 c^2 c^1}$
$= \sqrt{a^4}\sqrt{b^4}\sqrt{b}\sqrt{c^2}\sqrt{c}$
$= (a^2)(b^2)(\sqrt{b})(c)(\sqrt{c}) = a^2 b^2 c\sqrt{bc}$

Practice Problems 4.1

1. Multiply and simplify $2\sqrt{18} \cdot 6\sqrt{2}$.

   A) 72   B) 48   C) $12\sqrt{6}$

   D) $8\sqrt{2}$ E) 36

2. Find $(3\sqrt{3})^3$.

   A) $27\sqrt{3}$ B) $81\sqrt{3}$ C) 81   D) $9\sqrt{3}$ E) 243

3. Combine $\sqrt{80} + \sqrt{45} - \sqrt{20}$.

   A) $9\sqrt{5}$ B) $5\sqrt{5}$ C) $-\sqrt{5}$ D) $3\sqrt{5}$
   E) $-2\sqrt{5}$

4. $(3 + \sqrt{5})(3 - \sqrt{5}) =$

   A) 4   B) $4 - 6\sqrt{5}$ C) -16
   D) -1 E) 6

5. Simplify $\sqrt{20a^7b^3c^8}$.

   A) $10a^3bc^4\sqrt{ab}$ B) $2abc\sqrt{abc}$
   C) $5a^3bc^4\sqrt{2ab}$ D) $2abc^4\sqrt{5a^3b}$
   E) $2a^3bc^4\sqrt{5ab}$

6. $\dfrac{\sqrt{32b^3}}{\sqrt{8b}} =$

   A) $2\sqrt{b}$ B) $\sqrt{2b}$ C) $2b$
   D) $\sqrt{2b^2}$ E) $b\sqrt{2b}$

7. $(\tfrac{1}{2}\sqrt{2})(\sqrt{6} + \tfrac{1}{2}\sqrt{2}) =$

   A) $\sqrt{3} + \tfrac{1}{2}$ B) $\tfrac{1}{2}\sqrt{3}$ C) $\sqrt{6} + 1$
   D) $\sqrt{6} + \tfrac{1}{2}$ E) $\sqrt{6} + 2$

8. Combine $\tfrac{1}{2}\sqrt{180} + \tfrac{1}{3}\sqrt{45} - \tfrac{2}{5}\sqrt{20}$.

   A) $3\sqrt{10} + \sqrt{15} + 2\sqrt{2}$ B) $\tfrac{16}{5}\sqrt{5}$ C) $\sqrt{97}$
   D) $\tfrac{24}{5}\sqrt{5}$ E) 7

Solutions on page 47

4.2 Rationalizing the Denominator

To **rationalize** a fraction with a radical in the denominator is to write an equivalent fraction with the denominator being rational. This is done by multiplying by the identity element expressed in a useful form.

Form of Denominator to be Rationalized	Form of Identity Element to be Used for Rationalizing
$\sqrt{x}$	$\sqrt{x}$
$a\sqrt{x}$	$\sqrt{x}$
$a \pm \sqrt{x}$	$a \mp \sqrt{x}$
$a \pm b\sqrt{x}$	$a \mp b\sqrt{x}$
$\sqrt{x} \pm \sqrt{y}$	$\sqrt{x} \mp \sqrt{y}$

The expressions $a + b\sqrt{x}$ and $a - b\sqrt{x}$ or $a\sqrt{x} + b\sqrt{y}$ and $a\sqrt{x} - b\sqrt{y}$ are called conjugate pairs.

Examples:

1. Rationalize $\dfrac{1}{\sqrt{7}}$.

2. Rationalize $\dfrac{3}{4-\sqrt{5}}$.

3. Find the reciprocal of $2 + 2\sqrt{3}$.

4. Divide $\sqrt{5}$ by $\sqrt{7}$.

Solutions:

1. The denominator is of the form $\sqrt{x}$. Use $\sqrt{x}$ to rationalize.

$$\frac{1}{\sqrt{7}}\left(\frac{\sqrt{7}}{\sqrt{7}}\right)=\frac{\sqrt{7}}{7}$$

2. The denominator is of the form $a - \sqrt{x}$. Use $a + \sqrt{x}$ to rationalize.

$$\frac{3}{4-\sqrt{5}}\left(\frac{4+\sqrt{5}}{4+\sqrt{5}}\right)=\frac{3(4)+3\sqrt{5}}{4(4)+4\sqrt{5}-4\sqrt{5}-\sqrt{5}\sqrt{5}}$$

$$=\frac{12+3\sqrt{5}}{16-5}=\frac{12+3\sqrt{5}}{11}$$

3. The reciprocal is 1 over the number.

$$\text{Reciprocal} = \frac{1}{2+2\sqrt{3}}$$

Rationalizing:

$$\frac{1}{2+2\sqrt{3}}\left(\frac{2-2\sqrt{3}}{2-2\sqrt{3}}\right)=\frac{2-2\sqrt{3}}{4-4(3)}$$

$$=\frac{2-2\sqrt{3}}{4-12}=\frac{2(1-\sqrt{3})}{-8}=-\frac{(1-\sqrt{3})}{4}$$

4. $\dfrac{\sqrt{5}}{\sqrt{7}}\cdot\dfrac{(\sqrt{7})}{(\sqrt{7})}=\dfrac{\sqrt{35}}{7}$

Practice Problems 4.2

1. Rationalize $\dfrac{2}{\sqrt{11}}$.

A) $\dfrac{11\sqrt{11}}{2}$ B) $\dfrac{2\sqrt{11}}{11}$ C) $\dfrac{2+\sqrt{11}}{2}$

D) $\dfrac{2-\sqrt{11}}{2}$ E) $\dfrac{4}{11}$

2. Rationalize $\dfrac{\sqrt{3}}{1+\sqrt{2}}$.

A) $\dfrac{\sqrt{3}-\sqrt{6}}{-1}$ B) $\sqrt{6}-\sqrt{3}$ C) -1

D) $1-\sqrt{2}$ E) $\sqrt{2}-1$

3. Find the reciprocal of $\dfrac{2\sqrt{3}}{3}$.

A) $\dfrac{\sqrt{3}}{6}$ B) 9 C) $\dfrac{3}{2}$

D) $\dfrac{\sqrt{3}}{2}$ E) $\dfrac{\sqrt{6}}{2}$

4. The expression $\dfrac{1}{1+\sqrt{3}}$ is equivalent to

A) 1 B) $\dfrac{-\sqrt{3}}{2}$ C) $\dfrac{1-\sqrt{3}}{-2}$

D) $1+\sqrt{3}$ E) $\dfrac{1+\sqrt{3}}{2}$

Solutions on page 48

Cumulative Review

1. Find the numerical value of

   $|-3| + |7| - |-7|.$

   A) -3 B) 3 C) 11 D) -11 E) 17

2. Which is an irrational number?

   A) $\sqrt{9}$ B) $\sqrt{2} \cdot \sqrt{8}$ C) $\frac{2}{3}$ D) 0
   C) $\sqrt{2}$

3. The sum of $3\sqrt{3}$ and $\sqrt{12}$ is

   A) $5\sqrt{3}$ B) $7\sqrt{3}$ C) $3\sqrt{15}$
   D) $\sqrt{21}$ E) 18

4. If $h(x) = x^2 + 1$ and $k(x) = 3x$, then $(h - k)(x) =$

   A) $-2x + 1$ B) $x^2 - 2$ C) $x^2 - 3x + 1$
   D) $-x^2 + 3x - 1$ E) $9x^2 + 1$

5. The reciprocal of $\dfrac{\sqrt{3}}{2}$ is

   A) $\dfrac{3}{4}$ B) $\dfrac{3}{2}$ C) $\dfrac{2\sqrt{2}}{3}$

   D) $\dfrac{2\sqrt{3}}{3}$ E) $\dfrac{3\sqrt{2}}{2}$

6. If $f(x) = 3x + 2$, then $f^{-1}(x) =$

   A) $\dfrac{3}{x} + 2$ B) $2x + 3$ C) $x - 2$

   D) $\dfrac{x - 2}{3}$ E) $\dfrac{x}{3} - 2$

7. Which of the following graphs represents a function?

   A)   B)   C)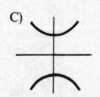

D)    E)

8. What type of symmetry does the equation $y = 2x^7 + 3x^5 - 2x$ have?

   A) x-axis B) y-axis C) Origin
   D) $x = y$ E) None

9. Given that $f(x) = x^2 - 1$ and $g(x) = x + 1$ find $(f \cdot g)(x)$.

   A) $x^2 + 2x$ B) x^2 C) $x^3 + x^2 - x - 1$
   D) $x^3 - 1$ E) $x^3 - x^2 + x - 1$

10. Which graph represents the equation $y = |x - 1| + 3$?

    A) 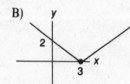   B)   C)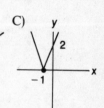

    D) $(-1,3)$ E) 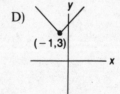 $(1,3)$

11. Rationalize $\dfrac{\sqrt{3}}{\sqrt{15} - 3}$.

    A) $\dfrac{\sqrt{15} + 3}{6}$ B) $\dfrac{\sqrt{15} + 3}{2}$ C) $\dfrac{\sqrt{5} + \sqrt{3}}{2}$

    D) $\dfrac{\sqrt{45} + \sqrt{9}}{6}$ E) $\dfrac{4\sqrt{3}}{3}$

12. If $f(x) = x^2 - 3$, find

$$\frac{f(x + h) - f(x)}{h}, h \neq 0.$$

A) $2x + h$ B) $2x$ C) $2xh + h^2$
D) $2x^2$ E) $\dfrac{2x^2 + 2xh + h^2 - 6}{h}$

13. Simplify $\sqrt{49a^3b} - \sqrt{a^3b} + 2a\sqrt{ab}$.

A) $8a^2\sqrt{ab}$ B) $8a\sqrt{ab}$ C) $8a^3\sqrt{b}$
D) $8a^3b$ E) $11a\sqrt{ab}$

14. The graphic representation of $|x + 1| < 3$ is:

A) ———o———o———   B) ◄——o———o——►

C) ———●———●———   D) ———o———o———

C) ◄——o———o——►

15. If $f(x) = 3x^2 - 2$ and $g(x) = x + \sqrt{x}$, find $f(g(4))$.

A) $46 + \sqrt{46}$ B) 10 C) $22 + \sqrt{22}$
D) 106 E) 46

16. Simplify $(2 - \sqrt{3})(2 + \sqrt{3})$.

A) $4 - \sqrt{3}$ B) $4 + \sqrt{3}$ C) $1 + 4\sqrt{3}$
D) 7 E) 1

17. If $f = \{(0, 1), (2, 7), (-1, 9), (-3, -3)\}$, find f^{-1}.

A) $\{(0, -1), (2, -7), (-1, -9), (-3, 3)\}$
B) $\{(0, 1), (-2, 7), (1, 9), (3, -3)\}$
C) $\{(1, 0), (7, 2), (9, -1), (-3, -3)\}$
D) $\{(-1, 0), (-7, 2), (-9, -1), (3, -3)\}$
E) $\{(0, 1), (2, 9), (-1, 8), (-3, 0)\}$

18. Positive integers are synonymous with

A) whole numbers   B) even numbers
C) fractions   D) natural numbers
E) rational numbers

19. An odd function is synonymous with

A) x-axis symmetry   B) y-axis symmetry
C) origin symmetry   D) $x = y$ symmetry
E) a constant function

20. The number $\sqrt[3]{-\frac{1}{8}}$ is

A) real and rational
B) real and irrational
C) imaginary   D) complex
E) an improper fraction

Solutions on page 48

Diagnostic Chart for Cumulative Review Test

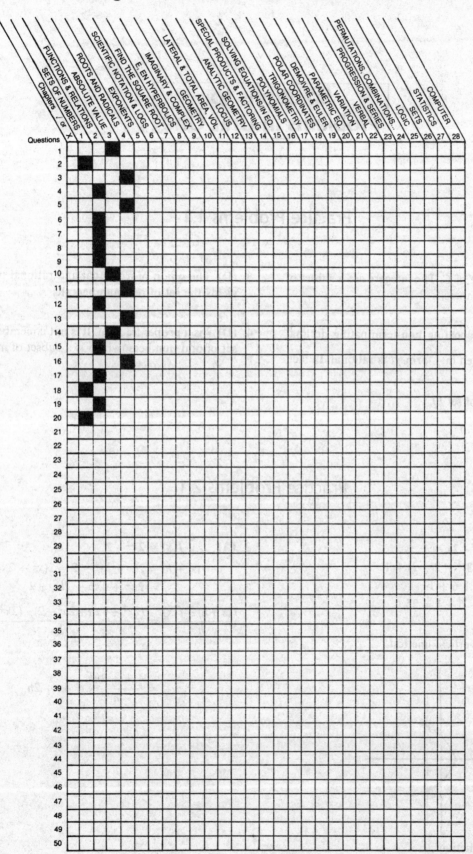

Solutions to Problems in Unit One

Practice Problems 1.1

1. **(D)** J is a subset of C. The integers are a subset of the complex numbers.

2. **(D)** $.166\overline{6}$ is a repeating non-terminating decimal and can be written in $\dfrac{a}{b}$ form. It is rational. In fact, $.166\overline{6} = \frac{1}{6}$.
   It is a member of set Q.

3. **(D)** The union of irrational and rational numbers yields the set of real numbers (R).

4. **(D)** $\frac{5}{7}$ is a proper fraction. It is not a member of the set of odd numbers (which is a subset of integers).

Practice Problems 2.1

1. **(E)**
$$f(x) = 2x^2 - 2x + 1$$
$$= 2(-3)^2 - 2(-3) + 1$$
$$= 2(9) + 6 + 1$$
$$= 18 + 6 + 1 = 25$$

2. **(E)** Using the vertical line test,

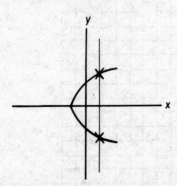

"E" is not a function

3. **(A)**
$$f(x) = 2x^2 - 2$$
$$f(x + h) = 2(x + h)^2 - 2 = 2(x^2 + 2xh + h^2) - 2$$
$$= 2x^2 + 4xh + 2h^2 - 2$$
$$\frac{f(x+h) - f(x)}{h} = \frac{(2x^2 + 4xh + 2h^2 - 2) - (2x^2 - 2)}{h}$$

$$= \frac{4xh + 2h^2}{h} = 4x + 2h$$

44

4. **(B)** A simple sketch is very useful for determining domain and range.

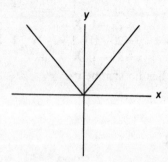

Domain: All real numbers (all x are good values)
Range: $y \geqslant 0$ (only values of y greater than or equal to zero are possible)

5. **(A)** $f(x) = -x^2 - 3x - 2$
$\qquad = -(-3)^2 - 3(-3) - 2$
$\qquad = -9 + 9 - 2 = -2$

6. **(A)** A simple sketch is very useful. This is a circle with center $(0,0)$ and radius $= 3$.

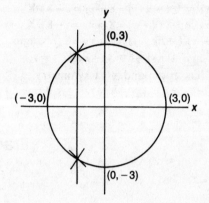

Using the vertical line test this *is not* a function.

Practice Problems 2.2

1. **(B)** $y = |x| + 3$

   Test Points

$\quad (x, y) \qquad\qquad y = |x| + 3$
$\quad (-x, y) \qquad\quad y = |-x| + 3 = |x| + 3 \;\checkmark$ y-axis

$\quad (x, -y) \qquad\; -y = |x| + 3$ **X**
$\quad (-x, -y) \quad -y = |-x| + 3 = |x| + 3$ **X**
$\quad (y, x) \qquad\qquad x = |y| + 3$ **X**

Has y-axis symmetry only.

2. **(B)** $y = 7$

   Test Points

$\quad (x, y) \qquad\qquad y = 7$
$\quad (-x, y) \qquad\quad y = 7$ y-axis $\checkmark$
$\quad (x, -y) \qquad\; -y = 7$ **X**
$\quad (-x, -y) \quad -y = 7$ **X**
$\quad (y, x) \qquad\qquad x = 7$ **X**

Has y-axis symmetry only.

3. **(D)** $x^2 - y^2 = 4$

Test Points

$\quad (x, y) \qquad\quad x^2 - y^2 = 4$
$\quad (-x, y) \qquad (-x)^2 - y^2 = 4 \Rightarrow x^2 - y^2 = 4 \;\checkmark$ y-axis

$\quad (x, -y) \qquad x^2 - (-y)^2 = 4 \Rightarrow x^2 - y^2 = 4 \;\checkmark$ x-axis
$\quad (-x, -y) \quad (-x)^2 - (-y)^2 = 4 \Rightarrow x^2 - y^2 = 4 \;\checkmark$ Origin
$\quad (y, x) \qquad\quad (y)^2 - (x)^2 = 4 \Rightarrow y^2 - x^2 = 4$ **X**

Has x, y and origin symmetry.

4. **(B)** $y = \sin x$

Test Points

$\quad (x, y) \qquad\qquad y = \sin x$
$\quad (-x, y) \qquad\quad y = \sin(-x) = \sin x$ **X**
$\quad (-x, -y) \quad -y = \sin(-x) = -\sin x \;\checkmark$ Origin

Origin symmetry is odd.

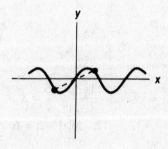

$y = \sin x$ is a function
$\therefore y = \sin x$ is an odd function

5. **(C)** $xy = 8$
   Test Points

(x, y)	$xy = 8$
$(-x, y)$	$(-x)(y) = 8 \Rightarrow xy = -8$ **X**
$(x, -y)$	$(x)(-y) = 8 \Rightarrow xy = -8$ **X**
$(-x, -y)$	$(-x)(-y) = 8 \Rightarrow xy = 8$ $\sqrt{}$ origin
(y, x)	$(y)(x) = 8 \Rightarrow xy = 8$ $\sqrt{}$ $x = y$

   $xy = 8$ has origin and $x = y$ symmetry.

6. **(A)** $y = 2x - 1$
   Test Points

(x, y)	$y = 2x - 1$
$(-x, y)$	$y = 2(-x) - 1 \Rightarrow -2x - 1$ **X**
$(x, -y)$	$-y = 2x - 1$ **X**
$(-x, -y)$	$-y = 2(-x) - 1 = -2x - 1$ **X**
(y, x)	$x = 2y - 1$ **X**

   $y = 2x - 1$ has no symmetry.

Practice Problems 2.3

1. **(B)**
$$(f + g)x = f(x) + g(x) = (3x - 3) + (2x + 1)$$
$$= 5x - 2$$

2. **(D)**
$$(f \cdot g)x = f(x) \cdot g(x) = (3x - 3)(2x + 1)$$
$$= 6x^2 - 3x - 3$$

3. **(E)**
$$(f \circ g)x = f(g(x)) = f(2x + 1) = 3(2x + 1) - 3$$
$$= 6x$$

4. **(C)**
$$(g \circ f)x = g(f(x)) = g(3x - 3) = 2(3x - 3) + 1$$
$$= 6x - 6 + 1 = 6x - 5$$

5. **(D)** $\quad f(x) = \dfrac{x - 1}{x + 1}$

   Step 1 $\quad y = \dfrac{x - 1}{x + 1}$

Step 2 $\quad x = \dfrac{y - 1}{y + 1}$

Step 3 $\quad xy + x = y - 1$

$$xy - y = -x - 1$$

$$y(x - 1) = -(x + 1)$$

$$y = \frac{-(x + 1)}{(x - 1)}$$

Step 4 $\quad f^{-1}(x) = \dfrac{-(x + 1)}{(x - 1)}$

6. **(C)** To be the graph of the inverse the relation should represent the reflection of $f(x)$ about the line $y = x$.

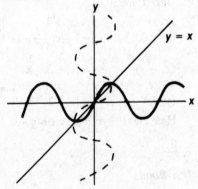

Practice Problems 3.1

1. **(D)** $\left|\dfrac{-6}{-2}\right| = |3| = 3, \quad \dfrac{|-6|}{|-2|} = \dfrac{6}{2} = 3$

   $|6(-2)| = |-12| = 12, \quad |6| \cdot |-2| = 6 \cdot 2 = 12$

   $\left|\dfrac{-6}{-2}\right| + \dfrac{|-6|}{|-2|} + |6(-2)| + |6| \cdot |-2|$

   $\qquad\qquad = 3 + 3 + 12 + 12 = 30$

2. **(B)** $|x| \leqslant 3 \Rightarrow x \leqslant 3$ and $-x \leqslant 3 \Rightarrow x \geqslant -3$

   $$-3 \leqslant x \leqslant 3$$

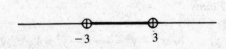

3. (A) The critical value of $y = |2x - 6| + 5$ occurs when $2x - 6 = 0$ . . $x = 3$

x	y
0	11
1	9
2	7
3	5
4	7
5	9
6	11

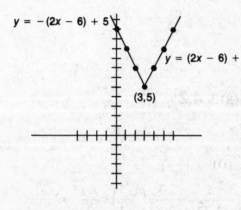

The answer is $y = -(2x - 6) + 5$

4. (E) The critical value is at $x + 3 = 0$
$\therefore x = -3$
$y = |x + 3| - 1$

x	y
0	2
−1	1
−2	0
−3	−1
−4	0
−5	1
−6	2

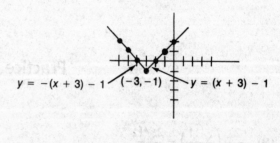

Practice Problems 4.1

1. (A) $2\sqrt{18} \cdot 6\sqrt{2} = 2 \cdot 6\sqrt{18 \cdot 2}$
$$= 12\sqrt{36} = 12(6) = 72$$

2. (B) $(3\sqrt{3})^3 = (3\sqrt{3})(3\sqrt{3})(3\sqrt{3})$
$$= 27\sqrt{3}\sqrt{3}\sqrt{3}$$
$$= 27 \cdot 3 \cdot \sqrt{3}$$
$$= 81\sqrt{3}$$

3. (B) $\sqrt{80} + \sqrt{45} - \sqrt{20}$
$$= \sqrt{16 \cdot 5} + \sqrt{9 \cdot 5} - \sqrt{4 \cdot 5}$$
$$= \sqrt{16}\sqrt{5} + \sqrt{9}\sqrt{5} - \sqrt{4}\sqrt{5}$$
$$= 4\sqrt{5} + 3\sqrt{5} - 2\sqrt{5}$$
$$= 5\sqrt{5}$$

4. (A) $(3 + \sqrt{5})(3 - \sqrt{5})$
$$= 3(3) - 3\sqrt{5} + 3\sqrt{5} - \sqrt{5}\sqrt{5}$$
$$= 9 \qquad\qquad -5$$
$$= 4$$

5. (E) $\sqrt{20a^7b^3c^8}$
$$= \sqrt{4 \cdot 5a^6a^1b^2b^1c^8}$$
$$= \sqrt{4}\sqrt{5}\sqrt{a^6}\sqrt{a}\sqrt{b^2}\sqrt{b}\sqrt{c^8}$$
$$= 2\sqrt{5}\ a^3\sqrt{a}\ b\ \sqrt{b}\ c^4$$
$$= 2a^3bc^4\sqrt{5ab}$$

6. (C) $\dfrac{\sqrt{32b^3}}{\sqrt{8b}} = \sqrt{\dfrac{32b^3}{8b}} = \sqrt{4b^2}$
$$= \sqrt{4}\sqrt{b^2} = 2b$$

7. **(A)** $(\frac{1}{2}\sqrt{2})(\sqrt{6} + \frac{1}{2}\sqrt{2})$

$$= \frac{1}{2}\sqrt{2}\sqrt{6} + \frac{1}{2}\sqrt{2}\cdot\frac{1}{2}\sqrt{2}$$

$$= \frac{1}{2}\sqrt{12} + \frac{1}{4}\sqrt{2}\sqrt{2}$$

$$= \frac{1}{2}\sqrt{4\cdot 3} + \frac{1}{4}(2)$$

$$= \frac{1}{2}\sqrt{4}\sqrt{3} + \frac{2}{4}$$

$$= \frac{1}{2}(2)\sqrt{3} + \frac{2}{4} = \sqrt{3} + \frac{1}{2}$$

8. **(B)**

$$\frac{1}{2}\sqrt{180} + \frac{1}{3}\sqrt{45} - \frac{2}{5}\sqrt{20}$$

$$= \frac{1}{2}\sqrt{36\cdot 5} + \frac{1}{3}\sqrt{9\cdot 5} - \frac{2}{5}\sqrt{4\cdot 5}$$

$$= \frac{1}{2}\sqrt{36}\sqrt{5} + \frac{1}{3}\sqrt{9}\sqrt{5} - \frac{2}{5}\sqrt{4}\sqrt{5}$$

$$= \frac{1}{2}(6)\sqrt{5} + \frac{1}{3}(3)\sqrt{5} - \frac{2}{5}(2)\sqrt{5}$$

$$= 3\sqrt{5} + \sqrt{5} - \frac{4}{5}\sqrt{5}$$

$$= 4\sqrt{5} - \frac{4}{5}\sqrt{5}$$

$$= \frac{16}{5}\sqrt{5}$$

Practice Problems 4.2

1. **(B)** $\dfrac{2}{\sqrt{11}}\left(\dfrac{\sqrt{11}}{\sqrt{11}}\right) = \dfrac{2\sqrt{11}}{11}$

2. **(A)**

$$\frac{\sqrt{3}}{1+\sqrt{2}}\left(\frac{1-\sqrt{2}}{1-\sqrt{2}}\right) = \frac{\sqrt{3} - \sqrt{3}\sqrt{2}}{1(1) - \sqrt{2} + \sqrt{2} - \sqrt{2}\sqrt{2}}$$

$$= \frac{\sqrt{3} - \sqrt{6}}{1-2} = \frac{\sqrt{3} - \sqrt{6}}{-1}$$

3. **(D)**

$$\frac{1}{\dfrac{2\sqrt{3}}{3}} = \frac{3}{2\sqrt{3}}\left(\frac{\sqrt{3}}{\sqrt{3}}\right) = \frac{3\sqrt{3}}{2(3)} = \frac{\sqrt{3}}{2}$$

4. **(C)**

$$\frac{1}{1+\sqrt{3}}\left(\frac{1-\sqrt{3}}{1-\sqrt{3}}\right) = \frac{1-\sqrt{3}}{1-\sqrt{3}+\sqrt{3}-3}$$

$$= \frac{1-\sqrt{3}}{1-3} = \frac{1-\sqrt{3}}{-2} \text{ or } \frac{\sqrt{3}-1}{2}$$

Cumulative Review Test

1. **(B)** $|-3| = 3 \quad |7| = 7 \quad |-7| = 7$

$|-3| + |7| - |-7|$

$= 3 + 7 - 7 = 3$

2. **(E)** An irrational number cannot be written in $\dfrac{a}{b}$

form.

ans. $\sqrt{2}$

3. **(A)** $3\sqrt{3} + \sqrt{12}$

$$= 3\sqrt{3} + \sqrt{4\cdot 3} = 3\sqrt{3} + \sqrt{4}\sqrt{3}$$

$$= 3\sqrt{3} + 2\sqrt{3}$$

$$= 5\sqrt{3}$$

4. **(C)**

$$(h - k)(x) = h(x) - k(x) = (x^2 + 1) - (3x)$$

$$= x^2 - 3x + 1$$

5. **(D)** $\dfrac{2}{\sqrt{3}}\dfrac{\sqrt{3}}{(\sqrt{3})} = \dfrac{2\sqrt{3}}{3}$

6. **(D)**

$$f(x) = 3x + 2$$
$$y = 3x + 2$$
$$x = 3y + 2 \quad \text{interchange } x \text{ and } y$$
$$y = \frac{x-2}{3} \quad \text{solve for } y$$
$$f^{-1}(x) = \frac{x-2}{3} \quad \text{replace } y \text{ with } f^{-1}(x)$$

7. **(E)** Using the vertical line test

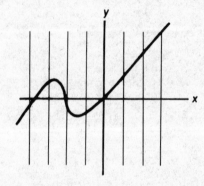

8. **(C)** $y = 2x^7 + 3x^5 - 2x$

Test Points

$$(x, y) \Rightarrow \quad y = 2x^7 + 3x^5 - 2x$$
$$(-x, y) \Rightarrow \quad y = -2x^7 - 3x^5 + 2x$$
$$(x, -y) \Rightarrow -y = 2x^7 + 3x^5 - 2x$$
$$(-x, -y) \Rightarrow -y = -2x^7 - 3x^5 + 2x$$
$$\therefore y = 2x^7 + 3x^5 - 2x \ \sqrt{}$$

There is origin symmetry.

9. **(C)**

$$(f \cdot g)(x) = f(x) \cdot g(x) = (x^2 - 1)(x + 1)$$
$$= x^3 + x^2 - x - 1$$

10. **(E)** The critical value occurs when $x - 1 = 0$
$\therefore x = 1.$

When $x = 1 \quad y = 3$

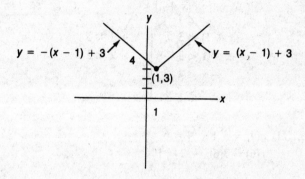

11. **(C)**

$$\frac{\sqrt{3}}{\sqrt{15} - 3} \cdot \frac{\sqrt{15} + 3}{(\sqrt{15} + 3)} = \frac{\sqrt{3}(\sqrt{15} + 3)}{15 - 9}$$

Note:
$$\sqrt{45} = \sqrt{9}\sqrt{5}$$
$$= 3\sqrt{5}$$

$$= \frac{\sqrt{45} + 3\sqrt{3}}{6}$$

$$= \frac{3\sqrt{5} + 3\sqrt{3}}{6} = \frac{3(\sqrt{5} + \sqrt{3})}{6}$$

$$= \frac{\sqrt{5} + \sqrt{3}}{2}$$

12. **(A)**

$$f(x) = x^2 - 3$$
$$f(x + h) = (x + h)^2 - 3$$
$$= x^2 + 2xh + h^2 - 3$$
$$\frac{f(x + h) - f(x)}{h} = \frac{(x^2 + 2xh + h^2 - 3) - (x^2 - 3)}{h}$$
$$= \frac{2xh + h^2}{h} = 2x + h$$

13. **(B)**

$$\sqrt{49a^3b} = \sqrt{49a^2a^1b^1} = \sqrt{49}\sqrt{a^2}\sqrt{a}\sqrt{b}$$
$$= 7a\sqrt{a}\sqrt{b}$$
$$= 7a\sqrt{ab}$$
$$\sqrt{a^3b} = \sqrt{a^2a^1b^1} = \sqrt{a^2}\sqrt{a}\sqrt{b}$$
$$= a\sqrt{ab}$$

$$\sqrt{49a^3b} - \sqrt{a^3b} + 2a\sqrt{ab}$$
$$= 7a\sqrt{ab} - a\sqrt{ab} + 2a\sqrt{ab}$$
$$= 8a\sqrt{ab}$$

14. **(A)**

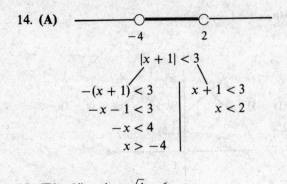

$$|x + 1| < 3$$

$$-(x + 1) < 3 \qquad x + 1 < 3$$
$$-x - 1 < 3 \qquad x < 2$$
$$-x < 4$$
$$x > -4$$

15. **(D)** $g(4) = 4 + \sqrt{4} = 6$

$f(6) = 3(6)^2 - 2 = 106$

16. **(E)**

$$(2 - \sqrt{3})(2 + \sqrt{3}) = 2(2) + 2\sqrt{3} - 2\sqrt{3} - 3$$
$$= 1$$

17. **(C)** The inverse involves interchange of the x and y coordinates:

$$f = \{(0, 1), (2, 7), (-1, 9), (-3, -3)\}$$
$$f^{-1} = \{(1, 0), (7, 2), (9, -1), (-3, -3)\}$$

18. **(D)** Positive integers are also called natural numbers.

19. **(C)** An odd function has origin symmetry.

20. **(A)** $\sqrt[3]{-\frac{1}{8}} = \dfrac{\sqrt[3]{-1}}{\sqrt[3]{8}} = \dfrac{-1}{2}$

This is real and rational.

UNIT TWO

5. Exponents

5.1 Essentials

$x^0 = 1 \; (x \neq 0)$	$x^a \cdot x^b = x^{a+b}$	$(xy)^a = x^a y^a$
$x^1 = x$	$\dfrac{x^a}{x^b} = x^{a-b} \quad (x \neq 0)$	$\left(\dfrac{x}{y}\right)^a = \dfrac{x^a}{y^a}$
$x^{-a} = \dfrac{1}{x^a} \quad (x \neq 0)$ $x^{1/a} = \sqrt[a]{x}$ $x^{b/a} = (\sqrt[a]{x})^b = \sqrt[a]{x^b}$	$(x^a)^b = x^{ab}$	If $x^a = x^b$, then $a = b \quad (x \neq 0, 1)$.

 Common error: $3^2 \cdot 3^3 = 9^{2+3} = 9^5$. Should be: $3^2 \cdot 3^3 = 3^{2+3} = 3^5$. When multiplying exponential terms the base must be the same and remains the same; add the exponents.

Examples:

1. Evaluate $x^0 + x^{1/2} + x^{-2}$ when $x = 9$.

2. Simplify $\dfrac{3^7 \cdot 3^4}{3^5}$.

3. Simplify $\dfrac{n^{a+2} \cdot n^{2a-1}}{n^{2a+1}}$.

Solutions:

1. Substitute $x = 9$
$$9^0 + 9^{1/2} + 9^{-2}$$
$$= 1 + \sqrt{9} + \frac{1}{9^2}$$
$$= 1 + 3 + \frac{1}{81}$$
$$= 4\tfrac{1}{81}$$

2. $\dfrac{3^{7+4}}{3^5} = \dfrac{3^{11}}{3^5} = 3^{11-5} = 3^6$

3. $\dfrac{n^{(a+2)+(2a-1)}}{n^{2a+1}} = \dfrac{n^{3a+1}}{n^{2a+1}}$
$$= n^{(3a+1)-(2a+1)} = n^a$$

To **solve an equation which has the variable as an exponent**, express each side of the equation as a power of the same base, then set the exponents equal to each other.

Examples:

4. Solve $2^x = 32$.

5. Solve for x: $3^{x+1} = 243$.

6. Solve for y: $2^y = 4^{y+1}$.

7. Solve for r: $9^{r+3} = 27^{r-1}$.

8. Solve for x: $125^x = \dfrac{1}{25}$.

9. Simplify $(x^2 y^3)^5$.

10. Simplify $\dfrac{(x^2 y^7)^3}{x^4 y}$.

Solutions:

4. Write 32 as a power of 2:

$$2^x = 2^5$$

Since the bases are equal, equate the exponents:

$$x = 5$$

5. $3^{x+1} = 3^5$
$$x + 1 = 5$$
$$x = 4$$

6. Develop a common base:

$$2^y = (2^2)^{y+1}$$
$$2^y = 2^{2(y+1)} = 2^{2y+2}$$
$$y = 2y + 2$$
$$y = -2$$

7. Convert to a common base:
$$(3^2)^{r+3} = (3^3)^{r-1}$$
$$3^{2(r+3)} = 3^{3(r-1)}$$
$$2(r + 3) = 3(r - 1)$$
$$2r + 6 = 3r - 3$$
$$r = 9$$

8. Write both sides of the equation in base 5:

$$(5^3)^x = \frac{1}{5^2} = 5^{-2}$$
$$5^{3x} = 5^{-2}$$
$$3x = -2$$
$$x = \frac{-2}{3}$$

9. $(x^2 y^3)^5 = x^{10} y^{15}$

10. $\dfrac{(x^2 y^7)^3}{x^4 y} = \dfrac{x^6 y^{21}}{x^4 y^1} = x^2 y^{20}$

To solve an equation of the form: $x^{-2/3} = 4$ or $2x^{1/3} = 3$, first rearrange the equation with only the variable term, with coefficient 1, on the left side, then raise each side of the equation to the reciprocal (up-side-down) power of the variable. The objective is to create the term $x^1 = x$.

Examples:

11. Solve $x^{1/2} = 8$.

12. Solve for x: $x^{-2/3} = 4$.

13. Solve for x: $2x^{1/3} = 3$.

Solutions:

11. $x^{1/2} = 8$

Raise each side of the equation to the $\frac{2}{1}$ power:

$(x^{1/2})^{2/1} = 8^{2/1}$

$x^1 = 8^2 \implies x = 64$

12. $(x^{-2/3})^{-3/2} = (4)^{-3/2}$

$x^1 = 4^{-3/2} = \dfrac{1}{(\sqrt{4})^3} \implies x = \dfrac{1}{8}$

13. $x^{1/3} = \dfrac{3}{2}$

$(x^{1/3})^3 = \left(\dfrac{3}{2}\right)^3$

$x^1 = \dfrac{3^3}{2^3} \implies x = \dfrac{27}{8}$

Practice Problems 5.1

1. The value of $2^0 + 4^{1/2} + (\frac{1}{2})^{-1}$ is

A) $1\frac{1}{2}$ B) $3\frac{1}{2}$ C) 4 D) 5 E) 8

2. Simplify $\dfrac{a^{x+1} \cdot a^{2x-1}}{a^{-x+1}}$.

A) a^{4x-1} B) a^{2x+1} C) a^{-x+1}
D) a^{2x-1} E) a^{4x+1}

3. Solve for m: $4^{2m} = 2^{m+1}$.

A) 4 B) 2 C) 1 D) $\frac{1}{3}$ E) $\frac{1}{15}$

4. Solve for v: $\frac{1}{16} = 2^{v-1}$.

A) -3 B) -1 C) 0
D) 3 E) 17

5. Solve for x: $x^{-2} = 16$.

A) 18 B) -8 C) 4
D) -4 E) $\pm\frac{1}{4}$

6. Simplify $\dfrac{2^3 \cdot 2^4}{(2^3)^2}$.

A) $\frac{1}{2}$ B) 2 C) 4 D) $2^{7/6}$ E) 2^6

7. Simplify $(5^{2a})(5)(5^{a+1})$.

A) 5^{3a+1} B) 5^{3a+2} C) 15^{3a+1}
D) 25^{3a+1} E) 125^{3a+2}

8. Evaluate $3x^0 + (3x)^0 + x^{-1} + 2x$
when $x = \frac{1}{2}$.

A) -1 B) $1\frac{1}{2}$ C) $6\frac{1}{2}$ D) 7 E) 10

Solutions on page 68

6. Scientific Notation and Logarithms

6.1 Scientific Notation

Very large or very small numbers are often expressed in the form:

$$N \times 10^a, \text{ where } 1 \leqslant N < 10 \text{ and } a \text{ is an integer.}$$

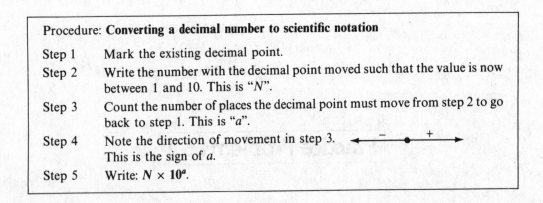

Procedure: **Converting a decimal number to scientific notation**

Step 1   Mark the existing decimal point.

Step 2   Write the number with the decimal point moved such that the value is now between 1 and 10. This is "N".

Step 3   Count the number of places the decimal point must move from step 2 to go back to step 1. This is "a".

Step 4   Note the direction of movement in step 3. This is the sign of a.

Step 5   Write: $N \times 10^a$.

Scientific notation and the laws of exponents may be used together in calculations.

Examples:

1. Express 7123 in scientific notation.

Solutions:

1. Step 1   7123.

   Step 2   7.123 $N = 7.123$

   Step 3   7.123 $a = 3$

   Step 4   to the right, a is "$+$"

   Step 5 $7.123 \times 10^{+3}$

2. Express .0048 in scientific notation.

2.   .0048   Step 1

       4.8   Step 2 $N = 4.8$

       004.8   Step 3 $a = 3$

       Step 4 a is "$-$"

       4.8×10^{-3} step 5

3. If 6,000,000 is written as 6×10^a, find the value of a.

3. 6,000,000.

6.000,000 $N = 6$

6.000,000 $a = 6$

a is "+"

$6,000,000 = 6 \times 10^{+6}$ $a = +6$

4. Using scientific notation, calculate $(250,000)(.000003)$.

4. $250,000 = 2.5 \times 10^5$

$.000003 = 3 \times 10^{-6}$

$(2.5 \times 10^5)(3 \times 10^{-6}) = (2.5)(3) \times 10^{5+(-6)}$

$= 7.5 \times 10^{-1} = .75$

Practice Problems 6.1

1. Express 278300 in scientific notation.

A) 2783×10^2 B) 2.783×10^5
C) $.2783 \times 10^{-4}$ D) 2.783×10^{-1}
E) 27.83×10^4

2. If .007932 is written as 7.932×10^a, the value of a is

A) 3   B) 2   C) 1   D) -2 E) -3

3. Calculate $\dfrac{12000}{.04}$ using scientific notation.

A) .00003   B) .003   C) .3
D) 300   E) 300,000

4. Write $\dfrac{1}{1000}$ in scientific notation.

A) 1×10^3 B) 1×10^2 C) 1×10^{-2}
D) 1×10^{-3} E) 1×10^{-4}

Solutions on page 69

6.2   Logarithms

$\log_b N = a \leftrightarrow b^a = N$	b = base (positive, real, $\neq 1$) N = number (positive, real) a = the logarithm, the exponent of b to get N

$\log_b 1 = 0$ $\quad$ $\log_b xy = \log_b x + \log_b y$

$\log_b b = 1$ $\quad$ $\log_b\left(\dfrac{x}{y}\right) = \log_b x - \log_b y$

$\begin{cases} \log_b b^x = x \\ b^{\log_b x} = x \end{cases}$ $\quad$ $\log_b x^y = y \log_b x$ *Note:* $\log_b \sqrt[y]{x} = \log_b x^{1/y}$

$\log_b \sqrt[y]{x} = \dfrac{\log_b x}{y} = \dfrac{1}{y} \log_b x$

$\log(N \times 10^a) = \log N + a$

☞ $\log_{10}$ is frequently written as log. "log" which is a logarithm to the base 10 is called a common logarithm.

☞ • $\log_e$ is frequently written as ln. "ln" which is a logarithm to the base e is called a natural logarithm.
• All other bases must be indicated.
• $e \approx 2.718281828459 \ldots$

Examples:

1. Express $\log_7 49 = 2$ in exponential form.

2. Express $8^{-2} = \frac{1}{64}$ in logarithmic form.

3. Find the logarithm of 10^2 in base 10.

4. Express $\log \dfrac{xy}{z}$ in expanded form.

5. Express $\log \dfrac{x^2 y^3 z}{\sqrt{m}}$ in expanded form.

6. If $\log 2.13 = t$, express $\log(2.13 \times 10^{-4})$ in terms of t.

7. If $\log 7.64 = m$, express the logarithm of 764 in terms of m.

8. If $\log_b 9 = 2$, find b.

9. If $\ln e^2 = a$, find a.

10. If $\log 2 = .3010$ and $\log 5 = .6990$, find $\log 40$.

Solutions:

1. $7^2 = 49$

2. $\log_8 \frac{1}{64} = -2$

3. $\log_{10} 10^2 = a$
   Change to exponential form: $10^a = 10^2$
   $a = 2$

4. $\log \dfrac{xy}{z} = \log x + \log y - \log z$

5. $\log \dfrac{x^2 y^3 z}{\sqrt{m}} = 2 \log x + 3 \log y + \log z$
   $- \dfrac{\log m}{2}$

6. $\log(N \times 10^a) = \log N + a$
   $\log(2.13 \times 10^{-4}) = \log 2.13 + (-4)$
   $\qquad\qquad\qquad = t - 4$

7. Express 764 in scientific notation:
   $\log 764 = \log(7.64 \times 10^2) = \log 7.64 + 2$
   $\qquad\qquad = m + 2 \qquad\qquad m$

8. $b^2 = 9 = 3^2$
   $b = 3$

   ☞ If $b^2 = 9$, $b = \pm 3$, however, in logarithms the base must be positive.

9. $\ln = \log_e$
   $\log_e e^2 = a$
   $e^a = e^2, \quad a = 2$

10. $\log 40 = \log(2^3 \cdot 5)$
    $\qquad\qquad = 3 \log 2 + \log 5$
    $\qquad\qquad = 3(.3010) + (.6990) = 1.6020$

11. If $\ln 7 = m$,
    express $\ln \frac{1}{7}$ in terms of m.

11. $\ln \frac{1}{7} = \ln 1 - \ln 7$
    $\ln 1 = 0 \quad (\log_b 1 = 0)$
    $\ln 7 = m$
    $\ln 1 - \ln 7 = 0 - m = -m$

Practice Problems 6.2

1. If $\log_m 125 = 3$, then $m =$

   A) 2 B) 3 C) 4 D) 5 E) 6

2. If $\log_{1/7} 49 = y$, then $y =$

   A) 7 B) 2 C) -2 D) $-\frac{1}{2}$
   E) 5

3. The expression $\ln \frac{a^2}{b^3}$ is equivalent to

   A) $2 \ln a - 3 \ln b$ B) $\frac{2}{3} \ln \frac{a}{b}$

   C) $\frac{(\ln a)^2}{(\ln b)^2}$ D) $\frac{2 \ln a}{3 \ln b}$

   E) $\frac{2}{3}(\ln a - \ln b)$

4. If $\log 3 = .4771$ and $\log 7 = .8451$, then $\log \sqrt{21} =$

   A) 1.1499 B) 2.6444 C) .6611
   D) .2016 E) .6350

5. $2 \log a + \dfrac{\log b}{2}$ is equivalent to

   A) $\log\left(2a + \dfrac{b}{2}\right)$ B) $\log 2a + \log \dfrac{b}{2}$

   C) $\log \dfrac{a^2 b}{2}$ D) $\log a^2 \sqrt{b}$

   E) $\log \sqrt{a^2 b}$

6. If $\log 9.17 = t$, express the log $.00917$ in terms of t.

   A) $3 - t$ B) $t - 3$ C) $3t$
   D) $2.17 + t$ E) $\dfrac{t}{3}$

7. Express $\log \sqrt{\dfrac{x^3}{zm}}$ in expanded form.

   A) $3 \log x - \log z + \log m$
   B) $\frac{1}{2}(3 \log x - \log z + \log m)$
   C) $\frac{1}{2}(3 \log x - \log z - \log m)$
   D) $\dfrac{3 \log x}{\log z + \log m}$
   E) $\dfrac{1}{2}\left(\dfrac{3 \log x}{\log z + \log m}\right)$

8. If $\log_8 x = \frac{1}{3}$, then $x =$

   A) $\frac{1}{8}$ B) 2 C) 24
   D) $\frac{8}{3}$ E) 3^8

Solutions on page 69

7. Finding the Square Root of a Number

7.1 Basics and Beyond

There are two basic techniques for finding the square root of a number. One is iterative, the other logarithmic.

Iterative Procedure:

Numerical Example: $\sqrt{8}$

Step 1   Pair off the digits under the square root sign in each direction from the decimal point. Each pair of digits will generate one digit of the square root.

$$\sqrt{\underset{\smile\;\;\smile}{08.0000}}$$

Step 2   Bring up the decimal point.

$$\sqrt{08.0000}$$

Step 3   Find the largest whole number square root that is less than or equal to the first pair of numbers. Place this number on the top and the square of the number under the first pair of digits.

$$\begin{array}{c} 2. \\ \sqrt{08.0000} \\ 4 \end{array}$$

Step 4   Subtract and bring down the next pair of digits.

$$\begin{array}{c} 2. \\ \sqrt{08.0000} \\ \underline{4} \\ 400 \end{array}$$

Step 5   Double the top number and place it outside to the left. Put a dash after this number (this becomes the divisor) and a dash on top over the next pair of digits (this becomes the multiplier).

$$\begin{array}{c} 2.\_ \\ \sqrt{08.0000} \\ \underline{4} \\ 4\_\overline{\smash{)}400} \end{array}$$

Step 6   Find the integer (0 through 9) which when placed on both dashes will provide the closest product, less than or equal to the required number. This is a trial and error process.

Try 7

$$\begin{array}{c} 2.7 \\ \sqrt{08.0000} \\ \underline{4} \\ 47\,\overline{\smash{)}400} \end{array}$$

329 ← [Too Small]

$7 \times 47 = 329$

60

Try 8

$$\begin{array}{r} 2.\underline{8} \\ \sqrt{08.0000} \\ 4 \end{array}$$

$$48\overline{)400}$$

$$384 \leftarrow \boxed{\text{OK}}$$

$$8 \times 48 = 384$$

Try 9

$$\begin{array}{r} 2.\underline{9} \\ \sqrt{08.0000} \\ 4 \end{array}$$

$$49\overline{)400}$$

$$441 \leftarrow \boxed{\text{Too Large}}$$

$$9 \times 49 = 441$$

Step 7   Write the digit that replaces the two dashes and show the product of the multiplication.

Step 8   Go back to step 4.

$$\begin{array}{r} 2.\underline{82} \\ \sqrt{08.0000} \\ 4 \end{array}$$

$$48\overline{)400}$$

$$\begin{array}{r} 384 \\ 562\overline{)1600} \\ 1124 \\ \hline 476 \end{array}$$

Logarithmic Procedure:

Step 1A   Create an equation.

Step 1B   Create a log equation.

find the $\sqrt{8}$

$x = \sqrt{8}$

$\log x = \log\sqrt{8} = \log(8^{1/2})$

Step 2   Use the rules of logarithms.

$\log x = \dfrac{\log 8}{2} = \dfrac{1}{2}\log 8$

Step 3   From the log tables, find log of the number.

$\log 8 = .9031$

Step 4   Substitute into the equation.

$\log x = \dfrac{.9031}{2}$

Step 5   Solve the equation for log x.

$\log x = .45155$

Step 6   Use the table to find the antilog.

$x \approx 2.83$

Examples:

1. Find the $\sqrt{15}$ to the nearest tenth using the iterative procedure.

Solutions:

1. $\sqrt{15.0000}$

$$\begin{array}{r} 3.\underline{87} \\ \sqrt{15.0000} \\ 9 \end{array}$$

$$68\overline{)600}$$

$$\begin{array}{r} 544 \\ 767\overline{)5600} \\ 5369 \\ \hline 231 \end{array}$$

Try:
$68 \times 8 = 544$
$69 \times 9 = 621$ Too large

Try:
$767 \times 7 = 5369$
$768 \times 8 = 6144$ Too large

$\sqrt{15} = 3.9$ to the nearest tenth

2. Find the $\sqrt{15}$ to the nearest tenth using logarithms.
   Note: log 1.5 = .1761

              log 3.87 = .5877

2. $x = \sqrt{15}$

    $\log x = \log \sqrt{15} = \log(15^{1/2})$

            $= \frac{1}{2} \log 15$

            $= \frac{1}{2} \log(1.5 \times 10^1)$ $\log(1.5 \times 10^1)$

            $= \frac{1}{2}(1.1761)$ $= \log 1.5 + \log 10^1$

            $= .5881$ $= .1761 + 1$

    $x = \text{antilog } .5881$ $= 1.1761$

    $x \approx 3.87$

    $x = 3.9$ to the nearest tenth

Practice Problems 7.1

1. Using the iterative procedure, find the $\sqrt{3}$ to the nearest tenth.

   A) 1.6   B) 1.68   C) 1.7   D) 1.73   E) 1.732

2. Find $\sqrt{12}$ to the nearest tenth using the iterative procedure.

   A) 3   B) 3.4   C) 3.46   D) 3.5   E) 3.464

3. Using logarithms, find the $\sqrt{3}$ to the nearest integer.

   A) 1   B) 2   C) 1.7   D) 1.73   E) 1.732

4. Find the $\sqrt{12}$ to the nearest tenth using logarithms.

   A) 3   B) 3.4   C) 3.46   D) 3.5   E) 3.464

Solutions on page 70

8. e, ln, Hyperbolics

8.1 Fundamental Facts

e   an irrational number $\approx 2.718281828459\ldots$

ln   natural logarithm $= \log_e$

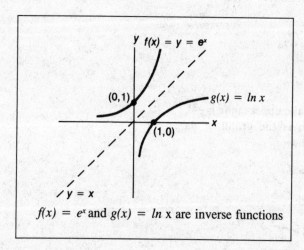

$f(x) = e^x$ and $g(x) = ln\ x$ are inverse functions

sinh x:	hyperbolic sin $x = \dfrac{e^x - e^{-x}}{2}$
cosh x:	hyperbolic cos $x = \dfrac{e^x + e^{-x}}{2}$
tanh x:	hyperbolic tan $x = \dfrac{e^x - e^{-x}}{e^x + e^{-x}}$

Examples:

1. Show that $\cosh^2 x - \sinh^2 x = 1$.

Solutions:

1. $\left(\dfrac{e^x + e^{-x}}{2}\right)^2 - \left(\dfrac{e^x - e^{-x}}{2}\right)^2 \overset{?}{=} 1$

$\left(\dfrac{(e^x + e^{-x}) \cdot (e^x + e^{-x})}{2} \cdot \dfrac{}{2}\right) - \left(\dfrac{(e^x - e^{-x}) \cdot (e^x - e^{-x})}{2} \cdot \dfrac{}{2}\right) \overset{?}{=} 1$

$\left(\dfrac{e^{2x} + 2e^0 + e^{-2x}}{4}\right) - \left(\dfrac{e^{2x} - 2e^0 + e^{-2x}}{4}\right) \overset{?}{=} 1$

63

$$\frac{2e^0 + 2e^0}{4} \overset{?}{=} 1$$

$$\frac{2+2}{4} \overset{?}{=} 1$$

$$1 = 1 \checkmark$$

Practice Problems 8.1

1. If ln 2 = .6931, find ln 16.

   A) $(.6931)^4$ B) $4(.6931)$ C) $2(.6931)$

   D) $(.6931)^2$ E) $\dfrac{1}{.6931}$

2. Express ln 9.974 = 2.3 in exponential form.

   A) $e^{9.974} = 2.3$ B) $e^{.026} = 2.3$

   C) $e^{2.3} = 9.974$ D) $\dfrac{1}{e^{2.3}} = 9.974$

   E) $e^{-2.3} = 9.974$

3. In the graphs shown below, if the upper curve is e^{2x}, then the lower curve, which is the graph of its inverse, is given by what equation?

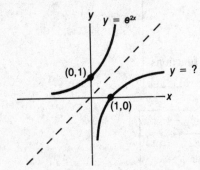

   A) $y = e^{-x}$ B) $y = e^{-2x}$ C) $y = \dfrac{1}{e^x}$

   D) $y = \ln 2$ E) $y = \ln x^{\frac{1}{2}}$

4. $2 \ln e^x =$

   A) x B) $2x$ C) x^2
   D) $x + 2$ E) $x - 2$

Solutions on page 70

Cumulative Review Test

1. Solve for x: $4^{3x} = \dfrac{1}{8^{x-3}}$.

   A) 1   B) $\dfrac{-3}{2}$ C) $\dfrac{-3}{5}$

   D) -3 E) -1

2. Express .03 in scientific notation.

   A) 3   B) $\dfrac{3}{100}$ C) 3×10^2

   D) 3×10^{-2} E) 3×10^{-3}

3. $\text{Log}(f^2 - g^2)$ equals

   A) $2 \log f - 2 \log g$

   B) $\log(f + g) + \log(f - g)$

   C) $\log 2f - \log 2g$ D) $\dfrac{2 \log f}{2 \log g}$

   E) $2 \log \dfrac{f}{g}$

4. The expression $\dfrac{2}{\sqrt{7} - 2}$ written as a fraction with a rational denominator is

   A) $2\sqrt{7} + 4$ B) $\dfrac{2\sqrt{7} + 4}{11}$ C) $\dfrac{4\sqrt{7}}{3}$

   D) $\dfrac{2\sqrt{7} + 4}{3}$ E) $\dfrac{2\sqrt{7} + 2}{3}$

5. Find $\sqrt{53}$ to the nearest tenth.

   A) 7.2   B) 7.28   C) 7.3

   D) 7.31   E) 7.4

6. Find the value of $2b^0 + b^{-1/2}$ when $b = 16$.

   A) -4 B) -2 C) -8

   D) $\dfrac{9}{4}$ E) $\dfrac{7}{4}$

7. The expression $r^n \div s^n$ is equivalent to

   A) $\left(\dfrac{r}{s}\right)^0$ B) $\dfrac{r}{s}$ C) $\dfrac{nr}{ns}$

   D) $\left(\dfrac{r}{s}\right)^n$ E) $\dfrac{r^n}{s^{-n}}$

8. The expression $\frac{1}{3}\log n - \frac{1}{2} \log l$ is equivalent to

   A) $\log(\sqrt[3]{n} - l^2)$ B) $\log(\sqrt[3]{n} - \sqrt{l})$

   C) $\log \dfrac{\sqrt[3]{n}}{\sqrt{l}}$ D) $\dfrac{1}{6} \log \dfrac{n}{l}$

   E) $\log \dfrac{n^3}{l^2}$

9. If $h(x) = \dfrac{x - 1}{x + 3}$, find $h(x - 3)$.

   A) $\dfrac{x - 4}{x}$ B) $\dfrac{x - 4}{x + 6}$ C) $\dfrac{x}{x - 4}$

   D) $\dfrac{x + 3}{x - 1}$ E) $2x + 2$

10. $.3 \times 10^{-4}$ is equivalent to

    A) .0003   B) 30×10^3
    C) $.03 \times 10^{-5}$ D) 3×10^{-3}
    E) 3×10^{-5}

11. The expression $-2(x)^3 + (-2x)^3 =$

    A) $4x^3$ B) $16x^3$ C) $16x^6$
    D) $-10x^3$ E) $-16x^3$

12. If $\log_8 x = \frac{2}{3}$, find the value of x.

    A) $\frac{16}{3}$ B) 4   C) $8\sqrt{2}$
    D) $\frac{4}{3}$ E) 16

13. Solve for x: $9^{3x} = 3^{x+5}$.

    A) -2 B) -1 C) 0
    D) 1   E) 2

14. Simplify $\dfrac{3^3 + 3^3 + 3^3}{3^4}$.

   A) $\frac{1}{3}$ B) 3^5 C) 3 D) 1 E) 9

15. Express $\dfrac{3.2 \times 10^2}{.32 \times 10^{-2}}$ in scientific notation.

   A) 10 B) 1×10^1 C) 1×10^4
   D) 1×10^5 E) 1×10^6

16. log 4.1 = .6128, log 6.40 = .8062 and log 6.41 = .8069. Find the $\sqrt{41}$ to the nearest tenth.

   A) 6.2 B) 6.3 C) 6.4
   D) 6.5 E) 6.6

17. Simplify $\left(\dfrac{x^{1/4}y^{1/2}}{z^{-1}}\right)^4$.

   A) xy^2z^4 B) $\dfrac{xy^2}{z^4}$ C) xy^2z

   D) $x^{17/4}y^{9/2}z^4$ E) $\dfrac{z}{xy^2}$

18. Evaluate $\dfrac{|-6|}{6} + \dfrac{|4|}{-4} + 2|-2| + (-2)|2|$.

   A) 8 B) 6 C) 0 D) 2 E) -10

19. $1 + \sinh^2 x =$

   A) $\cosh x$ B) $\dfrac{e^{2x} + e^{-2x}}{4}$

   C) $\dfrac{e^{2x} + 4 + e^{-2x}}{4}$ D) $\dfrac{e^x + e^{-x}}{2}$

   E) $\dfrac{e^{2x} + 2 + e^{-2x}}{4}$

20. $.35\overline{3535}$ is an example of a(n).

   A) imaginary number
   B) irrational number
   C) rational number
   D) improper fraction
   E) natural number

Solutions on page 71

Diagnostic Chart for Cumulative Review Test

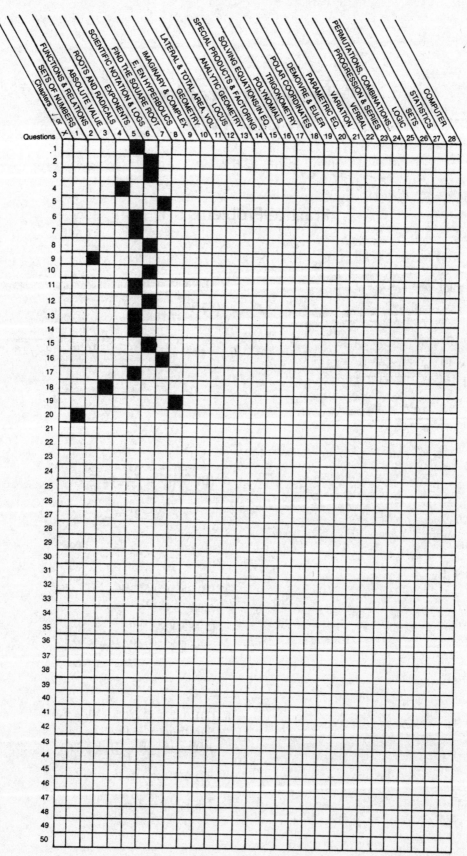

Solutions to Problems in Unit Two

Practice Problems 5.1

1. **(D)** $2^0 + 4^{1/2} + (\frac{1}{2})^{-1} = 1 + \sqrt{4} + \dfrac{1}{\frac{1}{2}}$

$$= 1 + 2 + 2 = 5$$

2. **(A)** $\dfrac{a^{x+1} \cdot a^{2x-1}}{a^{-x+1}} = \dfrac{a^{(x+1)+(2x-1)}}{a^{-x+1}}$

$$= \dfrac{a^{3x}}{a^{-x+1}} = a^{3x-(-x+1)}$$

$$= a^{4x-1}$$

3. **(D)** $4^{2m} = 2^{m+1}$

$(2^2)^{2m} = 2^{m+1}$

$2^{4m} = 2^{m+1}$

$4m = m + 1$

$3m = 1$

$m = \frac{1}{3}$

4. **(A)** $\dfrac{1}{16} = 2^{v-1}$

$\dfrac{1}{2^4} = 2^{v-1}$

$2^{-4} = 2^{v-1}$

$-4 = v - 1$

$v = -3$

5. **(E)** $x^{-2} = 16$

$(x^{-2})^{-1/2} = (16)^{-1/2}$

$$x^1 = 16^{-1/2} = \dfrac{1}{16^{1/2}} = \dfrac{1}{\pm\sqrt{16}}$$

$$= \pm\dfrac{1}{4}$$

6. **(B)** $\dfrac{2^3 \cdot 2^4}{(2^3)^2} = \dfrac{2^{3+4}}{2^{3 \cdot 2}} = \dfrac{2^7}{2^6}$

$$= 2^1 = 2$$

7. **(B)** $(5^{2a})(5)(5^{a+1})$

Note: $5 = 5^1$

$(5^{2a})(5^1)(5^{a+1})$

$$= 5^{(2a)+(1)+(a+1)} = 5^{3a+2}$$

8. **(D)** $3x^0 + (3x)^0 + x^{-1} + 2x$

Substitute $x = \frac{1}{2}$

$3(\frac{1}{2})^0 + (3 \cdot \frac{1}{2})^0 + (\frac{1}{2})^{-1} + 2(\frac{1}{2})$

$3(1) + 1 + \dfrac{1}{\frac{1}{2}} + 1$

$3 + 1 + 2 + 1 = 7$

Practice Problems 6.1

1. **(B)** 278300.

   2.78300 $N = 2.783$

   2.78300 $a = 5$

   $a = +$

   2.783×10^5

2. **(E)** .007932

   7.932 $N = 7.932$

   .007.932 $a = 3$

   a is "$-$"

   7.932×10^{-3}

   $a = -3$

3. **(E)**

   $12000 = 1.2 \times 10^4$

   $.04 = 4 \times 10^{-2}$

   $\dfrac{12000}{.04} = \dfrac{1.2 \times 10^4}{4 \times 10^{-2}} = \dfrac{1.2}{4} \times 10^{4-(-2)}$

   $= .3 \times 10^6$

   Note: .3 is *not* between 1 and 10. This is not in the correct format.

   $.3 \times 10^6 = 3. \times 10^5$

   $= 300,000$

4. **(D)** $\dfrac{1}{1000} = .001 = 1 \times 10^{-3}$

Practice Problems 6.2

1. **(D)** $\log_m 125 = 3 \Rightarrow m^3 = 125 = 5^3$

   $m = 5$

2. **(C)** $\log_{1/7} 49 = y \Rightarrow (\tfrac{1}{7})^y = 49$

   $(7^{-1})^y = 7^2$

   $7^{-y} = 7^2$

   $y = -2$

3. **(A)** $\ln \dfrac{a^2}{b^3} = \ln a^2 - \ln b^3$

   $= 2 \ln a - 3 \ln b$

4. **(C)** $\log \sqrt{21} = \dfrac{\log 21}{2} = \dfrac{\log 7 \cdot 3}{2}$

   $= \dfrac{\log 7 + \log 3}{2}$

   $= \dfrac{(.8451) + (.4771)}{2} = .6611$

5. **(D)** $2 \log a + \dfrac{\log b}{2}$

   $= \log a^2 + \log \sqrt{b}$

   $= \log a^2 \sqrt{b}$

6. **(B)** $\log .00917 = \log(9.17 \times 10^{-3})$

   $= \log 9.17 - 3$

   $= t - 3$

7. **(C)**

   $\log \sqrt{\dfrac{x^3}{zm}} = \dfrac{1}{2}[\log x^3 - \log zm]$

   $= \tfrac{1}{2}[3 \log x - (\log z + \log m)]$

   $= \tfrac{1}{2}[3 \log x - \log z - \log m]$

8. **(B)** $\log_8 x = \tfrac{1}{3} \Rightarrow 8^{1/3} = x$

   $2 = x$

Practice Problems 7.1

1. **(C)**

$$\begin{array}{r} 1.\ 7\ 3 \\ \sqrt{03.0000} \end{array}$$

$$\begin{array}{r} 1 \\ 27\overline{)\ 200} \\ 189 \\ 343\overline{)\ 1100} \end{array}$$

answer $1.73 = 1.7$ to the nearest tenth

Notes:

27	28	342	343
×7	×8	×2	×3
189	224	684	1029

 If the square root is required to a particular place value (i.e., tenths), work the calculation one place beyond. If the digit in this place is 0, 1, 2, 3 or 4, retain the required place value as is. If the digit in this place is 5, 6, 7, 8 or 9, round the required place up by 1.

$1.73 = 1.7$ to the nearest tenth
less than 5

$1.76 = 1.8$ to the nearest tenth
5 or higher

2. **(D)**

$$\begin{array}{r} 3.\ 4\ 6 \\ \sqrt{12.0000} \end{array}$$

$$\begin{array}{r} 9 \\ 64\overline{)\ 300} \\ 256 \\ 686\overline{)\ 4400} \\ 4116 \\ \overline{284} \end{array}$$

$3.46 = 3.5$ to the nearest tenth

3. **(B)**

$$x = \sqrt{3} \qquad \text{from the log table}$$
$$\log x = \log \sqrt{3} = \log(3^{\frac{1}{2}}) \qquad \log 3 = .4771$$
$$= \tfrac{1}{2} \log 3$$
$$= \tfrac{1}{2}(.4771)$$
$$\log x = .23855 \qquad \text{from the log table}$$
$$x \approx 1.73 \qquad \text{antilog } .2380 = 1.73$$
$$1.\underline{7}3 = 2. \text{ to the nearest integer}$$

4. **(D)**

$$x = \sqrt{12}$$
$$\log x = \log \sqrt{12} = \log (12^{1/2}) \quad \log 12 = \log(1.2 \times 10^1)$$
$$= \tfrac{1}{2} \log 12 \qquad\qquad\quad = \log 1.2 + \log 10^1$$
$$= \tfrac{1}{2}(1.0792)$$
$$= .5396 \qquad\qquad\quad \text{from the log table:}$$
$$x = \text{antilog } .5396 \qquad\qquad \log 1.2 = .0792$$
$$x \approx 3.46 \qquad\qquad\qquad \therefore \log 12 = \log 1.2 + \log 10^1$$
$$x = 3.5 \text{ to the nearest tenth} \qquad = .0792 + 1$$
$$\qquad\qquad\qquad\qquad\qquad = 1.0792$$

Practice Problems 8.1

1. **(B)** $\ln 16 = \ln 2^4 = 4 \ln 2$
$$= 4(.6931)$$
$$= 2.7724$$

2. **(C)** $\log_a b = c \Rightarrow a^c = b$
$$\ln 9.974 = 2.3$$
$$= \log_e 9.974 = 2.3 \Rightarrow e^{2.3} = 9.974$$

3. **(E)** The inverse of $y = e^{2x}$
$$\text{is } x = e^{2y}$$
$x = e^{2y} \Rightarrow \ln x = 2y$
$y = \frac{1}{2} \ln x$
$\quad = \ln x^{1/2}$

4. **(B)** Let $y = 2 \ln e^x$
$$y = 2 \log_e e^x$$
recall $2 \log a = \log a^2$
$$\therefore y = \log_e e^{2x}$$
changing to exponential form
$e^y = e^{2x}$
$\quad y = 2x$

Cumulative Review Test

1. **(A)** $4^{3x} = 8^{-(x-3)}$
$4^{3x} = 8^{-x+3}$
$(2^2)^{3x} = (2^3)^{3-x}$
$2^{6x} = 2^{9-3x}$
$6x = 9 - 3x$
$x = 1$

Since the hundredths place is greater than 5, the solution is 7.3 to the nearest tenth.

Note: $\quad$ 1445
$\qquad \underline{\times 5}$
$\qquad$ 7225

2. **(D)** $\quad$ .03 $\qquad$ Step 1
$\quad\quad$ 3.0 $\qquad$ Step 2 $\quad N = 3$
$\quad\quad$ 03. $\qquad$ Step 3 $\quad a = 2$
$\quad\quad \underset{\frown}{}$ $\qquad$ Step 4 $\quad a$ is "$-$"
$\quad\quad$ 3.0×10^{-2}

6. **(D)** $b^0 = 1 \quad b^{-1/2} = 16^{-1/2} = \frac{1}{4}$
$2b^0 + b^{-1/2} = 2(1) + \frac{1}{4} = \frac{9}{4}$

7. **(D)** $\left(\dfrac{a}{b}\right)^n = \dfrac{a^n}{b^n}$

$\therefore \dfrac{r^n}{s^n} = \left(\dfrac{r}{s}\right)^n \quad (s \neq 0)$

3. **(B)** $\log(f^2 - g^2) = \log[(f + g)(f - g)]$
$\qquad\qquad\qquad = \log(f + g) + \log(f - g)$

8. **(C)** $\frac{1}{3} \log n - \frac{1}{2} \log l$
$\qquad = \log \sqrt[3]{n} - \log \sqrt{l}$
$\qquad = \log \dfrac{\sqrt[3]{n}}{\sqrt{l}}$

4. **(D)**
$$\frac{2}{\sqrt{7} - 2}\left(\frac{\sqrt{7} + 2}{\sqrt{7} + 2}\right) = \frac{2(\sqrt{(7)} + 2)}{7 - 4} = \frac{2(\sqrt{7} + 2)}{3}$$
$$= \frac{2\sqrt{7} + 4}{3}$$

9. **(A)** $h(x) = \dfrac{x - 1}{x + 3}$

$h(x - 3) = \dfrac{(x - 3) - 1}{(x - 3) + 3} = \dfrac{x - 4}{x}$

5. **(C)**

$\quad\quad$ 7. 2 5
$\quad \sqrt{53.0000}$
$\quad\quad \underline{49}$
$\quad 142 \overline{\smash{)}400}$
$\quad\quad\quad \underline{284}$
$\quad 1445 \overline{\smash{)}11600}$

10. **(E)** $.3 \times 10^{-4} = .00003$
$\qquad\qquad = 3 \times 10^{-5}$

11. **(D)** $-2(x^3) + (-2x)^3$
$\qquad = -2x^3 + (-8x^3)$
$\qquad = -10x^3$

12. **(B)** $\log_8 x = \frac{2}{3} \Rightarrow 8^{2/3} = x$

$$x = (\sqrt[3]{8})^2$$
$$= 4$$

13. **(D)** $9^{3x} = 3^{x+5}$

$$(3^2)^{3x} = 3^{x+5}$$
$$3^{6x} = 3^{x+5}$$
$$6x = x + 5$$
$$x = 1$$

14. **(D)** $\dfrac{3^3 + 3^3 + 3^3}{3^4} = \dfrac{3(3^3)}{3^4} = \dfrac{3^4}{3^4} = 1$

15. **(D)** $\dfrac{3.2}{.32} = 10 \qquad \dfrac{10^2}{10^{-2}} = 10^{2-(-2)} = 10^4$

$$\frac{3.2 \times 10^2}{.32 \times 10^{-2}} = 10 \times 10^4 = 1 \times 10^5$$

16. **(C)** $\log 4.1 = .6128$

$$\log 41 = 1.6128$$
$$x = 41^{1/2} \quad \log x = \tfrac{1}{2} \log 41$$
$$= \frac{1.6128}{2} = .8064$$
$$\log x = .8064$$
$$x = 6.4$$

17. **(A)** $\left(\dfrac{x^{1/4}y^{1/2}}{z^{-1}}\right)^4 = \dfrac{x^1 y^2}{z^{-4}} = xy^2z^4$

18. **(C)** $|-6| = 6 \quad |4| = 4 \quad |-2| = 2 \quad |2| = 2$

$$\frac{|-6|}{6} + \frac{|4|}{-4} + 2|-2| + (-2)|2|$$
$$= \frac{6}{6} + \frac{4}{-4} + 2(2) + (-2)(2)$$
$$= 1 + (-1) + 4 + (-4) = 0$$

19. **(E)**

$$1 + \left(\frac{e^x - e^{-x}}{2}\right)^2 = 1 + \frac{e^{2x} - 2e^0 + e^{-2x}}{4}$$
$$= 1 + \frac{e^{2x} - 2 + e^{-2x}}{4}$$
$$= \frac{4 + e^{2x} - 2 + e^{-2x}}{4}$$
$$= \frac{e^{2x} + 2 + e^{-2x}}{4}$$
$$= \cosh^2 x$$

20. **(C)** $.35\overline{3535}$ is a *repeating* decimal. This is a rational number which can be expressed as a proper fraction.

UNIT THREE

9. Imaginary and Complex Numbers

9.1 Fundamental Facts

i The basic imaginary unit. It does not represent a variable. It equals $\sqrt{-1}$.

$a + bi$ or $x + yi$ The rectangular form of a complex number. The a is the real part, the bi is the imaginary part, where a and b are real numbers.

The successive non-negative integer powers of i follow a repeated sequence of

$$\mathbf{1, \, i, \, -1, \, -i}$$

$$i^0 = 1$$
$$i^1 = i$$
$$i^2 = -1$$
$$i^3 = -i$$

i^n The value of i^n, where n is a non-negative integer $= i^r$, where r is the remainder when n is divided by 4. For example:

$$i^{27} = i^3 = -i \qquad \text{because } \tfrac{27}{4} = 6r3.$$

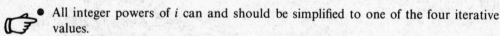

- All integer powers of i can and should be simplified to one of the four iterative values.
- Square roots of negative numbers should be put in $a + bi$ form prior to calculations.
- An expression with i in the denominator should be rationalized by using its conjugate.
- Every complex number can and should be simplified to $a + bi$ form.

VALUES OF i			
$i^0 = 1$ $i^4 = 1$ $i^8 = 1$			$i^n = i^r$
$i^1 = i$ $i^5 = i$ $i^9 = i$			
$i^2 = -1$ $i^6 = -1$ $\vdots$			n is a non-negative integer.
$i^3 = -i$ $i^7 = -i$ $\vdots$			r is the remainder when n is divided by 4.

COMPLEX NUMBERS (=, +, −, ∴, ÷)		
Equality	$a + bi = c + di$; then $a = c$ and $b = d$	Real terms are only equal to reals. Imaginary terms are only equal to imaginaries.
Sum	$(a + bi) + (c + di) = (a + c) + (b + d)i$	Reals add to reals. Imaginaries add to imaginaries.
Difference	$(a + bi) - (c + di) = (a - c) + (b - d)i$	Reals subtract from reals. Imaginaries subtract from imaginaries
Product	$(a + bi)(c + di) = (ac - bd) + (ad + bc)i$	Use FOIL. Do regular binomial multiplication. Recall: $i^2 = -1$
Conjugate Product	$(a + bi)(a - bi) = a^2 + b^2$	Middle terms drop out. $i^2 = -1$
Quotient	$\dfrac{(a + bi)}{(c + di)} = \dfrac{(a + bi)}{(c + di)}\dfrac{(c - di)}{(c - di)} = \dfrac{(ac + bd) + (bc - ad)i}{c^2 + d^2}$	Rationalize the denominator

Examples:

1. Add $\sqrt{-2} + \sqrt{-8}$.

2. Multiply $\sqrt{-2} \cdot \sqrt{-8}$.

3. Add $(3 + 2i) + (-2 - 6i)$.

4. Multiply $(2 + 3i)(2 - 3i)$.

Solutions:

1.
$$\sqrt{-2} = \sqrt{2}\sqrt{-1} = \sqrt{2}i$$
$$\sqrt{-8} = \sqrt{8}\sqrt{-1} = 2\sqrt{2}i$$
$$\sqrt{-2} + \sqrt{-8} = \sqrt{2}i + 2\sqrt{2}i = 3\sqrt{2}i$$

2.
$$\sqrt{-2} = \sqrt{2}i$$
$$\sqrt{-2}\sqrt{-8} = (\sqrt{2}i)(2\sqrt{2}i)$$
$$= 2(2)i^2$$
$$= 4i^2 = 4(-1) = -4$$

 $\sqrt{-2} \cdot \sqrt{-8} \neq \sqrt{16} = 4$
Use i notation when working the square root of negative numbers.
$(\sqrt{2}i)(\sqrt{8}i) = \sqrt{16}i^2 = 4(-1) = -4$

3. $(3 + 2i) + (-2 - 6i) = (3 - 2) + (2i - 6i)$
$$= (3 - 2) + (2 - 6)i$$
$$= 1 - 4i$$

4. $(2 + 3i)(2 - 3i) = 4 - 6i + 6i - 9i^2$
$$= 4 - 9i^2 \quad i^2 = -1$$
$$= 4 - 9(-1)$$
$$= 13$$

5. Simplify $\dfrac{2+i}{1-3i}$.

5. $\dfrac{(2+i)}{(1+3i)} \dfrac{(1+3i)}{(1+3i)} = \dfrac{2+6i+i+3i^2}{1+3i-3i-9i^2} = \dfrac{2+7i+3i^2}{1-9i^2}$

$$= \dfrac{2+7i+3(-1)}{1-9(-1)} = \dfrac{-1+7i}{10}$$

6. Find the reciprocal of i.

6. $\dfrac{1}{i} \cdot \dfrac{i}{i} = \dfrac{i}{i^2} = \dfrac{i}{-1} = -i$

7. Subtract $3 + 2i$ from $2 - i$.

7. $(2-i) - (3+2i) = 2 - i - 3 - 2i$

$$= -1 - 3i$$

☞ The $\sqrt{-8}$ can be written as $2i\sqrt{2}$ or $2\sqrt{2}i$. The former, $2i\sqrt{2}$, is used to make clear that the i is not under the radical sign. The latter, $2\sqrt{2}i$, is used because of its association to $a + bi$ form. Both notations are used depending on the author.

Practice Problems 9.1

1. Find the reciprocal of $\dfrac{4+3i}{25}$.

   A) $4 - 3i$ B) $3 - 4i$ C) $\dfrac{25(4-3i)}{7}$

   D) $\dfrac{4-3i}{25}$ E) $4 + 25i$

2. Find the simplified value of i^{25}.

   A) 0   B) 1   C) i
   D) -1 E) $-i$

3. Add $(3 + 2i) + (3 + 6i) + (-4 - 4i)$.

   A) 2   B) $4i$ C) $8i$ D) $2 + 4i$
   E) $4 + 2i$

4. Express $\sqrt{-3} + 3\sqrt{-27} - \sqrt{-12}$ in $a + bi$ form.

   A) $8\sqrt{3} + i$ B) $0 + 8\sqrt{3}i$ C) $8\sqrt{3} + 0i$
   D) $1 + 8\sqrt{3}i$ E) $8\sqrt{3} + 1i$

5. Simplify $\dfrac{3+2i}{i}$.

   A) $2 - 3i$ B) $2 + 3i$ C) $3 - 2i$
   D) $3 + 2i$ E) 5

6. Multiply $(3 + \sqrt{2}i)(3 - \sqrt{2}i)$.

   A) $11 - 6\sqrt{2}i$ B) $11 + 6\sqrt{2}i$ C) 9
   D) $9 - 6\sqrt{2}i$ E) 11

7. If $a + 3i = 7 - bi$ then $b =$

   A) 7   B) -3 C) -7
   D) 3   E) -4

8. $(3 + 7i)(2 - i) =$

   A) $13 + 11i$ B) $-1 + 11i$ C) $13 + 9i$
   D) $-1 + 9i$ E) $3i$

Solutions on page 127

9.2 Graphing of Complex Numbers

$a + bi$ can be represented as an ordered pair (a, b) on rectangular coordinates whose abscissa is the real axis and whose ordinate is the imaginary axis.

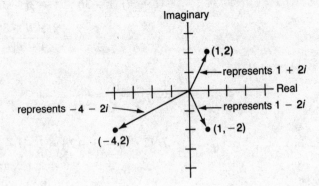

To add two complex numbers graphically, find the ordered pair associated with the diagonal of the parallelogram obtained using each complex number as one of the sides. For example, add $(2 + 3i) + (1 - i)$ graphically.

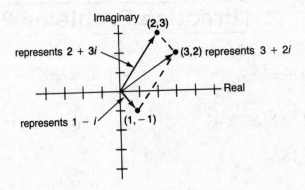

To subtract two complex numbers graphically, find the ordered pair associated with the diagonal of the parallelogram obtained using the first complex number and the negative of the complex number being subtracted. For example:

$$(2 + 3i) - (1 - i) \Rightarrow (2 + 3i) + (-1 + i)$$

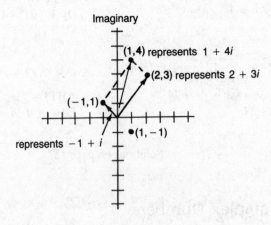

Practice Problems 9.2

1. Add $(-2 + 3i) + (3 - 2i)$ graphically.

2. Subtract $(-2 + 3i)$ from $(6 + 2i)$ graphically.

Solutions on page 128

10. Geometry

10.1 Angles

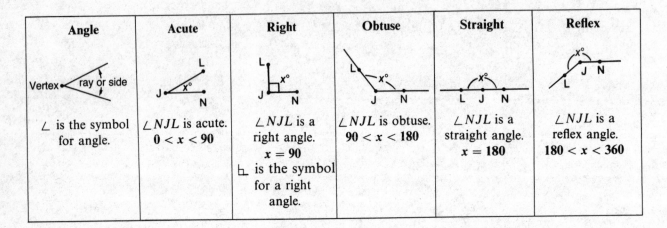

Angle	Acute	Right	Obtuse	Straight	Reflex
$\angle$ is the symbol for angle.	$\angle NJL$ is acute. $0 < x < 90$	$\angle NJL$ is a right angle. $x = 90$ ∟ is the symbol for a right angle.	$\angle NJL$ is obtuse. $90 < x < 180$	$\angle NJL$ is a straight angle. $x = 180$	$\angle NJL$ is a reflex angle. $180 < x < 360$

☞ Within this text:

$m\angle A$ indicates the measure of angle A as measured in degrees. Hence it is not necessary to use the word 'degrees' or the symbol (°) when referring to the measure of an angle.

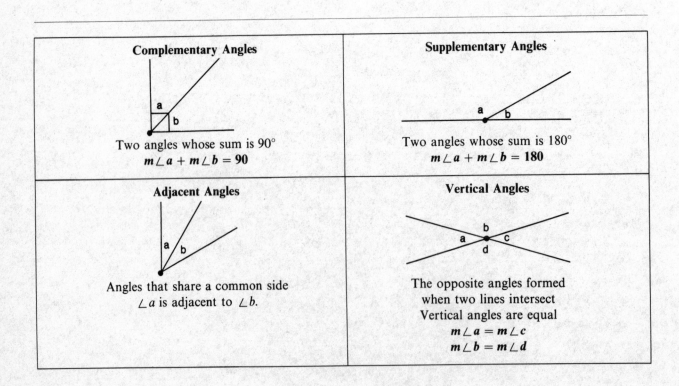

Complementary Angles

Two angles whose sum is 90°
$m\angle a + m\angle b = 90$

Supplementary Angles

Two angles whose sum is 180°
$m\angle a + m\angle b = 180$

Adjacent Angles

Angles that share a common side
$\angle a$ is adjacent to $\angle b$.

Vertical Angles

The opposite angles formed when two lines intersect
Vertical angles are equal
$m\angle a = m\angle c$
$m\angle b = m\angle d$

Perpendicular Lines	Parallel Lines
Lines which intersect at right ∠s. The symbol ⊥ indicates perpendicular. $n \perp l$	 Lines which do not meet. The symbol ‖ indicates parallel. $n \parallel l$
Transversal 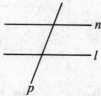 A line which intersects parallel lines. Line *p* is a transversal.	Concurrent 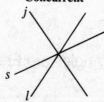 Lines are concurrent if they share a common intersection. Line *j*, *s* and *l* are concurrent.

Parallel Lines

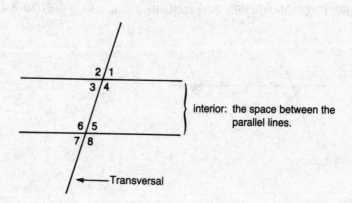

interior: the space between the parallel lines.

Transversal

When a transversal cuts two parallel lines a series of relationships exists among the eight angles formed.

Adjacent Angles are Supplementary

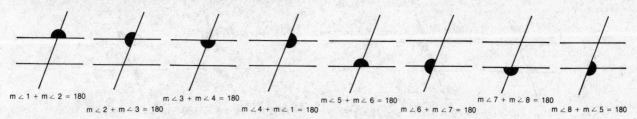

m∠1 + m∠2 = 180

m∠2 + m∠3 = 180

m∠3 + m∠4 = 180

m∠4 + m∠1 = 180

m∠5 + m∠6 = 180

m∠6 + m∠7 = 180

m∠7 + m∠8 = 180

m∠8 + m∠5 = 180

Vertical Angles are Equal

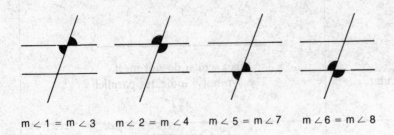

m∠1 = m∠3   m∠2 = m∠4   m∠5 = m∠7   m∠6 = m∠8

Angle Locations

A ——————— B
     2/1
     3/4

C ——————— D
     6/5
     7/8

AB ‖ CD

Corresponding Angles are Equal

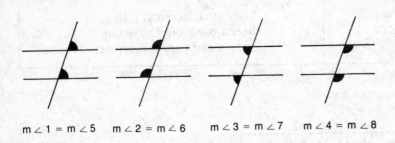

m∠1 = m∠5   m∠2 = m∠6   m∠3 = m∠7   m∠4 = m∠8

Alternate Interior Angles are Equal

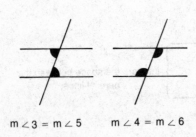

m∠3 = m∠5   m∠4 = m∠6

Alternate Exterior Angles are Equal

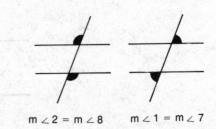

m∠2 = m∠8   m∠1 = m∠7

Interior Angles on the Same Side of the Transversal are Supplementary

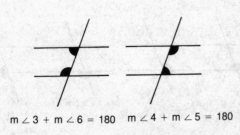

m∠3 + m∠6 = 180   m∠4 + m∠5 = 180

Exterior Angles on the Same Side of the Transversal are Supplementary

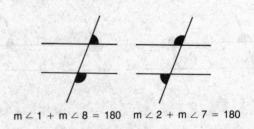

m∠1 + m∠8 = 180   m∠2 + m∠7 = 180

Examples:

1. If $AB \parallel CD$ and $m\angle 1 = 35$,

   A) find $m\angle 5$
   B) find $m\angle 8$
   C) find $m\angle 6$

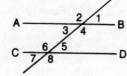

Solutions:

1. A) $\angle 1$ and $\angle 5$ are corresponding angles.
   $$m\angle 1 = m\angle 5 = 35$$
   B) $\angle 8 + \angle 1$ are exterior angles on the same side of the transversal. They are supplementary.
   $$m\angle 8 + m\angle 1 = 180$$
   $$m\angle 8 + 35 = 180$$
   $$m\angle 8 = 145$$
   Note: $\angle 8$ is also supplementary to $\angle 5$.
   C) $\angle 6$ is a vertical angle with $\angle 8$.
   $$m\angle 6 = m\angle 8 = 145$$

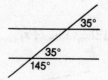

2. If AB is parallel to CD and $m\angle 1 = 3x + 1$ and $m\angle 2 = 2x - 6$, find the value of x.

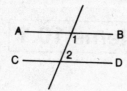

2. $\angle 1$ and $\angle 2$ are interior angles on the same side of the transversal.
   $$m\angle 1 + m\angle 2 = 180$$
   $$(3x + 1) + (2x - 6) = 180$$
   $$5x - 5 = 180$$
   $$5x = 185$$
   $$x = 37$$

3. The measures of two supplementary angles are represented by $7x - 27$ and $2x$. Find the measure of the smaller angle.

3.

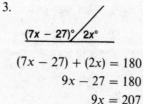

$$(7x - 27) + (2x) = 180$$
$$9x - 27 = 180$$
$$9x = 207$$
$$x = 23$$
$$2x = 2(23) = 46$$
$$7x - 27 = 7(23) - 27 = 134$$

4. If $AB \parallel CD$, $m\angle A = 35$ and $m\angle C = 45$, find the $m\angle AEC$.

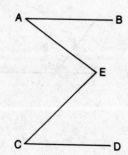

4.

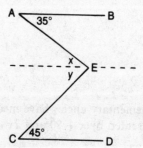

Construct a line parallel to AB and CD through E.
The $m\angle x = m\angle A = 35$, alternate interior angles.
The $m\angle y = m\angle C = 45$, alternate interior angles.
$$m\angle AEC = m\angle x + m\angle y = 35 + 45 = 80$$

5. If $\angle a$ and $\angle b$ are in the ratio of $2:3$, find the measure of angle a.

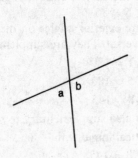

5. $m\angle a + m\angle b = 180$ Let $m\angle a = 2x$
$$2x + 3x = 180 \qquad m\angle b = 3x$$
$$5x = 180$$
$$x = 36$$
$$m\angle a = 2x = 2(36) = 72$$

6. Find the value of x.

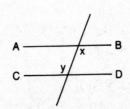

$2x°$ $(3x - 12)°$

6. $2x = 3x - 12$ Vertical angles
$$x = 12$$

Practice Problems 10.1

1. If $AB \parallel CD$, find $m\angle x$ when $m\angle y = 20$.

A) 20   B) 25   C) 70
D) 110   E) 160

A ———— x ———— B

C ———— y ———— D

2. Two complementary angles have measures, in degrees, represented by $x + 20$ and $2x + 1$. Find x.

A) 21   B) 23   C) 37
D) 53   E) 113

3. If parallel lines l and m are cut by a transversal and angles a and b are in the ratio of $4:5$, find the measure of $\angle b$.

A) 10   B) 20   C) 50
D) 80   E) 100

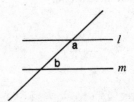

4. In the diagram, $AB \perp BC$. If XBY is a straight line and $m\angle XBC = 37$, find $m\angle ABY$.

A) 37   B) 53   C) 63   D) 127
E) 143

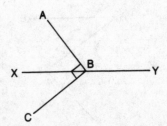

Solutions on page 128

10.2 Triangles

A triangle is a closed three sided figure. Triangles are classified by sides and/or angles.

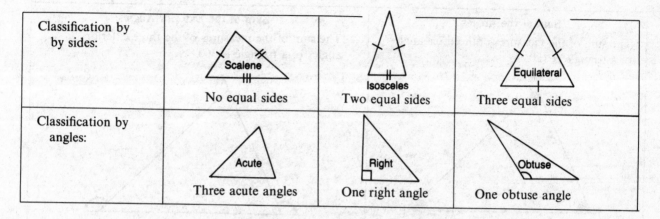

Classification by by sides:	Scalene No equal sides	Isosceles Two equal sides	Equilateral Three equal sides
Classification by angles:	Acute Three acute angles	Right One right angle	Obtuse One obtuse angle

Parts of a Triangle

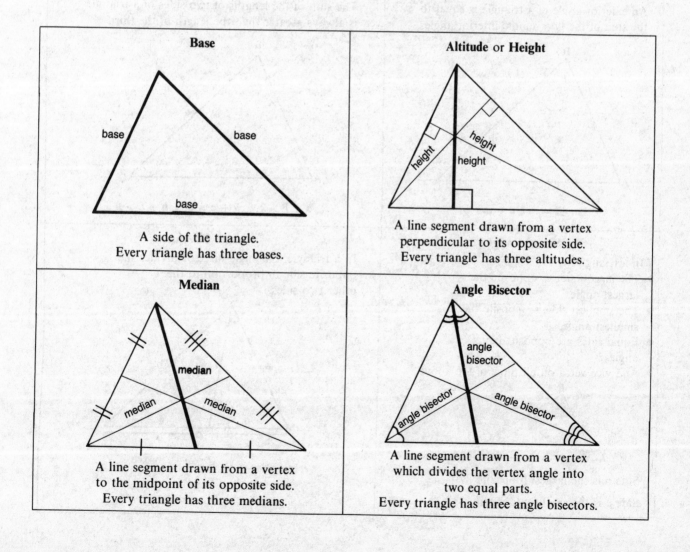

Base

A side of the triangle.
Every triangle has three bases.

Altitude or Height

A line segment drawn from a vertex
perpendicular to its opposite side.
Every triangle has three altitudes.

Median

A line segment drawn from a vertex
to the midpoint of its opposite side.
Every triangle has three medians.

Angle Bisector

A line segment drawn from a vertex
which divides the vertex angle into
two equal parts.
Every triangle has three angle bisectors.

General Properties

Sum of the Angles

The sum of the measures of the three angles of a triangle is 180.

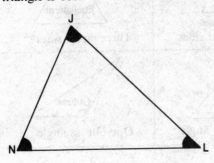

$$m\angle N + m\angle J + m\angle L = 180$$

Sum of the Exterior Angles

The sum of the measures of the three exterior angles of a triangle is 360.

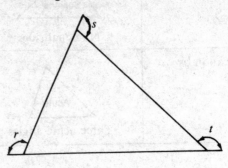

$$m\angle r + m\angle s + m\angle t = 360$$

Exterior Angle

An exterior angle of a triangle is equal to the sum of the two remote interior angles.

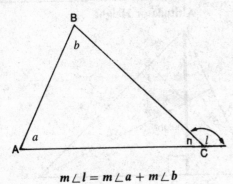

$$m\angle l = m\angle a + m\angle b$$

Lengths of Sides

The sum of the lengths of two sides of a triangle is always greater than the length of the third side.

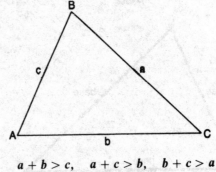

$$a + b > c, \quad a + c > b, \quad b + c > a$$

Sides vs Angles

In a triangle:
- The largest side is opposite the largest angle.
- The smallest side is opposite the smallest angle.
- Equal sides are opposite equal angles.
- And vice versa on all three of the above.

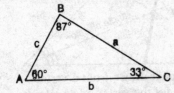

Side b is the largest (opposite 87°)

Side c is the smallest (opposite 33°)

Angle Bisector

In a triangle, the angle bisector divides the opposite side in proportion to the other two sides.

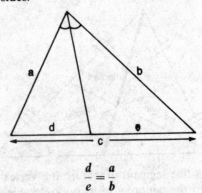

$$\frac{d}{e} = \frac{a}{b}$$

Area of a Triangle

The area of a triangle is $\frac{1}{2}$(base)(height). $A = \frac{1}{2}bh$

The base is any side of the triangle.
The height is the altitude drawn perpendicular to the base from the vertex not on the base.

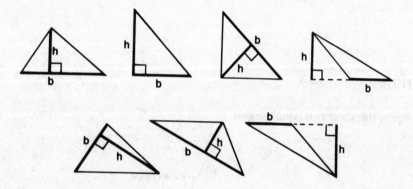

☞ In an obtuse triangle the height does not have to be in the triangle. The requirements to be an altitude are: a) be perpendicular to the base or its extension; b) be drawn from the vertex not on the base.

The area of a triangle is given by the equation: $A = \sqrt{s(s-a)(s-b)(s-c)}$

s is the semi-perimeter ($\frac{1}{2}$ the perimeter)
a, b, c are the sides of the triangle

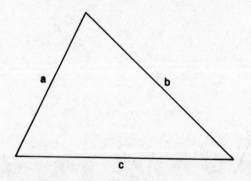

$$s = \frac{a+b+c}{2}$$

The area of a triangle is given by the equations: $A = \frac{1}{2}ab \sin C$
$= \frac{1}{2}bc \sin A$
$= \frac{1}{2}ac \sin B$

Area = half the product of two sides of a triangle multiplied by the Sin of the included angle.

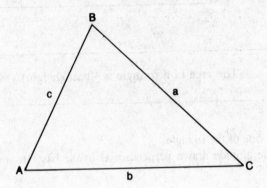

Isosceles Triangle:

Has **two equal sides** and **two equal angles**.

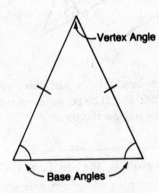

The equal angles are opposite the equal sides and are called base angles. The vertex angle is opposite the non-equal side.

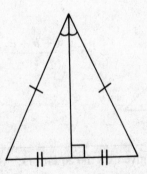

In an isosceles triangle, the altitude from the vertex angle is also the median and angle bisector.

Equilateral Triangle:

Has **three equal sides** and **three equal angles**.
Each angle is 60°.

 An easy way to calculate the area of an equilateral triangle is:

$$A = \frac{(\text{side})^2}{4}\sqrt{3}$$

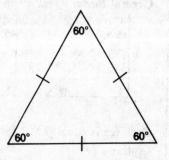

Right Triangle:

Pythagorean Theorem

$$(\text{leg 1})^2 + (\text{leg 2})^2 = (\text{hypotenuse})^2$$

This is the **Pythagorean theorem** and is a property of every right triangle.

A Pythagorean triple is a set of **integer values** which represent the sides of a right triangle, e.g., $7^2 + 24^2 = 25^2$.

Any multiple or division of a triple which results in integral values is also a triple, e.g., multiplying the 3, 4, 5 triple by 2⇒6, 8, 10:

$$6^2 + 8^2 = 10^2$$

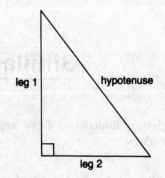

Common Pythagorean Triples		
leg	leg	hypotenuse
3	4	5
5	12	13
7	24	25
8	15	17
9	40	41
11	60	61

examples: $8^2 + 15^2 = 17^2$
$11^2 + 60^2 = 61^2$

Special Right Triangles

$45° - 45° - 90°$	$30° - 60° - 90°$
General form: 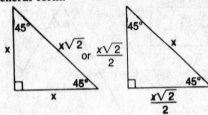	**General form:** 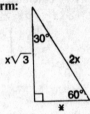
Each leg is equal. The hypotenuse is leg times $\sqrt{2}$.	The leg opposite the 30° angle is half the hypotenuse. The leg opposite the 60° angle is half the hypotenuse times $\sqrt{3}$.
Numerical master: 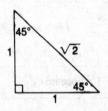	**Numerical master:** 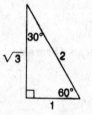

Similar Triangles

Two triangles are similar ($\sim$) if two angles of one triangle are equal to two angles of a second triangle.

☞ **Similar** ($\sim$) means the same shape, not necessarily the same size

$\triangle \sim \triangle$ $\square \sim \square$ $\odot \sim \odot$

Equal ($=$) means the same size (i.e., area), not necessarily the same shape

A = 8 A = 8

Congruent ($\cong$) means the same size and shape (identical)

If two figures are similar, their corresponding parts are proportional.

Line Ratio $= \dfrac{a}{x} = \dfrac{b}{y} = \dfrac{c}{z} = \dfrac{\text{any line in } \triangle I}{\text{any corresponding line in } \triangle II}$

$= \dfrac{\text{perimeter } \triangle I}{\text{perimeter } \triangle II}$

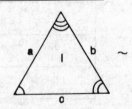

Perimeter is made of lines and follows the line ratio.

$$\text{Area Ratio} = (\text{line ratio})^2 = \left(\frac{a}{x}\right)^2 = \left(\frac{b}{y}\right)^2 = \left(\frac{c}{z}\right)^2 \cdots$$

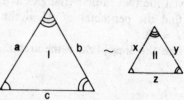

 Area ratio is the square of the line ratio, area is *not* made of lines.

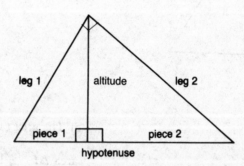

 All parts of the diagram are named after the large right triangle. All three right triangles are similar.

Rule 1:	**(altitude)^2 = (piece 1)(piece 2)**
Rule 2:	**(leg)^2 = (whole hypotenuse)(piece next to leg)** **(leg 1)^2 = (hypotenuse)(piece 1)** **(leg 2)^2 = (hypotenuse)(piece 2)**
Rule 3:	**Pythagorean Theorem:** **(leg 1)^2 + (leg 2)^2 = (hypotenuse)^2** **(piece 1)^2 + (altitude)^2 = (leg 1)^2** **(piece 2)^2 + (altitude)^2 = (leg 2)^2**

Examples:

1. The legs of a right triangle are 5 and 8. Find the length of the hypotenuse.

2. If a rectangle of width 10 has a diagonal of length 26, find the length of the rectangle.

Solutions:

1. By the Pythagorean theorem
$$(\text{leg})^2 + (\text{leg})^2 = (\text{hyp.})^2$$
$$5^2 + 8^2 = x^2$$
$$25 + 64 = x^2$$
$$89 = x^2$$
$$x = \sqrt{89}$$

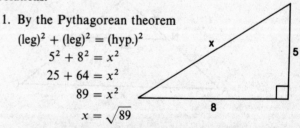

2. $(\text{leg})^2 + (\text{leg})^2 = (\text{hyp.})^2$
$$10^2 + x^2 = 26^2$$
$$100 + x^2 = 676$$
$$x^2 = 576$$
$$x = \sqrt{576} = 24$$

3. Given the two similar triangles as shown below,
   a) find the perimeter of △I, if the perimeter of △II = 12.
   b) find the area of △I if the area △II = 12.

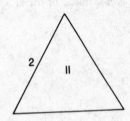

3.

a) The line ratio $= \dfrac{\text{side } \triangle I}{\text{Corresponding side } \triangle II} = \dfrac{1}{2}$

$$\frac{1}{2} = \frac{\text{perim. } \triangle I}{\text{perim. } \triangle II} = \frac{x}{12}$$

$$2x = 12$$

$$x = 6$$

b) The area ratio $= (\text{line ratio})^2 = \left(\dfrac{1}{2}\right)^2$

$$\frac{\text{area } \triangle I}{\text{area } \triangle II} = \left(\frac{1}{2}\right)^2 = \frac{1}{4}$$

$$\frac{x}{12} = \frac{1}{4}$$

$$4x = 12$$

$$x = 3$$

4. In △ABC, NL ∥ AC, find the length of AC.

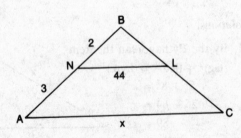

4. △NBL ~ △ABC

$$\frac{2}{5} = \frac{4}{x}$$

$$2x = 20$$

$$x = 10$$

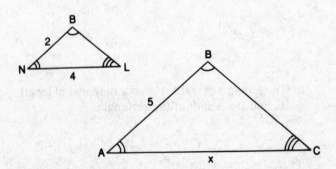

5. The measure of the base angles of an isosceles triangle are $3x - 1$ and $2x + 16$. Find the measure of the vertex angle.

5.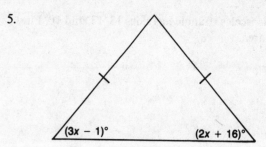

The measure of the base angles of an isosceles triangle are equal.

$3x - 1 = 2x + 16$

$x = 17$

The measure of the base angles are:

$3x - 1 = 3(17) - 1 = 50$

$2x + 16 = 2(17) + 16 = 50$

The vertex angle is the non-equal angle. Since the measures of the three angles must add up to 180 and the two base angles total 100, the measure of the vertex angle = $180 - 100 = 80$.

6. Three angles of a triangle are represented by $3x - 1$, $10x + 11$ and $x + 2$. Find the measure of the smallest angle.

6.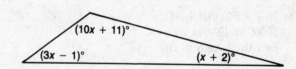

The sum of the measures of the angles is 180.

$(3x - 1) + (10x + 11) + (x + 2) = 180$

$14x + 12 = 180$

$14x = 168$

$x = 12$

The measures of the three angles are:

$3x - 1 = 3(12) - 1 = 35$

$10x + 11 = 10(12) + 11 = 131$

$x + 2 = 12 + 2 = 14$

Ans.: 14

7. An isosceles triangle has sides 13, 13, and 10. Find the area.

7. The altitude is also the median. By the 5-12-13 triple $x = 12$.

Area $= \frac{1}{2}bh = \frac{1}{2}(12)(10) = 60$

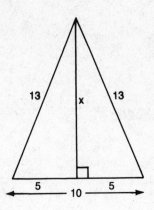

Alternative Solution:

Semi-perimeter $= \dfrac{13 + 13 + 10}{2} = 18$

$A = \sqrt{s(s - a)(s - b)(s - c)}$

$\quad = \sqrt{18(18 - 13)(18 - 13)(18 - 10)}$

$\quad = \sqrt{18(5)(5)(8)}$

$\quad = 5\sqrt{(18)(8)} = 5\sqrt{9 \cdot 2 \cdot 2 \cdot 4} = 5 \cdot 3 \cdot 2 \cdot 2$

$\quad = 60$

8. In $\triangle ABC$, $BD \perp AC$.
If $AC = 13$ and $AD = 4$,
find the length of BD.

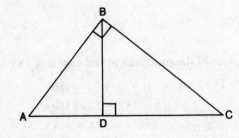

8.

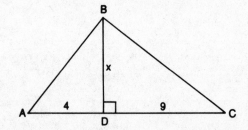

(altitude)^2 = (piece 1)(piece 2)

$\quad x^2 = (4)(9) = 36$

$\quad x = 6$

9. Find the area of an equilateral triangle whose side is 10.

9. $A = \dfrac{s^2}{4}\sqrt{3} = \dfrac{10^2}{4}\sqrt{3}$

$\quad = 25\sqrt{3}$

10. In the given $\triangle NJL$, find the lengths of j and l.

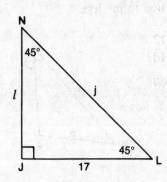

10. Using the numerical master of the 45°-45°-90°

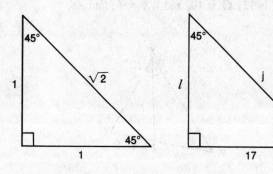

$$\text{line ratio} = \frac{\text{master}}{\text{problem}} = \frac{1}{17} = \frac{1}{l}$$

$$l = 17$$

$$\frac{\text{master}}{\text{problem}} = \frac{1}{17} = \frac{\sqrt{2}}{j}$$

$$j = 17\sqrt{2}$$

11. The angles of a triangle are in the ratio of $2:3:4$. What is the classification of the triangle (by angles and by sides)?

11. $2x + 3x + 4x = 180$

$$9x = 180$$

$$x = 20$$

$$2x = 2(20) = 40$$

$$3x = 3(20) = 60$$

$$4(x) = 4(20) = 80$$

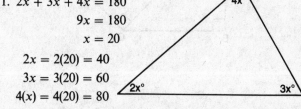

The triangle: is acute (all three angles are acute)

is scalene (no equal angles implies no equal sides)

12. In $\triangle ABC$, shown, find the length of side a.

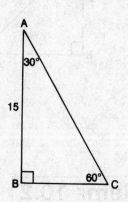

12. Using the numerical master:

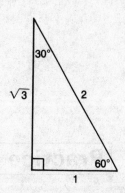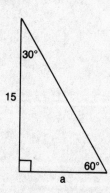

$$\text{line ratio} = \frac{\text{master}}{\text{problem}} = \frac{\sqrt{3}}{15} = \frac{1}{a}$$

$$a\sqrt{3} = 15$$

$$a = \frac{15}{\sqrt{3}}\left(\frac{\sqrt{3}}{\sqrt{3}}\right) = \frac{15\sqrt{3}}{3} = 5\sqrt{3}$$

13. In the figure, XZ is the angle bisector of $< X$. If WX is 12, XY is 10, and $WZ = 6$, find ZY.

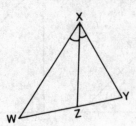

13. The angle bisector divides the opposite sides in proportion to the legs.

$$\frac{6}{a} = \frac{12}{10}$$

$$12a = 60$$

$$a = 5$$

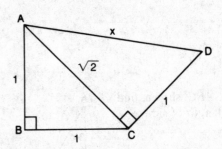

14. Find x, the length of AD.

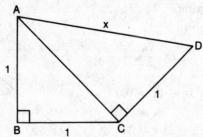

14. In $\triangle ABC$

$$(\text{leg})^2 + (\text{leg})^2 = (\text{hyp.})^2$$
$$1^2 + 1^2 = (AC)^2$$
$$2 = (AC)^2$$
$$AC = \sqrt{2}$$

In $\triangle ACD$
$$(\text{leg})^2 + (\text{leg})^2 = (\text{hyp.})^2$$
$$(\sqrt{2})^2 + 1^2 = x^2$$
$$2 + 1 = x^2$$
$$3 = x^2$$
$$x = \sqrt{3}$$

Practice Problems 10.2

1. The measures of the angles of a triangle are in the ratio $1:5:6$. This triangle is

   A) acute   B) right   C) obtuse
   D) equilateral   E) isosceles

2. The measure of the vertex angle of an isosceles triangle is 50. Find the number of degrees in each base angle.

   A) 50   B) 60   C) 65
   D) 55   E) 130

3. In △*ABC*, the measure of angle *A* is 6 more than 3 times the measure of angle *B*. The measure of the exterior angle at *C* is 106. Find the degree measure of angle *A*.

A) 25 B) 30 C) 74
D) 81 E) 106

4. In △*ABC*, *AB* = *BC*. If $m\angle A = 4x - 30$ and $m\angle C = 2x + 10$, find the number of degrees in ∠*B*.

A) 20 B) 40 C) 50
D) 80 E) 100

5. Find the altitude of an equilateral triangle whose side is 20.

A) 10 B) $20\sqrt{3}$ C) $10\sqrt{3}$
D) $20\sqrt{2}$ E) $10\sqrt{2}$

6. In △*RST*, *UV* ∥ *RS*, find *x*.

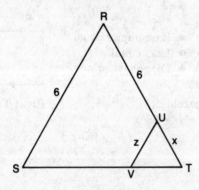

A) 2 B) 3 C) 4 D) 8 E) 9

7. In the figure, what is the value of *x*?

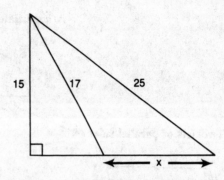

A) 8 B) 10 C) 12 D) 15 E) 20

8. Which of the following sets of numbers can represent the sides of a triangle?

A) {7, 8, 9} B) {3, 5, 8} C) {3, 3, 7}
D) {3, 10, 6} E) {4, 5, 10}

Solutions on page 129

10.3 Quadrilaterals

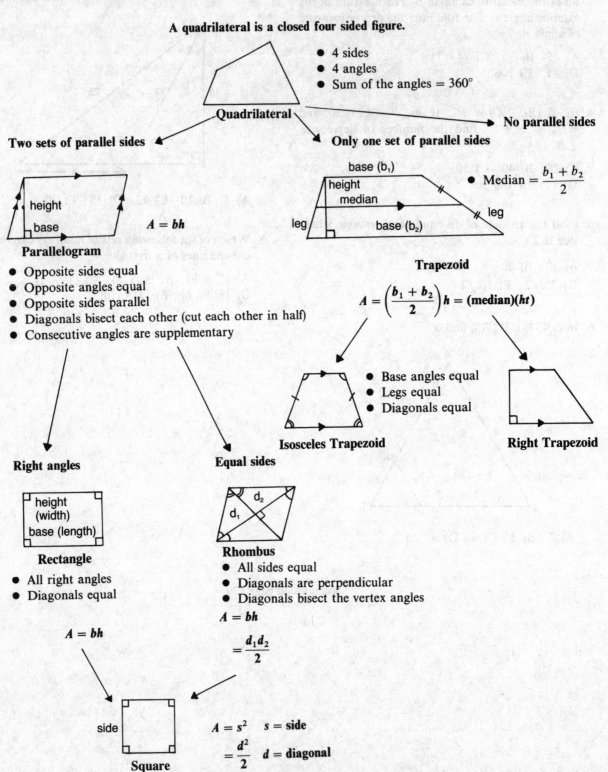

A quadrilateral is a closed four sided figure.

- 4 sides
- 4 angles
- Sum of the angles = 360°

Quadrilateral

No parallel sides

Two sets of parallel sides

Only one set of parallel sides

height
base

Parallelogram

$A = bh$

- Opposite sides equal
- Opposite angles equal
- Opposite sides parallel
- Diagonals bisect each other (cut each other in half)
- Consecutive angles are supplementary

base (b_1)
height
median
leg
base (b_2)
leg

- Median $= \dfrac{b_1 + b_2}{2}$

Trapezoid

$$A = \left(\dfrac{b_1 + b_2}{2}\right) h = (\text{median})(ht)$$

- Base angles equal
- Legs equal
- Diagonals equal

Isosceles Trapezoid

Right Trapezoid

Right angles

Equal sides

height
(width)
base (length)

Rectangle

- All right angles
- Diagonals equal

$A = bh$

d_2
d_1

Rhombus

- All sides equal
- Diagonals are perpendicular
- Diagonals bisect the vertex angles

$A = bh$

$= \dfrac{d_1 d_2}{2}$

side

Square

$A = s^2$ $s = \text{side}$

$= \dfrac{d^2}{2}$ $d = \text{diagonal}$

 Each figure in this tree has all the properties of the figures above it. For example, a rhombus has diagonals which bisect each other because its a parallelogram, and has 360° since it is a quadrilateral

☞ □ or Ⓟ is the symbol for a parallelogram.
□*ABCD* means label the vertices of the parallelogram *consecutively A, B, C, D*

Examples:

1. In □*ABCD*, if $m \angle A = 4x - 10$ and $m \angle C = 100 - 6x$, find x.

2. In quadrilateral *ABCD* the angles are in the ratio of $1:3:5:6$. Find the measure of each angle.

3. The diagonals of a rhombus are 6 and 10. Find a side of the rhombus.

Solutions:

1. Opposite angles are equal.

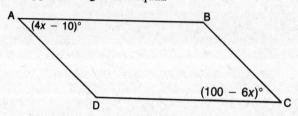

$$4x - 10 = 100 - 6x$$
$$10x = 110$$
$$x = 11$$

2. The sum of the angles of a quadrilateral is 360°.

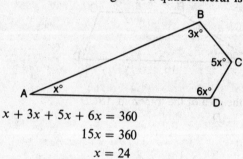

$$x + 3x + 5x + 6x = 360$$
$$15x = 360$$
$$x = 24$$

$$m \angle A = x = 24$$
$$m \angle B = 3x = 72$$
$$m \angle C = 5x = 120$$
$$m \angle D = 6x = 144$$

3. ● Diagonals bisect each other (because it is a parallelogram).
   ● Diagonals are perpendicular.

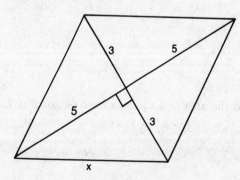

$$3^2 + 5^2 = x^2$$
$$9 + 25 = x^2$$
$$x^2 = 34$$
$$x = \sqrt{34}, \text{ all four sides are equal}$$

4. In rectangle $WXYZ$, $WY = 3x - 1$ and $XZ = 5x - 13$. Find the length of the diagonals.

4. The diagonals of a rectangle are equal.

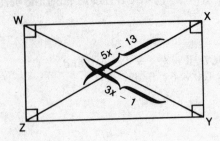

$$3x - 1 = 5x - 13$$
$$12 = 2x$$
$$6 = x$$

Each diag $= 5(6) - 13 = 17$
$$= 3(6) - 1 = 17$$

5.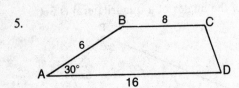

Find the area of the trapezoid $ABCD$ if $BC \parallel AD$

5.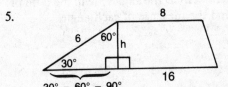

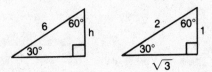

$$\text{line ratio} = \frac{\text{master}}{\text{problem}} = \frac{2}{6} = \frac{1}{h}$$
$$2h = 6$$
$$h = 3$$
$$A = \left(\frac{b_1 + b_2}{2}\right) h$$
$$= \left(\frac{8 + 16}{2}\right)(3) = 12(3) = 36$$

6. Find the area of a square whose diagonal is 7.

6.

$$A = \frac{d^2}{2} = \frac{7^2}{2} = \frac{49}{2}$$

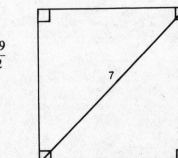

7. Find the area of a rhombus whose diagonals are 10 and 12.

7. $A_{\text{rhombus}} = \dfrac{d_1 d_2}{2} = \dfrac{10(12)}{2} = 60$

8. If two consecutive sides of a square are $7x + 4$ and $13x - 18$, find the area of the square.

8.
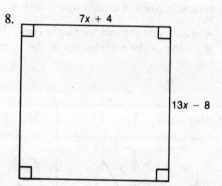

Sides of a square are equal.

$$7x + 4 = 13x - 8$$
$$12 = 6x$$
$$2 = x$$
∴ Each side = $7(2) + 4 = 18$
or $13(2) - 8 = 18$

Area = $s^2 = 18^2 = 324$

Practice Problems 10.3

1.
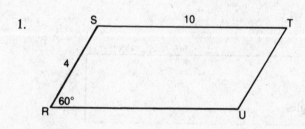

In $\square RSTU$, $ST = 10$, $RS = 4$ and $m\angle R = 60$. Find the area of the parallelogram.

A) 20 B) $20\sqrt{3}$ C) 40
D) $40\sqrt{3}$ E) 60

2. The diagonals of a rhombus are 2 and 4. Find the perimeter of the rhombus.

A) 4 B) $\sqrt{5}$ C) $4\sqrt{5}$
D) 8 E) 16

3. If the consecutive angles of a parallelogram are represented by $4t + 6$ and $5t - 15$, then the parallelogram could be a

I) rectangle II) rhombus III) square
IV) trapezoid

A) I only B) II only C) I, II, III only
D) II and III only
E) I and III only

4. In $\square ABCD$, $m\angle 1 = 2x + 12$ and the $m\angle 2 = 7x - 8$. Find x.

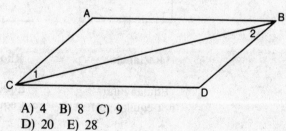

A) 4 B) 8 C) 9
D) 20 E) 28

Solutions on page 130

10.4 Polygons

A **convex polygon** is a closed figure with straight sides and

- Each side intersects exactly two other sides, one at each end point.
- None of the vertices lies in the interior of the polygon.

No. of Sides	3	4	5	6
Graphic				
Name	**Triangle**	**Quadrilateral**	**Pentagon**	**Hexagon**

7	8	10	12	No. of Sides
				Graphic
Heptagon	**Octagon**	**Decagon**	**Dodecagon**	Name

Equiangular: All equal angles
Equilateral: All equal sides
Regular: All equal sides and all equal angles

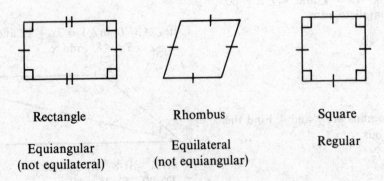

Rectangle	Rhombus	Square
Equiangular (not equilateral)	Equilateral (not equiangular)	Regular

	n = number of sides		
Any Convex Polygon	Sum of the measures of the exterior angles	=	**360**
	an interior angle + adjacent exterior angle = 180°		
	Sum of the measures of the interior angles	=	**(n − 2)(180)**
Regular Polygons Only	The measure of each exterior angle	=	$\dfrac{360}{n}$
	The measure of each interior angle	=	$\dfrac{(n - 2)(180)}{n}$

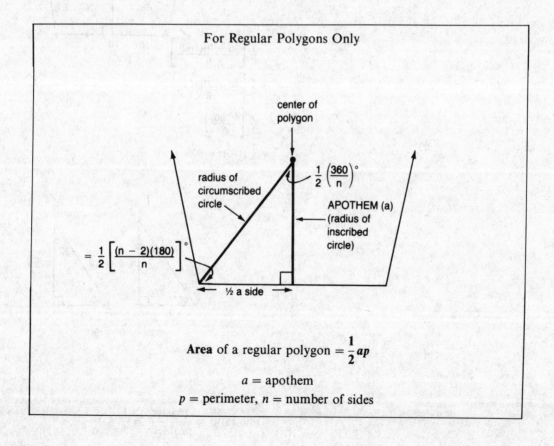

Bottom view of a polygon

For Regular Polygons Only

center of polygon

radius of circumscribed circle

$\frac{1}{2}\left(\dfrac{360}{n}\right)^{\circ}$

APOTHEM (a) (radius of inscribed circle)

$= \frac{1}{2}\left[\dfrac{(n - 2)(180)}{n}\right]^{\circ}$

½ a side

Area of a regular polygon $= \dfrac{1}{2}ap$

a = apothem

p = perimeter, *n* = number of sides

Examples:

1. Given a polygon with 22 sides, find the sum of the measures of the exterior angles.

Solutions:

1. 360. The sum of the measures of the exterior angles of any polygon is 360.

2. Given a polygon with 22 sides, find the sum of the measures of the interior angles.

2. $(n - 2)(180) = (22 - 2)(180) = 3600$

3. Find the measure of an interior angle of a regular pentagon.

3. The measure of each interior

$$= \frac{(n - 2)(180)}{n} = \frac{(5 - 2)(180)}{5}$$

$$= \frac{3(180)}{5} = 108$$

4. Find the area of a regular hexagon whose side is 10.

4. $A = \frac{1}{2}ap$

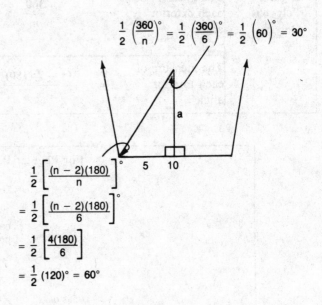

$$\frac{1}{2}\left(\frac{360}{n}\right)^{\circ} = \frac{1}{2}\left(\frac{360}{6}\right)^{\circ} = \frac{1}{2}\left(60\right)^{\circ} = 30^{\circ}$$

$$\frac{1}{2}\left[\frac{(n - 2)(180)}{n}\right]^{\circ}$$

$$= \frac{1}{2}\left[\frac{(n - 2)(180)}{6}\right]^{\circ}$$

$$= \frac{1}{2}\left[\frac{4(180)}{6}\right]$$

$$= \frac{1}{2}(120)^{\circ} = 60^{\circ}$$

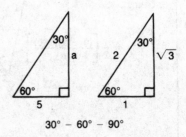

$30^{\circ} - 60^{\circ} - 90^{\circ}$

$$\text{line ratio} = \frac{\text{master}}{\text{problem}} = \frac{1}{5} = \frac{\sqrt{3}}{a}$$

$$a = 5\sqrt{3}$$

perimeter of hexagon (six sides) $= 6 \cdot 10 = 60$

Area $= \frac{1}{2}ap = \frac{1}{2}(5\sqrt{3})(60) = 150\sqrt{3}$

Practice Problems 10.4

1. Find the measure of each exterior angle of a regular decagon.

   A) 36   B) 144   C) 180
   D) 360   E) 720

2. Find the sum of the measures of the interior angles of a convex polygon of 12 sides.

   A) 30   B) 36   C) 360
   D) 1800   E) 2160

3. If a regular polygon has an exterior angle of 72°, how many sides does it have?

   A) 4   B) 5   C) 6   D) 7   E) 8

4. If the sum of the measures of the interior angles of a regular polygon is equal to the sum of the measures of the exterior angles, how many sides does the polygon have?

   A) 4   B) 6   C) 8   D) 10   E) 12

Solutions on page 131

10.5 Circles

A circle is the set of points in a plane equidistant from a given point called the center.

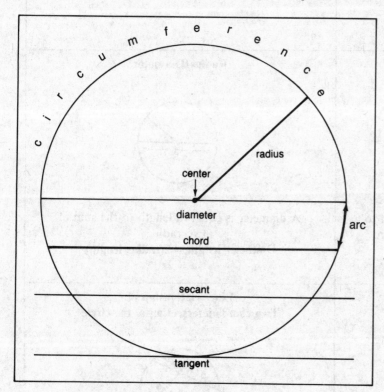

circumference	The boundary of the circle.
radius:	A line segment from the center to the circumference.
diameter:	A line segment from the circumference to the circumference through the center.
chord:	A line segment from a point on the circumference to any other point on the circumference
secant:	The line determined by a chord. A line that intersects the circle at two points.
tangent:	A line outside the circle which touches the circle at one point.
arc:	A piece of the circumference.

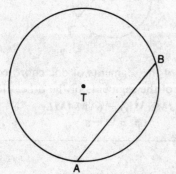

Notation

A circle is named for its center.
⊙*T* is shown on the left.
Chord AB is represented as **AB**
Arc AB is represented as **A͡B**

Basic Circle Rules

The number of degrees of arc in a circle is **360**.

Central Angle

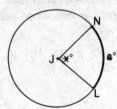

The measure of a central angle is equal to the measure of its arc.

$$m\angle J = m\widehat{NL}$$
$$x = a$$

Radii

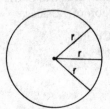

r = radius

All radii of a circle are equal in length.

Inscribed Angle

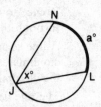

The measure of an inscribed angle is equal to one-half the measure of its intercepted arc.

$$m\angle J = \tfrac{1}{2}m\widehat{NL}$$
$$x = \tfrac{1}{2}a \quad \text{or} \quad a = 2x$$

Radius/Diameter

A diameter is equal in length to the sum of two radii.

Diameter length = 2(radius length)
$$d = 2r$$

Angle formed by:
Two chords intersecting in the circle

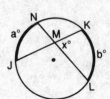

$m\angle KML$ is one-half the sum of the measures of the intercepted arcs.

$$m\angle KML = \tfrac{1}{2}(m\widehat{NJ} + m\widehat{KL})$$
$$x = \tfrac{1}{2}(a + b)$$

Line relationship of:
Two chords intersecting in the circle

The product of the segments of one chord equals the product of the segments of the other chord.

$$(JM)(MK) = (NM)(ML)$$
$$p \cdot q = r \cdot s$$

Angle formed by:

- **Two tangents intersecting outside the circle**

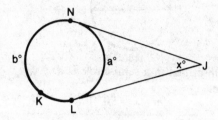

$$m \angle J = \tfrac{1}{2}(m\widehat{NKL} - m\widehat{NL})$$
$$x = \tfrac{1}{2}(a - b)$$

- **A tangent and secant intersecting outside the circle**

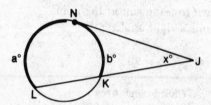

$$m \angle J = \tfrac{1}{2}(m\widehat{NL} - m\widehat{NK})$$
$$x = \tfrac{1}{2}(a - b)$$

- **Two secants intersecting outside the circle**

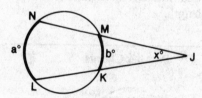

$$m \angle J = \tfrac{1}{2}(m\widehat{NL} - m\widehat{MK})$$
$$x = \tfrac{1}{2}(a - b)$$

The measure of the external angle at *J* equals one-half the difference of its intercepted arcs.

Line relationship of:

- **Two tangents intersecting outside the circle**

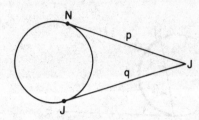

$$\therefore \overline{NJ} = \overline{LJ}$$
$$p = q$$

- **A tangent and secant intersecting outside the circle**

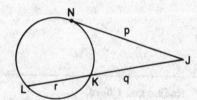

$$(NJ)(NJ) = (LJ)(KJ)$$
$$p \cdot p = (r + q)(q)$$

- **Two secants intersecting outside the circle**

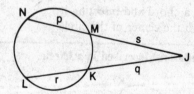

$$(NJ)(MJ) = (LJ)(KJ)$$
$$(p + s)s = (r + q)q$$

The whole secant or tangent times the external segment equals the whole other secant or tangent times its external segment.

Angle formed by:
A tangent and a chord drawn to the point of contact

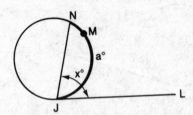

$m\angle NJL$ is one-half the measure of its intercepted arc.

$$m\angle NJL = \tfrac{1}{2}m\widehat{NMJ}$$
$$x = \tfrac{1}{2}a$$

Angle inscribed in a semi-circle

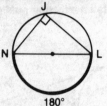

An angle inscribed in a semi-circle is a right angle.

$$m\angle J = 90$$

Radius drawn to a tangent at the point of contact

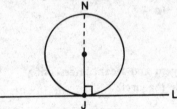

A radius drawn to a tangent at the point of contact forms a right angle with the tangent.

$$m\angle NJL = 90$$

Radius and Chord

- A radius drawn ⊥ to a chord bisects the chord and its arc.
- A radius which bisects a chord is ⊥ to the chord.
- A line ⊥ to a chord and bisecting it goes through the center of the circle.

Chords and Arcs

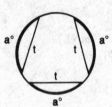

- Chords of equal length intercept arcs of equal measure.
- Arcs of equal measure determine chords of equal length.

Quadrilateral Inscribed in a Circle

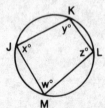

If a quadrilateral is inscribed in a circle, the opposite angles are supplementary.

- $m\angle M + m\angle K = 180$
  $$w + y = 180$$
- $m\angle J + m\angle L = 180$
  $$x + z = 180$$

Parallel Chords

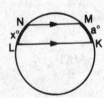

Parallel chords intercept arcs of equal measure.

$$m\widehat{NL} = m\widehat{MK}$$
$$x = a$$

Circumference length	**Area**
$C = 2\pi r$ $\;\;\;= \pi d$	$A = \pi r^2$ $\;\;\;= \dfrac{\pi d^2}{4}$
The length of the circumference of a circle $\;\;= 2\pi r$ or πd.	The area of a circle $= \pi r^2$ or $\dfrac{\pi d^2}{4}$.

Arc Length (degrees)	**Area of a Sector** (degrees)
$\text{Arc length} = \dfrac{x}{360} \cdot 2\pi r$	$\text{Area of sector} = \dfrac{x}{360} \cdot \pi r^2$
The arc length is $\dfrac{\text{central angle}}{360}$ of the whole circumference.	A sector contains $\dfrac{\text{central angle}}{360}$ of the whole circle area.
The curvature of the arc is measured in degrees. It is equal to the measure of its central angle. The length of the arc is measured in linear units such as inches, centimeters, etc., and is a piece of the circumference length.	

Arc Length (radians)	**Area of Sector** (radians)
r is the radius θ is the central angle in radians S is the arc length $S = r\theta$	r is the radius θ is the central angle in radians $A_{\text{sector}} = \dfrac{1}{2}r^2\theta$

Examples:

1. In $\odot N$, $m\angle ANB = 40$. Find the $m\angle C$.

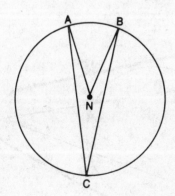

Solutions:

1. $\angle ANB$ is a central angle.
   $m\widehat{AB} = m\angle ANB = 40$

   $\angle C$ is an inscribed angle.
   $m\angle C = \frac{1}{2}m\widehat{AB} = \frac{1}{2}(40) = 20$

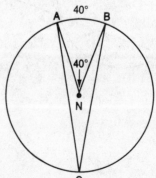

2. Find the value of x.

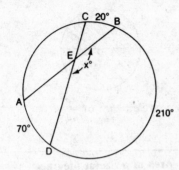

2. $m\angle BED = \frac{1}{2}(m\overset{\frown}{CA} + m\overset{\frown}{BD})$

$x = \frac{1}{2}(m\overset{\frown}{CA} + 210)$

However, we need $m\overset{\frown}{CA}$.

$m\overset{\frown}{CA} + 20 + 210 + 70 = 360$

$m\overset{\frown}{CA} = 60$

$\therefore x = \frac{1}{2}(60 + 210) = 135$

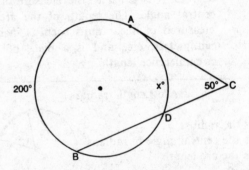

3. The $m\overset{\frown}{AB} = 200$, $m\angle C = 50$. Find the $m\overset{\frown}{AD}$.

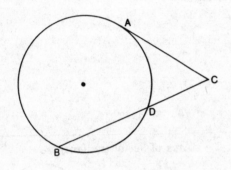

3. $m\angle C = \frac{1}{2}(m\overset{\frown}{AB} - m\overset{\frown}{AD})$

$50 = \frac{1}{2}(200 - x)$

$100 = 200 - x$

$x = 100$

4. $ED = 5$, $DC = 4$.
   If $AC = 12$, find the length of BC.

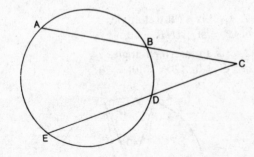

4. The whole secant times the segment outside equals the whole other secant times the segment outside.

$(EC)(DC) = (AC)(BC)$

$(9)(4) = (12)x$

$36 = 12x$

$x = 3$

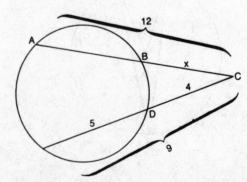

5. A chord of length 12 intercepts an arc of 60°. Find the radius of the circle.

5. First construct the radii as shown.

$\angle B$ is a central angle

$m\angle B = m\overset{\frown}{AC} = 60$

Triangle ABC is isosceles because $AB = BC$ (both are radii)

$\therefore m\angle A = m\angle C$

$m\angle A + m\angle B + m\angle C = 180$

$\qquad x + 60 + x = 180$

$\qquad\qquad 2x = 120$

$\qquad\qquad x = 60$

Triangle ABC is equilateral

$\therefore AB = BC = AC = 12$

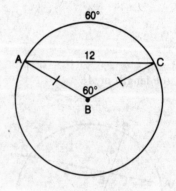

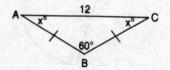

6. Find the area of a circle whose circumference is 18π.

6. $C = 2\pi r = 18\pi$

$\qquad 2r = 18$

$\qquad r = 9$

$A = \pi r^2 = \pi(9)^2 = 81\pi$

7. Find the area of a sector of a circle whose central angle is 72° and whose radius is 5.

7. $A = \frac{72}{360}\cdot\pi(5)^2$

$\quad = \frac{1}{5}\cdot 25\pi = 5\pi$

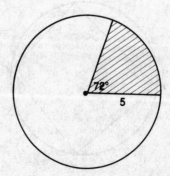

8. If a circle of radius 3 has a central angle of $\frac{\pi}{6}$ radians, find the arc length associated with the central angle.

8. $S = r\theta$

$$S = \frac{\pi}{6}(3) = \frac{3\pi}{6} = \frac{\pi}{2}$$

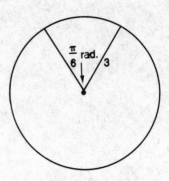

9. Find x, the length of AE.

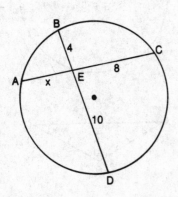

9. The product of the lengths of the segments of one chord equals the product of the lengths of the segments of the other chord.

$$(AE)(EC) = (BE)(ED)$$
$$x(8) = 4(10)$$
$$x = 5$$

10. In the Figure, $m\angle B = 6x - 20$ and $m\angle D = 4x + 10$
Find the value of x.

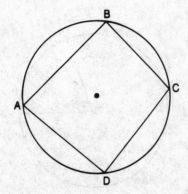

10. If a quadrilateral is inscribed in a circle, the opposite angles are supplementary.

$$m\angle B + m\angle D = 180$$
$$(6x - 20) + (4x + 10) = 180$$
$$10x - 10 = 180$$
$$10x = 190$$
$$x = 19$$

Practice Problems 10.5

1. In the figure, what is the value of x.

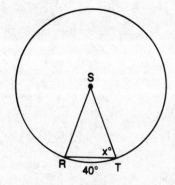

A) 40 B) 60 C) 70
D) 80 E) 140

2. In the figure, what is the value of x.

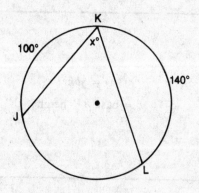

A) 30 B) 60 C) 90
D) 120 E) 150

3. $m\widehat{AB}:m\widehat{BC}:m\widehat{CNA} = 2:3:7.$
Find $m\angle P.$

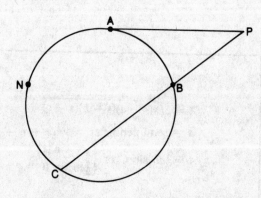

A) $22\frac{1}{2}$ B) 30 C) 60 D) 75 E) 150

4. The area of a circle is 12π. Find the length of the circumference.

A) 6π B) 12π C) 36π D) $6\pi\sqrt{2}$ E) $4\pi\sqrt{3}$

5. A square is inscribed in a circle of radius 4. Find the shaded area.

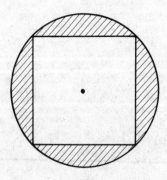

A) 16π B) $8\pi - 32$ C) $8\pi - 64$
D) $16\pi - 32$ E) $16\pi - 64$

6. AC is 12 and B is the midpoint of AC.
$\widehat{ANB}$ and $\widehat{BMC}$ are semi-circles. Find the length of the curved path $ANBMC$.

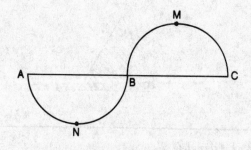

A) 6π B) 10π C) 12π D) 18π E) 24π

7. A chord of length 4 is 6 units from the center of a circle. Find the radius of the circle.

A) $10\sqrt{2}$ B) $2\sqrt{10}$ C) 8 D) $2\sqrt{13}$ E) 6

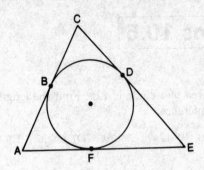

8. *AC*, *CE* and *AE* are tangents. If *CD* = 4, *DE* = 5 and *AE* = 9, find the length of *AC*.

   A) 6 B) 7 C) 8 D) 9 E) 10

Solutions on page 132

10.6 Area

Triangle:	
 (triangle figures with height *h* and base *b*)	$A = \frac{1}{2}bh$ *b* = base *h* = height
Equilateral Triangle: (equilateral triangle with 60° angles and sides *s*)	$A = \frac{s^2}{4}\sqrt{3}$ *s* = side
Any Triangle: (triangle *ABC* with sides *a*, *b*, *c*)	$A = \sqrt{s(s-a)(s-b)(s-c)}$ *s* = semi-perimeter = ½ (a + b + c) $A = \frac{1}{2}ab \sin C$ or $\begin{cases} \frac{1}{2}bc \sin A \\ \frac{1}{2}ac \sin B \end{cases}$

Any Triangle:

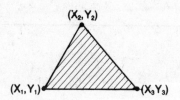

$$A = \tfrac{1}{2}|(x_1 y_2 + x_2 y_3 + x_3 y_1) - (y_1 x_2 + y_2 x_3 + y_3 x_1)|$$

Parallelogram:

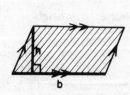
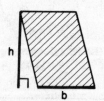

$$A = bh$$
b = base h = height

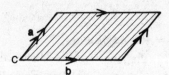

$$A = ab \sin C$$

Rectangle:

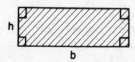

$$A = bh$$
b = base h = height

Rhombus:

$$A = \frac{d_1 d_2}{2}$$

d_1 = diagonal 1 d_2 = diagonal 2

Square:

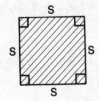

$$A = s^2$$
$$s = \text{side}$$

$$A = \frac{d^2}{2}$$

$$d = \text{diagonal}$$

Trapezoid:

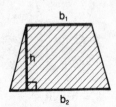 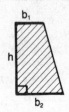

$$A = \left(\frac{b_1 + b_2}{2}\right) h$$

$$b_1 = \text{base 1} \quad b_2 = \text{base 2} \quad h = \text{height}$$

Any Quadrilateral:

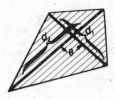

$$A = \tfrac{1}{2}(d_1)(d_2) \sin \theta$$
$$d_1 = \text{diagonal 1} \quad d_2 = \text{diagonal 2}$$
$$\theta = \text{angle between the diagonals}$$

Any Regular Polygon:

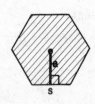

$$A = \tfrac{1}{2}ap$$
$$a = \text{apothem} \quad p = \text{perimeter}$$

$$A = \tfrac{1}{4}ns \cot \left(\frac{180}{n}\right)^{\circ}$$
$$n = \text{number of sides}$$
$$s = \text{length of a side}$$

Circle:

$$A = \pi r^2 \quad r = \text{radius}$$

$$A = \frac{\pi d^2}{4} \quad d = \text{diameter}$$

Practice Problems 10.6

1.

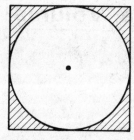

A circle of radius 6 is inscribed in a square. Find the area of the shaded region.

A) $36\pi - 144$ B) $36\pi - 36$ C) $144 - 12\pi$
D) $144\pi - 36$ E) $144 - 36\pi$

2. Find the area of triangle *NJL*.

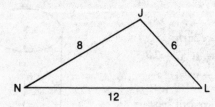

A) 455 B) 24 C) 12 D) $\sqrt{455}$

E) $\dfrac{\sqrt{455}}{2}$

3.

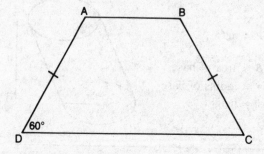

In isosceles trapezoid *ABCD*, $m\angle D = 60$. If $AB = 10$ and $AD = 8$, find the area of the trapezoid.

A) 40 B) $56\sqrt{2}$ C) $56\sqrt{3}$ D) 80 C) 140

4.

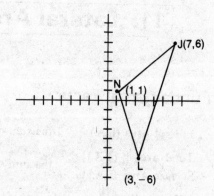

Find the area of triangle *NJL*.

A) $\dfrac{\sqrt{53}\sqrt{61}}{2}$ B) 26 C) 20 D) 13 E) $17\frac{1}{2}$

Solutions on page 133

11. Lateral Area, Total Area, Volume

11.1 Fundamental Facts

Lateral Area (L.A.)   the sum of the areas of the lateral faces

Total Area   (T.A.)
Surface Area (S.A.)   the sum of the areas of all the faces (lateral and bases)

Volume (V)   the cubic content of the solid

Right Prism

base

$\text{L.A.} = ph$
$\text{T.A.} = \text{L.A.} + 2B$
$V = Bh$

B = area of base
p = perimeter of base
h = height

lateral edge

lateral face

Right Cylinder

$\text{L.A.} = 2\pi rh$
$\text{T.A.} = \text{L.A.} + 2B$
$V = \pi r^2 h$

B = area of base = πr^2
r = radius of base
h = height

Oblique Prism

base

$V = Bh$

B = area of base
p = perimeter of base
h = height

lateral edge

lateral face

Oblique Cylinder

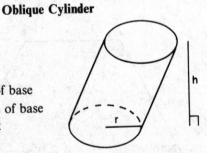

$V = \pi r^2 h$

B = area of base
r = radius of base
h = height

Regular Pyramid

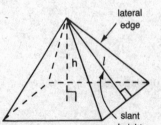

$\text{L.A.} = \frac{1}{2}pl$
$\text{T.A.} = \text{L.A.} + B$
$V = \frac{1}{3}Bh$

l = slant height
B = area of base
p = perimeter of base
h = height

lateral edge

slant height

Right Cone

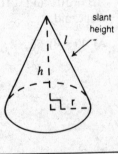

$\text{L.A.} = \pi rl$
$\text{T.A.} = \text{L.A.} + B$
$V = \frac{1}{3}\pi r^2 h$

l = slant height
B = area of base = πr^2
r = radius of base
h = height

slant height

Sphere	Cube
T.A. $= 4\pi r^2$ $V = \frac{4}{3}\pi r^3$ r = radius of sphere 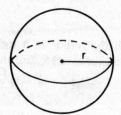	**L.A.** $= 4e^2$ **T.A.** $= 6e^2$ $V = e^3$ e = edge of cube 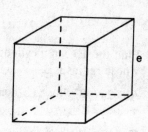

Examples:

1. If the total area of a cube is 36, find its volume.

2. The bases of a right prism, whose height is 8, are equilateral triangles with side 10. Find the lateral area, total area and volume of the prism.

3. A regular square pyramid has a base edge of 8 and a height of 4. Find the lateral area, total area and volume of the pyramid.

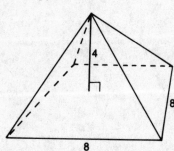

Solutions:

1. T.A. $= 6e^2 = 36$ $V = e^3$
   $e^2 = 6$ $V = (\sqrt{6})^3 = 6\sqrt{6}$
   $e = \sqrt{6}$

2. L.A. $= ph$
   perimeter of base $= 30$
   height $= 8$
   L.A. $= 30(8) = 240$
   T.A. $=$ L.A. $+ 2B$

   $B =$ area of base $= \frac{S^2}{4}\sqrt{3} = \frac{10^2}{4}\sqrt{3}$
   $= 25\sqrt{3}$
   T.A. $= 240 + 2(25\sqrt{3})$
   $= 240 + 50\sqrt{3}$

   $V = Bh = (25\sqrt{3})8 = 200\sqrt{3}$

3.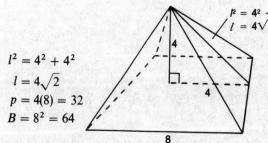

   $l^2 = 4^2 + 4^2$
   $l = 4\sqrt{2}$
   $p = 4(8) = 32$
   $B = 8^2 = 64$

   L.A. $= \frac{1}{2}pl = \frac{1}{2}(32)(4\sqrt{2}) = 64\sqrt{2}$
   T.A. $=$ L.A. $+ B = 64\sqrt{2} + 64$
   $V = \frac{1}{3}Bh = \frac{1}{3}(64)(4) = \frac{256}{3}$

Practice Problems 11.1

1. Find the T.A. of a cylinder whose height is 10 and whose radius is 3.

   A) 69π B) 78π C) 60π D) 18π E) 9π

2. If the height and radius of a cylinder are of equal length and the volume is 125π, find the radius of the base.

   A) 5   B) 10   C) 25   D) 5π E) $62\frac{1}{2}$

3. Find the volume of the given sphere.

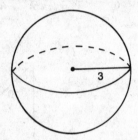

   A) 36π B) 9π C) 18π D) 27π E) 64π

4. If the radius of a right cone is 8 and the height (altitude) is 15, find the lateral area.

   A) 8π B) 15π C) 120π D) 136π E) 17π

Solutions on page 134

12. Locus

12.1 Basics and Beyond

Locus: The region formed by the set of all points satisfying a given condition.
Loci: The plural of locus

The locus of points (in a plane) **a distance r from a given point P**	The locus of points (in a plane) **a distance d from a given line l_1**	The locus of points (in a plane) **equidistant from two parallel lines**
	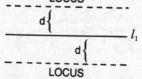	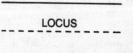
A circle with center at P and radius r	**Two lines** each parallel to the given line at a distance d	**A line** parallel to the two given lines and mid-way between them
The locus of points (in a plane) **equidistant from two fixed points A and B**	The locus of points (in a plane) **equidistant from the ends of a line segment**	The locus of points (in a plane) **equidistant from the rays of a given angle ABC**
		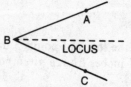
The **perpendicular bisector** of the line segment connecting points A and B.	The **perpendicular bisector** of the given line segment	The **angle bisector** of angle ABC
The locus of points (in a plane) **equidistant from two intersecting lines l_1 and l_2**	The locus of points (in a plane) **equidistant from the vertices of a triangle**	The locus of points (in a plane) **equidistant from the sides of a triangle**
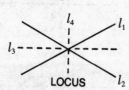	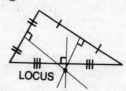	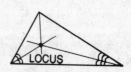
Two lines l_3 and l_4, the angle bisectors of the angles formed between l_1 and l_2	**A point.** The intersection of the perpendicular bisectors of the sides of the triangle	**A point.** The intersection of the angle bisectors of the angles of the triangle.

The locus of points (in a plane) such that **the distance from a fixed point F is equal to the distance to a fixed line l**

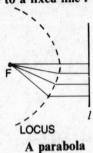

LOCUS
A parabola

The locus of points (in a plane) such that **the sum of the distances to each of two fixed points F_1 and F_2 is a positive constant**

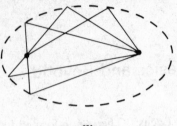

An ellipse

The locus of points in a plane such that **the absolute value of the difference of the distances from two fixed points F_1 and F_2 is constant**

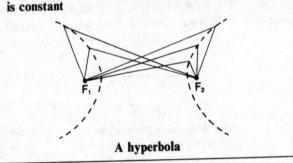

A hyperbola

☞ **Compound locus:** Do each piece of the problem separately. The intersection of the individual solutions is the solution set or locus for the compound example.

Examples:

1. Find the locus of points 3 units from a given line and 4 inches from a given point on the line.

Solutions:

1. Do each piece of the problem separately.

3 units from a given line

4 inches from a given point on the line

Composite

Four points (shown by Xs)

2. Given two points N and L 8 units apart, what is the locus of points 6 units from N and 3 units from L?

2. Do each piece of the problem separately.

6 units from point N

3 units from point L

Composite

Two points (designated by the Xs)

Practice Problems 12.1

1. In a plane, parallel lines l and m are 6 units apart and P is a point on line l. The total number of points in the plane that are equidistant from l and m and also 3 units from P is

A) 0   B) 1   C) 2   D) 3   E) 4

2. The equation of the locus of points in a plane equidistant from the points $(7, 3)$ and $(-3, 3)$ is

A) $x^2 + y^2 = 100$ B) $x = 5$ C) $x = 2$
D) $y = 5$ E) $y = 2$

3. In a plane the locus of points 4 units from the point $N(-2, 3)$ is

A) $x = -2$ B) $y = 3$ C) $x^2 + y^2 = 16$
D) $(x - 2)^2 + (y + 3)^2 = 16$
E) $(x + 2)^2 + (y - 3)^2 = 16$

4. In a plane, how many points are equidistant from two parallel lines and also equidistant from two points on one of these lines?

A) 0   B) 1   C) 2   D) 3   E) 4

Solutions on page 135

Cumulative Review
Chapters 1-12

1. If $\log a = p$, express $\log 10a^2$ in terms of p.

   A) $2p$ B) p^2 C) $1 + 2p$ D) 1^{2p}
   E) $1 + p^2$

2. The square of $(2 - 2i)$ is

   A) 0 B) $-8i$ C) $4 - 4i$ D) 4
   E) $8 - 8i$

3. In the accompanying diagram, secant ADB and tangent AC are drawn to circle O from external point A. If $AD = 3$ and $DB = 9$, what is the length of AC?

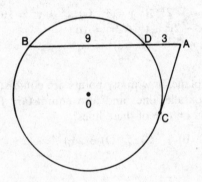

   A) 6 B) $\sqrt{6}$ C) 27 D) $\sqrt{27}$ E) $\sqrt{12}$

4. In the accompanying diagram $AB \parallel CD$, EF is a transversal and $m\angle 1 = 110$. What is the $m\angle 2$?

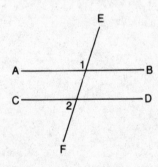

   A) 110 B) 70 C) 50 D) 80 E) 20

5. Given point P on a line. In a plane containing the line, the total number of points which are at a distance of 4 units from P and also a distance of 3 units from the given line is

   A) 0 B) 1 C) 2 D) 3 E) 4

6. $i^{219} = 7$.

   A) 0 B) 1 C) i D) -1 E) $-i$

7. A cylindrical pail has a radius of 3 units and a height of 6 units. Approximately how many gallons will it hold if there are 27 cubic units to a gallon?

   A) 1 B) 3 C) π D) 2π E) π^2

8. In the accompanying diagram, right triangle JSL has altitude h drawn to the hypotenuse, dividing it into segments of 8 and 10. What is the value of h?

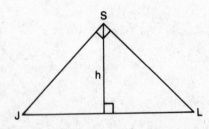

   A) $4\sqrt{5}$ B) 80 C) $3\sqrt{2}$ D) $\sqrt{2}$ E) $2\sqrt{10}$

9. The range of the function shown in the given graph is

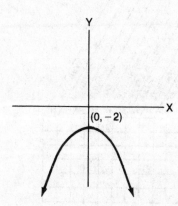

(0, −2)

A) all real numbers   B) imaginary   C) $x > 0$
D) $y \geqslant -2$ E) $y \leqslant -2$

10. Find the value of $(a + 1)^0 + 4(a)^{-1/2}$ when $a = 1$.

A) 6   B) 5   C) −2   D) −1   E) $1\frac{1}{2}$

11. What is the measure of the supplement of an angle whose complement is 70°

A) 20   B) 110   C) 160   D) 140   E) 130

12. Express $\log \dfrac{a}{b\sqrt{c}}$ in terms of the sums and differences of logarithms.

A) $\dfrac{\log a}{\log b + \frac{1}{2}\log c}$ B) $\log a - \log b + \frac{1}{2}\log c$

C) $\log a - \dfrac{\log b}{\log \sqrt{c}}$

D) $\log a - (\log b + \frac{1}{2}\log c)$

E) $\log a + \log b - \frac{1}{2}\log c$

13. In a given plane, the total number of points which are equidistant from two intersecting lines and

also 5 units from their point of intersection is

A) 0   B) 1   C) 2   D) 3   E) 4

14. What is the smallest integral value of x for which $\sqrt{2 - x}$ is imaginary?

A) −1   B) 0   C) 1   D) 2   E) 3

15. The base of a rectangular tank is 6 units by 10 units and its height is 12 units. Find the number of cubic units in the tank when it is $\frac{1}{4}$ empty.

A) 270   B) 180   C) 150   D) 540   E) 720

16. Find the value of $\sqrt{19}$ to the nearest tenth.

A) 4.2   B) 4.3   C) 4.4   D) 4.42   E) 4.5

17. Simplify $\left(\dfrac{49m^{-2}}{16r^{-2}}\right)^{-1/2}$

A) $\dfrac{49m}{16r}$ B) $\dfrac{4m}{7r}$ C) $\dfrac{4r}{7m}$

D) $\dfrac{7m}{4r}$ E) $\dfrac{7m}{2r}$

18. In triangle ABC, $m\angle C = 25$ and $BC = 13$. If the measure of an exterior angle at $B = 50$, find AB.

A) 13   B) 26   C) $7\frac{1}{2}$ D) $6\frac{1}{2}$ E) 130

19. The sum of $-3 - 2i$ and $2 - 3i$ is a complex number which, when graphically represented, lies in quadrant

A) I   B) II   C) III   D) IV
E) on the x-axis

20. Which set of numbers could represent the lengths of the sides of a right triangle?

A) $\{1, 7, 8\}$ B) $\{7, 8, 12\}$ C) $\{7, 9, 11\}$
D) $\{8, 15, 17\}$ E) $\{6, 8, 9\}$

Solutions on page 136

Diagnostic Chart for Cumulative Review Test

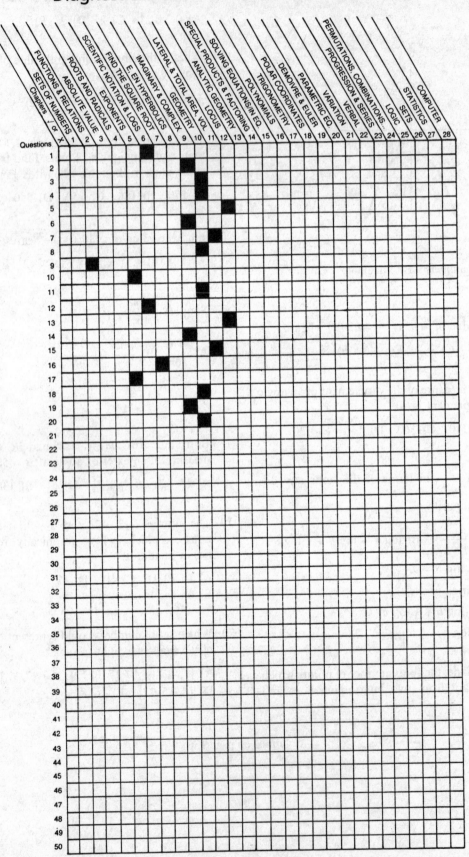

Solutions to Problems in Unit Three

Practice Problems 9.1

1. **(A)**

$$\frac{(25)}{(4 + 3i)} \frac{(4 - 3i)}{(4 - 3i)} = \frac{25(4 - 3i)}{16 - 9i^2} = \frac{25(4 - 3i)}{25}$$

$$= 4 - 3i$$

2. **(C)** $i^{25} = i^1 = i$

   Note: $\frac{25}{4} = 6r1$

3. **(D)**

$$(3 + 2i) + (3 + 6i) + (-4 - 4i)$$
$$= (3 + 3 - 4) + (2 + 6 - 4)i$$
$$= 2 + 4i$$

4. **(B)**

$$\sqrt{-3} = \sqrt{3}i$$
$$3\sqrt{-27} = 3(3\sqrt{3}i)$$
$$\sqrt{-12} = 2\sqrt{3}i$$

$$\sqrt{-3} + 3\sqrt{-27} - \sqrt{-12}$$
$$= \sqrt{3}i + 9\sqrt{3}i - 2\sqrt{3}i$$
$$= 8\sqrt{3}i$$
$$= 0 + 8\sqrt{3}i$$

5. **(A)**

$$\frac{(3 + 2i)}{i} \cdot \frac{i}{i} = \frac{3i + 2i^2}{i^2} = \frac{3i + 2(-1)}{-1}$$

$$= \frac{3i - 2}{-1} = 2 - 3i$$

6. **(E)**

$$(3 + \sqrt{2}i)(3 - \sqrt{2}i)$$
$$= 9 - 3\sqrt{2}i + 3\sqrt{2}i - 2i^2$$
$$= 9 - 2i^2 = 9 + 2$$
$$= 11$$

7. **(B)**   reals = reals

   imaginary = imaginary

$$a + 3i = 7 - bi$$
$$\therefore a = 7$$
$$b = -3$$

8. **(A)**

$$(3 + 7i)(2 - i) = 6 - 3i + 14i - 7i^2$$
$$= 6 + 11i - 7(-1)$$
$$= 13 + 11i$$

Practice Problems 9.2

1.

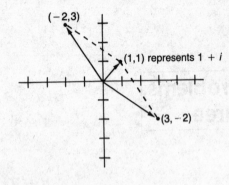

From the graph, the solution is $1 + i$.

2. $(6 + 2i) - (-2 + 3i)$ becomes
$(6 + 2i) + (2 - 3i)$

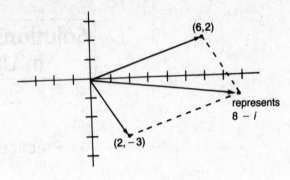

From the graph, the solution is $8 - i$.

Practice Problems 10.1

1. **(A)** $m \angle x = 20$.
   Alternate interior angles

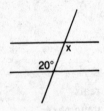

2. **(B)** Complementary angles add to 90°.
   $(x + 20) + (2x + 1) = 90$
   $3x + 21 = 90$
   $3x = 69$
   $x = 23$

3. **(E)** Angles a and b are supplementary.
   $m \angle a = 4x$
   $m \angle b = 5x$

$4x + 5x = 180$
$9x = 180$
$x = 20$
$m \angle b = 5x = 5(20) = 100$

4. **(D)**
   $\angle XBC$ and $\angle XBA$ are complementary.
   $37 + m \angle XBA = 90$
   $m \angle XBA = 53$
   $\angle XBA$ and $\angle ABY$ are supplementary.
   $53 + m \angle ABY = 180$
   $m \angle ABY = 127$

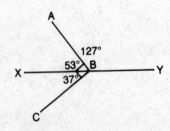

Practice Problems 10.2

1. (B) $1x + 5x + 6x = 180$

$$12x = 180$$
$$x = 15$$

$1x = 15$
$5x = 5(15) = 75$
$6x = 6(15) = 90$

The triangle is a right triangle.

2. (C) The base angles are equal.

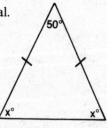

$$x + x + 50 = 180$$
$$2x + 50 = 180$$
$$2x = 130$$
$$x = 65$$

3. (D)

$$x + 3x + 6 = 106$$
$$4x + 6 = 106$$
$$4x = 100$$
$$x = 25$$

$$m\angle A = 3x + 6 = 3(25) + 6 = 81$$

4. (D) $4x - 30 = 2x + 10$

$$2x = 40$$
$$x = 20$$

$m\angle A = 4(20) - 30 = 50$
$m\angle C = 2(20) + 10 = 50$
$m\angle B = 180 - (50 + 50) = 80$

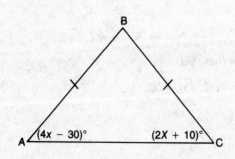

5. (C)

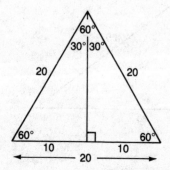

☞ An equilateral $\triangle$ is isosceles. The altitude is also the median and angle bisector.

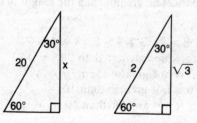

$$\text{line ratio} = \frac{\text{master}}{\text{problem}} = \frac{\sqrt{3}}{x} = \frac{2}{20}$$

$$2x = 20\sqrt{3}$$
$$x = 10\sqrt{3}$$

6. (B)

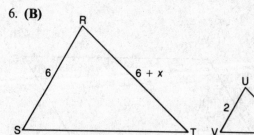

$$\triangle RST \sim \triangle UVT$$

$$\text{line ratio} = \frac{6}{2} = \frac{6 + x}{x}$$

$$6x = 12 + 2x$$
$$4x = 12$$
$$x = 3$$

7. **(C)** This is a 5 multiple of 3, 4, 5 ⇒ 15, 20, 25.

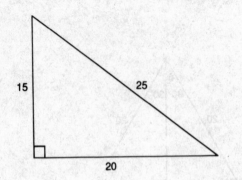

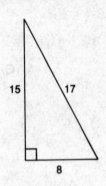

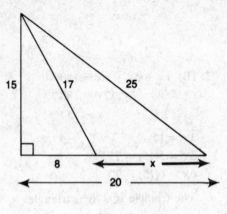

This is an 8, 15, 17
Pythagorean triple.

$x = 20 - 8 = 12$

8. **(A)** The sum of the lengths of two sides of a triangle must be greater than the length of the third side.

A) $7 + 8 > 9$ $7 + 9 > 8$ $8 + 9 > 7$
B) $3 + 5$ is not greater than 8
C) $3 + 3$ is not greater than 7
D) $3 + 6$ is not greater than 10
E) $4 + 5$ is not greater than 10

Practice Problems 10.3

1. **(B)**

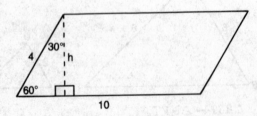

$$\text{line ratio} = \frac{\text{master}}{\text{problem}} = \frac{2}{4} = \frac{\sqrt{3}}{h}$$

$$2h = 4\sqrt{3}$$

$$h = 2\sqrt{3}$$

$$A = (\text{base})(ht)$$

$$= 10(2\sqrt{3}) = 20\sqrt{3}$$

2. (C)

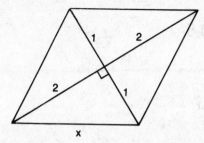

$$x$$

$$1^2 + 2^2 = x^2$$
$$5 = x^2$$
$$x = \sqrt{5} = \text{each side}$$
$$\text{perimeter} = 4 \times \sqrt{5} = 4\sqrt{5}$$

3. (E)

$(4t + 6)°$ $(5t - 15)°$

Consecutive angles of a parallelogram are supplementary.

$$(4t + 6) + (5t - 15) = 180$$
$$9t - 9 = 180$$

$$9t = 189$$
$$t = 21$$
$$4t + 6 = 4(21) + 6 = 90$$
$$5t - 15 = 5(21) - 15 = 90$$

A parallelogram with 90° angles is either a rectangle or a square.

4. (A)

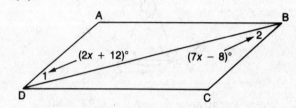

Since $AD \parallel BC$
$\angle 1$ and $\angle 2$ are alternate interior angles.

$$2x + 12 = 7x - 8$$
$$20 = 5x$$
$$4 = x$$

Practice Problems 10.4

1. (A) The measure of each exterior angle $= \dfrac{360}{n}$.

decagon = 10 sides $\therefore n = 10$

$$\frac{360}{10} = 36$$

2. (D) The sum of the measures of the interior angles $= (n - 2)180$

$$n = 12$$
$$(12 - 2)(180) = 1800$$

3. (B) The measure of each exterior $= \dfrac{360}{n}$.

$$72 = \frac{360}{n}$$
$$n = 5$$

4. (A)

The sum of the measures of the interior angles $=$ The sum of the measures of the exterior angles

$$(n - 2)(180) = 360$$
$$180n - 360 = 360$$
$$180n = 720$$
$$n = 4$$

Practice Problems 10.5

1. **(C)**

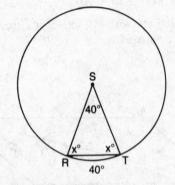

$\angle S$ is a central angle. $m\angle S = m\overset{\frown}{RT} = 40$

$\triangle RST$ is isosceles $\quad \therefore m\angle R = m\angle T = x$

$40 + x + x = 180$

$\qquad 2x = 140$

$\qquad x = 70$

2. **(B)**

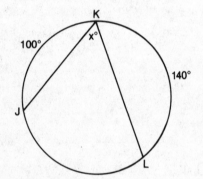

$\angle JKL$ is an inscribed angle

$m\angle JKL = x = \frac{1}{2}m\overset{\frown}{JL}$

$100 + 140 + m\overset{\frown}{JL} = 360$

$\qquad m\overset{\frown}{JL} = 120$

$\qquad x = \frac{120}{2} = 60$

3. **(D)**

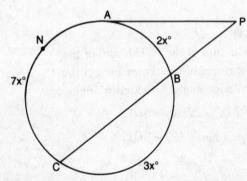

$m\overset{\frown}{AB} = 2x$

$m\overset{\frown}{BC} = 3x$

$m\overset{\frown}{CNA} = 7x$

$2x + 3x + 7x = 360$

$\qquad 12x = 360$

$\qquad x = 30$

$m\angle P = \dfrac{m\overset{\frown}{ANC} - m\overset{\frown}{AB}}{2} = \dfrac{7(30) - 2(30)}{2}$

$\qquad = \dfrac{210 - 60}{2}$

$\qquad = \dfrac{150}{2} = 75$

4. **(E)** $A = \pi r^2 = 12\pi$

$\qquad r^2 = 12$

$\qquad r = \sqrt{12}$

$C = 2\pi r = 2(\sqrt{12})\pi = 2\sqrt{12}\pi$

$\qquad = 2(2\sqrt{3})\pi$

$\qquad = 4\pi\sqrt{3}$

Note: $\sqrt{12} = \sqrt{4 \cdot 3}$

$\qquad = \sqrt{4} \cdot \sqrt{3}$

$\qquad = 2\sqrt{3}$

5. **(D)** Shaded area = Area $\odot$ − Area $\square$

Area circle = $\pi r^2 = \pi(4)^2 = 16\pi$

diagonal of square = 8

Area square = $\dfrac{d^2}{2} = \dfrac{8^2}{2} = 32$

Shaded area = $16\pi - 32$

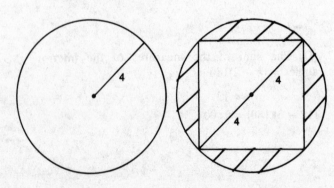

6. **(A)** The radius of each semi-circle is 3.

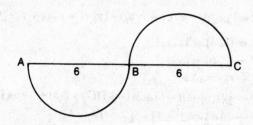

length semicircle path $= \frac{1}{2}$ circumference

length $\overset{\frown}{AB} = \frac{1}{2} \cdot 2\pi r$

$\qquad = \frac{1}{2}(2\pi(3))$

$\qquad = 3\pi$

length $BC = \frac{1}{2} \cdot 2\pi r$

$\qquad = \frac{1}{2}(2\pi(3))$

$\qquad = 3\pi$

$3\pi + 3\pi = 6\pi$

7. **(B)** A line from the center of a circle, perpendicular to a chord, bisects the chord. Draw the radius to complete a right triangle.

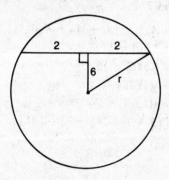

By the Pythagorean theorem

$2^2 + 6^2 = r^2$

$4 + 36 = r^2$

$r^2 = 40$

$r = \sqrt{40} = \sqrt{4 \cdot 10} = \sqrt{4}\sqrt{10} = 2\sqrt{10}$

8. **(C)**

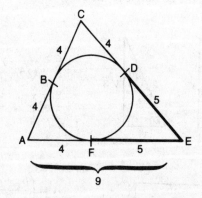

$DE = FE = 5$

Two tangents drawn from an outside point are equal in length.

$AF = 9 - FE \quad \therefore AF = 4$

$AB = AF = 4$

Two tangents drawn from an outside point are equal in length.

$BC = CD = 4$

Two tangents drawn from an outside point are equal in length.

$AC = AB + BC = 4 + 4 = 8$

Practice Problems 10.6

1. **(E)** Shaded area = area $\square$ minus area $\bigcirc$
Circle

$A_\square = S^2 = (12)^2 = 144$

$A_\bigcirc = \pi r^2 = \pi(6)^2 = 36\pi$

$\quad A = 144 - 36\pi$

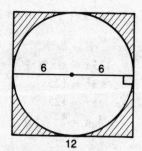

2. **(D)**

$A = \sqrt{s(s - a)(s - b)(s - c)}$

$s = \frac{1}{2}(6 + 8 + 12) = 13$

$A = \sqrt{13(13 - 6)(13 - 8)(13 - 12)}$

$\quad = \sqrt{13(7)(5)(1)}$

$\quad = \sqrt{455}$

3. **(C)**

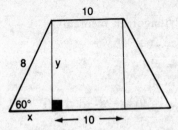

Using the 60° angle of the 30° − 60° − 90° triangle

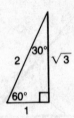
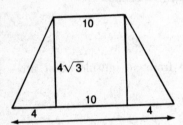

$$\frac{8}{2} = \frac{y}{\sqrt{3}} \quad y = 4\sqrt{3}$$

$$\frac{8}{2} = \frac{x}{1} \quad x = 4$$

$$A = \left(\frac{b_1 + b_2}{2}\right) h$$

$$= \left(\frac{10 + 18}{2}\right) 4\sqrt{3}$$

$$= 14(4\sqrt{3})$$

$$= 56\sqrt{3}$$

4. **(B)** Method 1:

$$A = \tfrac{1}{2}|(x_1 y_2 + x_2 y_3 + x_3 y_1) - (y_1 x_2 + y_2 x_3 + y_3 x_1)|$$

$N = (1, 1) = (x_1 y_1)$

$J = (7, 6) = (x_2 y_2)$

$L = (3, -6) = (x_3 y_3)$

$$= \tfrac{1}{2}|(1(6) + 7(-6) + 3(1)) - (1(7) + 6(3) + (-6)(1)|$$

$$= \tfrac{1}{2}|[6 - 42 + 3] - [7 + 18 - 6]|$$

$$= \tfrac{1}{2}|-33 - 19| = \tfrac{1}{2}|-52| = \tfrac{52}{2} = 26$$

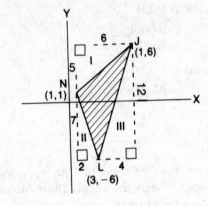

Method 2:

$$A_\triangle = A_{\text{rect}} - (A_\text{I} + A_\text{II} + A_\text{III})$$

$$A_{\text{rect}} = 12(6) = 72$$

$$A_\text{I} = \tfrac{1}{2}(5)(6) = 15$$

$$A_\text{II} = \tfrac{1}{2}(7)(2) = 7$$

$$A_\text{III} = \tfrac{1}{2}(12)(4) = 24$$

$$72 - [15 + 7 + 24] = 72 - [46] = 26$$

Practice Problems 11.1

1. **(B)**

$$\text{T.A.} = \text{L.A.} + 2B$$

$$= 2\pi rh + 2(\pi r^2)$$

$$= 2\pi(3)(10) + 2(\pi(3^2))$$

$$= 60\pi + 18\pi = 78\pi$$

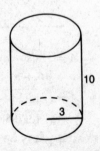

2. **(A)** Let x = height and radius.

$$V = \pi r^2 h$$

$$125\pi = \pi(x^2)(x)$$

$$\pi x^3 = 125\pi$$

$$x^3 = 125$$

$$x = 5$$

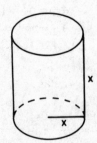

3. **(A)**

$V = \frac{4}{3}\pi r^3$
$= \frac{4}{3}\pi(3)^3$
$= 36\pi$

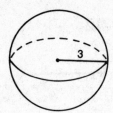

4. **(D)**

L.A. $= \pi rl$
L.A. $= \pi(8)(17)$
$= 136\pi$

Note: $l^2 = 15^2 + 8^2$
$l^2 = 225 + 64$
$l^2 = 289$
$l = 17$

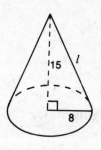

Practice Problems 12.1

1. **(B)** Do each part separately:

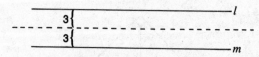

Equidistant from l and m

3 units from point P on l

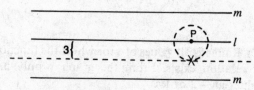

Composite

One point (indicated by the **X**)

2. **(C)**

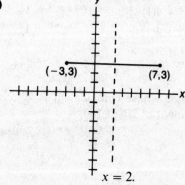

The locus in the perpendicular bisector of the line segment joining the two points.

3. **(E)**

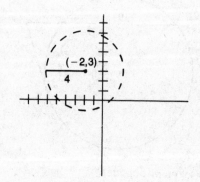

A circle with center $(-2, 3)$ and radius $= 4$
$(x - h)^2 + (y - k)^2 = r^2$
$(x + 2)^2 + (y - 3)^2 = 16$

4. **(B)** Do each part separately:

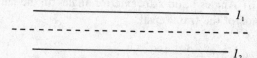

Equidistant from two parallel lines

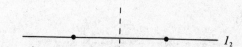

Equidistant from two points on one of the lines

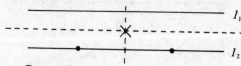

Composite

One point (indicated by the **X**)

Cumulative Review 1–12

1. **(C)**

$$\log 10a^2 = \underbrace{\log 10}_{1} + 2 \underbrace{\log a}_{p}$$

$$= 1 + 2p$$

2. **(B)**

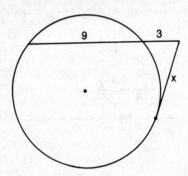

$$(2 - 2i)(2 - 2i) = 4 - 4i - 4i + 4i^2$$
$$= 4 - 8i + 4(-1)$$
$$= -8i$$

3. **(A)**

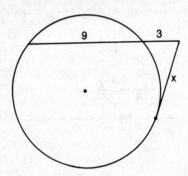

(whole)(segment outside)

$\qquad$ = (whole)(segment outside)

$$(9 + 3)(3) = x(x)$$
$$x^2 = 12 \cdot 3 = 36$$
$$x = 6$$

4. **(B)** Angles 1 and 2 are exterior angles on the same side of the transversal.

$$\underbrace{m \angle 1}_{110} + m \angle 2 = 180$$

$$m \angle 2 = 70$$

5. **(E)**

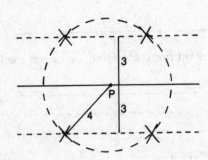

Four points (indicated by Xs)

6. **(E)**

$$\begin{array}{r} 54\,r3 \\ 4\overline{)219} \\ \underline{20} \\ 19 \\ \underline{16} \\ 3 \end{array} \qquad i^{219} = i^3 = -i$$

7. **(D)**

$$V = \pi r^2 h = \pi(3)^2(6) = 54\pi$$

$$\dfrac{54\pi \text{ units}^3}{27 \dfrac{\text{units}^3}{\text{gal}}} = 2\pi \text{ gal}$$

8. **(A)**

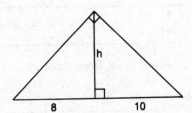

$$h^2 = 8 \cdot 10 = 80$$
$$h = \sqrt{80} = \sqrt{16}\sqrt{5} = 4\sqrt{5}$$

9. **(E)** Range is all values of y for which the function or relation exists. From the graph y only has values at -2 or less.

$$\therefore \text{ range: } y \leqslant -2$$

10. **(B)** $(a + 1)^0 + 4(a)^{-1/2}$

$\qquad (1 + 1)^0 + 4(1)^{-1/2} \qquad 1^{\text{any power}} = 1$

$\qquad\qquad 2^0 + 4(1) \qquad\qquad\qquad 2^0 = 1$

$\qquad\qquad 1 + 4 = 5$

11. **(C)** The angle whose complement is $70°$, is $20°$. The supplement of $20°$ is $160°$.

12. **(D)**

$$\log \dfrac{a}{b\sqrt{c}} = \log a - \log b\sqrt{c}$$

$$\log a - (\log b + \tfrac{1}{2} \log c)$$

13. **(E)**

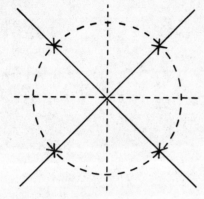

Four points (indicated by **X**s)

14. **(E)** The $\sqrt{2-x}$ is imaginary when
$$2 - x < 0 \quad \therefore x > 2.$$

The smallest integer value greater than 2, is 3.

15. **(D)** $\frac{1}{4}$ empty is $\frac{3}{4}$ full.

The water rises to $\frac{3}{4}(12) = 9$ units.

Volume = (base area)(height)
$$= (6 \cdot 10)(9)$$
$$= 540 \text{ cubic units}$$

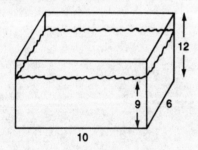

16. **(B)**

$$
\begin{array}{r}
4.\ 3\ 5 \\
\sqrt{19.0000} \\
\end{array}
$$

$$
\begin{array}{l}
16 \\
83\ |\ \overline{300} \\
\quad 249 \\
865\ |\ \overline{5100} \\
\end{array}
$$

Note: $\begin{array}{r} 865 \\ \times 5 \\ \hline 4325 \end{array}$

$\therefore$ to the nearest tenth $= 4.4$

17. **(B)**
$$\left(\frac{49m^{-2}}{16r^{-2}}\right)^{-1/2} = \frac{49^{-1/2}m^1}{16^{-1/2}r^1} = \frac{\frac{1}{7}m}{\frac{1}{4}r}$$
$$= \frac{4m}{7r}$$

18. **(A)**

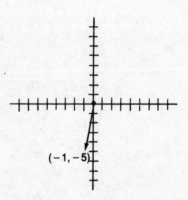

$m\angle ABC = 180 - 50 = 130$
$\therefore m\angle A = 180 - (25 + 130)$
$\qquad = 25$

$\therefore \triangle ABC$ is isosceles and $AB = BC = 13$

19. **(C)**
$$(-3 - 2i) + (2 - 3i) = -1 - 5i$$

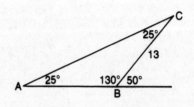

20. **(D)** To be a right triangle
$$(\text{leg})^2 + (\text{leg})^2 = (\text{hyp.})^2$$
$$8^2 + 15^2 = 17^2$$
$$64 + 225 = 289$$
$$289 = 289\ \sqrt{}$$

UNIT FOUR

13. Analytic Geometry

13.1 Basic Coordinate Geometry Equations

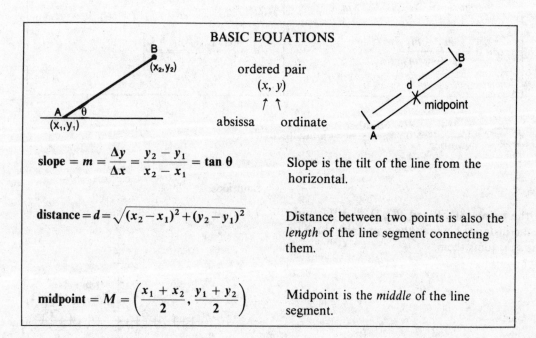

BASIC EQUATIONS

ordered pair
(x, y)

absissa   ordinate

$\text{slope} = m = \dfrac{\Delta y}{\Delta x} = \dfrac{y_2 - y_1}{x_2 - x_1} = \tan \theta$

Slope is the tilt of the line from the horizontal.

$\text{distance} = d = \sqrt{(x_2 - x_1)^2 + (y_2 - y_1)^2}$

Distance between two points is also the *length* of the line segment connecting them.

$\text{midpoint} = M = \left(\dfrac{x_1 + x_2}{2}, \dfrac{y_1 + y_2}{2} \right)$

Midpoint is the *middle* of the line segment.

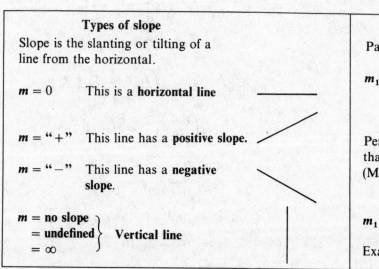

Types of slope

Slope is the slanting or tilting of a line from the horizontal.

$m = 0$ This is a **horizontal line**

$m = \text{``+''}$ This line has a **positive slope.**

$m = \text{``−''}$ This line has a **negative slope.**

$m = $ **no slope**
$= $ **undefined** $\Big\}$ **Vertical line**
$= \infty$

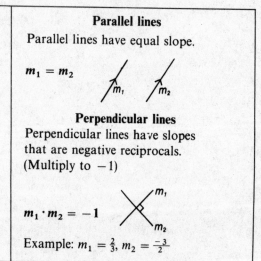

Parallel lines

Parallel lines have equal slope.

$m_1 = m_2$

Perpendicular lines

Perpendicular lines have slopes that are negative reciprocals.
(Multiply to -1)

$m_1 \cdot m_2 = -1$

Example: $m_1 = \frac{2}{3}$, $m_2 = \frac{-3}{2}$

☞ Colinear: Three points A, B and C are colinear (along the same line) if

$m_{AB} = m_{BC} = m_{AC}$

Miscellaneous Equations

Angle between two lines	**The distance from a point to a line**

Angle between two lines

If θ is the angle between two lines, θ is measured counter-clockwise.

$$\tan \theta = \frac{m_2 - m_1}{1 + m_1 m_2}$$

The distance from a point to a line

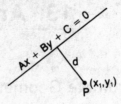

$$d = \left| \frac{Ax_1 + By_1 + C}{\sqrt{A^2 + B^2}} \right|$$

☞ $\dfrac{0}{\text{Number}} = 0,$ $\dfrac{\text{Number}}{0} = \text{Undefined}$

Examples:

1. Given the points $A(1, 3)$ and $B(7, 1)$, find the midpoint, the distance between them and the slope of the line that joins them.

Solutions:

1.

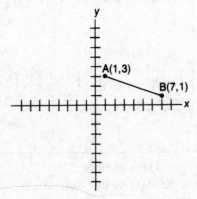

$$\text{midpoint} = M_{AB} = \left(\frac{x_1 + x_2}{2}, \frac{y_1 + y_2}{2} \right)$$

$$= \left(\frac{1 + 7}{2}, \frac{3 + 1}{2} \right) = (4, 2)$$

$$d = \sqrt{(x_2 - x_1)^2 + (y_2 - y_1)^2}$$

$$= \sqrt{(7 - 1)^2 + (1 - 3)^2}$$

$$= \sqrt{(6)^2 + (-2)^2} = \sqrt{36 + 4}$$

$$= \sqrt{40} = 2\sqrt{10}$$

$$m = \frac{\Delta y}{\Delta x} = \frac{y_2 - y_1}{x_2 - x_1} = \frac{1 - 3}{7 - 1} = \frac{-2}{6}$$

$$= -\frac{1}{3}$$

2. Given the points $C(-1, -5)$ and $D(1, -4)$, find the midpoint, the length of the line that connects them and the slope of the line.

2.

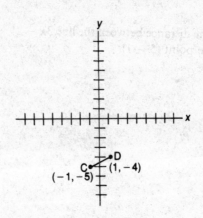

$$M_{AB} = \left(\frac{x_1 + x_2}{2}, \frac{y_1 + y_2}{2}\right)$$

$$= \left(\frac{-1 + 1}{2}, \frac{-5 + (-4)}{2}\right)$$

$$= \left(\frac{0}{2}, \frac{-9}{2}\right) = \left(0, -\frac{9}{2}\right)$$

$$d = \sqrt{(x_2 - x_1)^2 + (y_2 - y_1)^2}$$

$$= \sqrt{(-1 - 1)^2 + (-5 - (-4))^2}$$

$$= \sqrt{(-2)^2 + (-1)^2} = \sqrt{4 + 1} = \sqrt{5}$$

$$m = \frac{y_2 - y_1}{x_2 - x_1} = \frac{-4 - (-5)}{1 - (-1)}$$

$$= \frac{-4 + 5}{1 + 1} = \frac{1}{2}$$

3. Given the coordinates below, are the line segments AB and CD perpendicular?

   $A(-3, -1)$, $B(2, 2)$, $C(-3, 4)$, $D(3, -6)$

3.

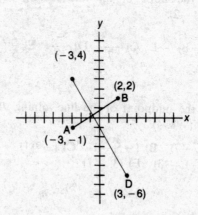

$$m_{AB} = \frac{y_2 - y_1}{x_2 - x_1} = \frac{2 - (-1)}{2 - (-3)} = \frac{2 + 1}{2 + 3} = \frac{3}{5}$$

$$m_{CD} = \frac{y_2 - y_1}{x_2 - x_1} = \frac{-6 - 4}{3 - (-3)} = \frac{-10}{6} = \frac{-5}{3}$$

m_{AB} and m_{CD} are negative reciprocals. The lines are perpendicular.

4. Find the distance between the line $3x - 2y - 2 = 0$ and the point $(5, -1)$.

4. $d = \left|\dfrac{Ax_1 + By_1 + C}{\sqrt{A^2 + B^2}}\right| = \left|\dfrac{3(5) - 2(-1) - 2}{\sqrt{3^2 + (-2)^2}}\right|$

$\left|\dfrac{15 + 2 - 2}{\sqrt{9 + 4}}\right| = \left|\dfrac{15}{\sqrt{13}}\right|$

$= \dfrac{15}{\sqrt{13}}\left(\dfrac{\sqrt{13}}{\sqrt{13}}\right) = \dfrac{15\sqrt{13}}{13}$

Practice Problems 13.1

1. Find the slope of the line joining the points $N(-3, -10)$ and $L(4, -4)$.

   A) $\dfrac{6}{7}$　B) $\dfrac{-6}{7}$　C) $\dfrac{-1}{14}$　D) $\dfrac{-14}{1}$　E) $\dfrac{7}{6}$

2. Find the length of the line joining the points $N(-7, 0)$ and $L(-4, -2)$

   A) 13　B) $\sqrt{13}$　C) $\sqrt{41}$　D) $\sqrt{97}$　E) $\sqrt{125}$

3. Find the slope of the line joining $R(5, 2)$ and $S(-5, -2)$.

   A) 1　B) -1　C) $\dfrac{-3}{7}$　D) $\dfrac{-7}{3}$　E) $\dfrac{2}{5}$

4. Find the midpoint of the line joining $J(-6, -2)$ and $K(-8, 4)$.

   A) $(1, -7)$　B) $(-5, -1)$　C) $(-7, 1)$
   D) $(-1, -3)$　E) $(3, -1)$

5. If the midpoint M of a line segment AB is $M(3, 2)$ and endpoint A is $(8, 3)$, find the coordinates of point B, the other endpoint.

   A) $(\tfrac{9}{2}, 3)$　B) $(-2, 1)$　C) $(\tfrac{5}{2}, \tfrac{1}{2})$
   D) $(-8, -3)$　E) $(1, -2)$

6. Are the points $N(-2, -11)$, $J(0, -7)$ and $L(4, 1)$ colinear?

   A) Yes　B) No

7. If the center of a circle has coordinates $C(2, 4)$ and a point $S(6, 1)$ is on the circumference, find the length of the radius.

   A) $\sqrt{5}$　B) $\sqrt{13}$　C) 7　D) 5　E) $\sqrt{89}$

8. If a line has a slope of $\tfrac{-5}{2}$, the slope of a perpendicular line is

   A) $\tfrac{-5}{2}$　B) $\tfrac{-2}{5}$　C) $\tfrac{5}{2}$　D) $\tfrac{2}{5}$　E) -1

Solutions on page 253

13.2 Lines

The equation of a line is a first degree equation.

☞ This means no x^2, y^2 or xy terms.

Format	Name	Comment
$Ax + By + C = 0$	General form	$AB \neq 0$ $m = -\dfrac{A}{B}$ If $A = 0$, line is horizontal. If $B = 0$, line is vertical. $b = -\dfrac{C}{B}$ If $C = 0$, line passes through origin. $m =$ slope, $b = y -$ intercept
$y = mx + b$	Slope-intercept	$m =$ slope, $b = y$-intercept
$(y - y_1) = m(x - x_1)$	Point-slope	$m =$ slope, (x_1, y_1) is a point on the line
$\dfrac{x}{a} + \dfrac{y}{b} = 1$	Intercept form	$a = x$-intercept, $b = y$-intercept
$x = c$	Vertical line	A vertical line at the value $x = c$
$y = c$	Horizontal line	A horizontal line at the value $y = c$

☞ **y-intercept: where the line crosses the y-axis.** At this point x has the value 0.

x-intercept: where the line crosses the x-axis. At this point y has the value 0.

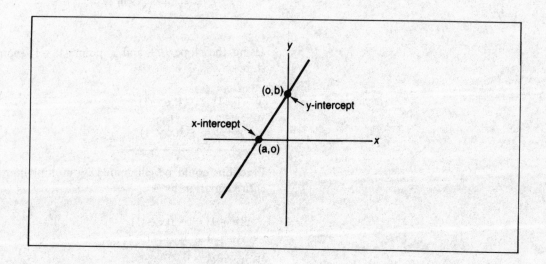

Examples:

1. Find the equation of the line passing through the points $N(-2, -1)$ and $L(1, 3)$.

2. Given the equation $3y = 6x - 9$, find the slope and y-intercept.

3. Write the equation of the line parallel to $2x + 6y = 11$ and passing through the point $V(1, -1)$.

Solutions:

1. Given two points, first find the slope:

$$m = \frac{y_2 - y_1}{x_2 - x_1} = \frac{3 - (-1)}{1 - (-2)} = \frac{4}{3}$$

Then use the point-slope form.

Using $m = \frac{4}{3}$ and *either* point, e.g., $L(1,3)$

$$y - y_1 = m(x - x_1)$$

$$\boxed{y - 3 = \tfrac{4}{3}(x - 1)}$$

2. Rearrange the equation into the slope-intercept form $y = mx + b$:

$$3y = 6x - 9$$

Divide by 3:

$$y = \frac{6x}{3} - \frac{9}{3} = 2x - 3$$

$$m = 2 \quad b = -3$$

3. To be parallel to $2x + 6y = 11$, the line we are looking for must have the same slope. First find the slope of the given line:

$$2x + 6y = 11$$
$$6y = -2x + 11$$
$$y = -\frac{1x}{3} + \frac{11}{3} \Rightarrow m = -\frac{1}{3}, b = \frac{11}{3}$$

Using the slope $-\frac{1}{3}$ and a point $(1, -1)$ (point-slope):

$$y - (-1) = -\tfrac{1}{3}(x - 1)$$

$$y + 1 = -\tfrac{1}{3}(x - 1)$$

Fractions could be eliminated by multiplying the entire equation by 3:

$$3(y + 1) = -1(x - 1)$$
$$3y + 3 = -x + 1$$
$$3y + x + 2 = 0$$

4. Find the x- and y-intercepts of $y = 3x - 7$.

4. The easiest way to find x-intercepts is to let $y = 0$.
The easiest way to find y-intercepts is let $x = 0$.

$$y = 3x - 7$$

Let $y = 0$.	Let $x = 0$.
$0 = 3x - 7$	
$x = \frac{7}{3}$	$y = 3(0) - 7$
	$y = -7$
The x-intercept is $\frac{7}{3}$.	The y-intercept is -7.
$(\frac{7}{3}, 0)$	$(0, -7)$

Graphing A Line

The three most common techniques are:

1) $y = mx + b$ Determine $m = $ slope $b = y$-intercept.
2) **intercepts**   Determine the x- and y-intercepts.
3) **table**   Create a table of coordinate values.

Graph $-3x + 2y = 6$ **using** $y = mx + b$

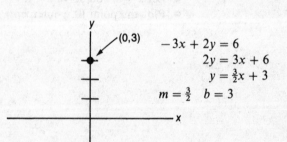

$$-3x + 2y = 6$$
$$2y = 3x + 6$$
$$y = \tfrac{3}{2}x + 3$$
$$m = \tfrac{3}{2} \quad b = 3$$

- Rearrange the equation into $y = mx + b$ form.
- Determine the slope (m) and y-intercept (b).
- On the graph, mark the y-intercept.
- The slope (m) should be written in fraction form.

$$\left(\frac{3}{2}, \frac{-3}{2}, \frac{2}{1}, \frac{-1}{5}, \frac{-6}{1} \right)$$

☞ $\quad m = \dfrac{\Delta y}{\Delta x} = \dfrac{\text{change in } y}{\text{change in } x} = \begin{cases} \text{a plus number means go up } (\uparrow) \\ \text{a minus number means go down } (\downarrow) \\ \text{a plus number means go right } (\rightarrow) \\ \text{a minus number means go left } (\leftarrow) \end{cases}$

$\dfrac{-3}{2}$ go 3 down ($\downarrow$) and then 2 to the right ($\rightarrow$)

$\dfrac{2}{5}$ go 2 up ($\uparrow$) then 5 to the right ($\rightarrow$)

- Beginning at the y-intercept use the slope to determine the next point. Repeat this step, progressing from one point to the next.
- Connect the points, and label the line.

Since $m = \frac{3}{2}$
go 3 up and 2 to the right

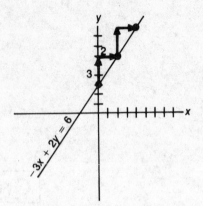

Graph $-3x + 2y = 6$ **using the x- and y-intercepts**

- Let $x = 0$ and solve for y.
- Plot the point (0, y-intercept).

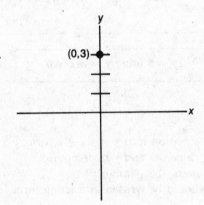

$$-3x + 2y = 6$$
$$-3(0) + 2y = 6$$
$$y = 3$$
$$\text{plot } (0,3)$$

- Let $y = 0$ and solve for x.
- Plot the point (x-intercept, 0).
- Connect the points and label the line.

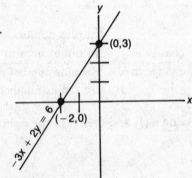

$$-3x + 2y = 6$$
$$-3x + 2(0) = 6$$
$$x = -2$$
$$\text{plot } (-2,0)$$

 If the line goes through the origin, that is, if both the x- and y-intercepts are (0,0), then determine a third point; **usually** either let $x = 1$ and solve for y or let $y = 1$ and solve for x.

Graph $-3x + 2y = 6$ **using a table (chart)**

- Prepare a chart with 3 to 5 values of x; these values are arbitrary. (Try to include 0.)
- Substitute each x into the equation and solve for the respective y value.
- Plot the points, join them, label the graph.

$$-3x + 2y = 6$$
$$-3(-1) + 2y = 6$$
$$2y = 3$$
$$y = \tfrac{3}{2}$$

$$-3(-2) + 2y = 6$$
$$2y = 0$$
$$y = 0$$

$$-3(0) + 2y = 6$$
$$2y = 3$$

$$-3(1) + 2y = 6$$
$$2y = 9$$
$$y = \tfrac{9}{2}$$

x	y
-1	$\tfrac{3}{2}$
-2	0
0	3
1	$\tfrac{9}{2}$

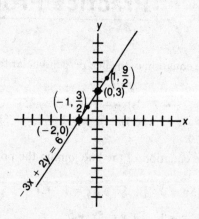

Special Line Graphs

Horizontal Lines
$$y = c$$

$y = 3$

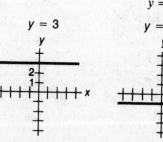

$y = -1$

$y = 0$

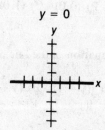

This is a horizontal
line at $y = 3$.

Note: this is a
graph of the
x-axis.

Vertical Lines

$$x = c$$

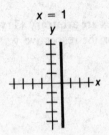
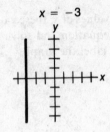
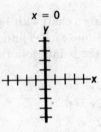

This is a vertical
line at $x = 1$.

Note: this is the
graph of the
y-axis.

Practice Problems 13.2

1. Write the equation of the line perpendicular to $y = \frac{1}{2}x - 7$ and passing through the point $(-1, 5)$.

   A) $y = -2x + 7$ B) $y + 1 = -2(x - 5)$ C) $y - 5 = -2(x - 1)$
   D) $y - 5 = -2(x + 1)$ E) $y + 1 = 2(x + 5)$

2. Write the equation of the line joining the points $N(5, 0)$ and $L(0, 5)$.

   A) $5x - 5y = 0$ B) $5y + 5x = 1$ C) $\frac{x}{5} + \frac{y}{5} = 1$
   D) $5x - 5y = 1$ E) $y = 5x + 5$

3. Find the slope of the line $7 + 3y = -4x$.

   A) -4 B) 7 C) $\frac{-4}{3}$ D) $\frac{-7}{3}$ E) $\frac{-3}{4}$

4. The coordinates of the x-intercept of the line passing through $N(0, 3)$ and $L(1, 4)$ are

   A) $(-3, 0)$ B) $(0, 0)$ C) $(1, 0)$ D) $(3, 0)$ E) $(4, 0)$

5. Write the equation of the line parallel to $\frac{x}{3} - \frac{y}{2} = 1$ and passing through the point $(1, 3)$.

   A) $(x - 3) = \frac{2}{3}(y - 1)$ B) $(y - 3) = \frac{2}{3}(x - 1)$ C) $(y - 1) = \frac{2}{3}(x - 3)$
   D) $(x - 1) = \frac{2}{3}(y - 3)$ E) $(y + 3) = \frac{2}{3}(x + 1)$

6. Each of the following points is on the line $2y = 4 - x$ except

   A) $(2, 1)$ B) $(-4, 4)$ C) $(-2, -3)$ D) $(0, 2)$ E) $(4, 0)$

7. If the points $N(1, 0)$, $J(3, 6)$ and $L(-4, k)$ are colinear, the value of k is

   A) 4   B) 2   C) -8 D) -10 E) -15

8. If $y = 3x + b$ and the point $(-3, 1)$ is on the line, find the value of b.

   A) -6 B) -8 C) -10 D) 10   E) 6

Solutions on page 254

13.3 Conic Sections—A Pictorial View

Conic sections are formed by the intersection of a plane with a right circular cone with two nappes.

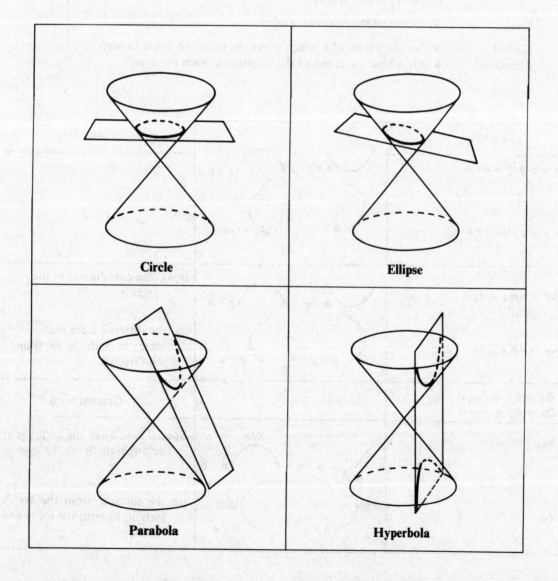

Circle	**Ellipse**
Parabola	**Hyperbola**

13.4 Parabola/Quadratic Equation

The equation of a parabola is called a quadratic equation. It is a second degree equation in which only one variable is squared and the other is linear.

Vertex: The turning point (minimum/maximum point). Has the coordinates (h, k).

Axis of Symmetry: A line through the middle of the parabola. The line containing the focus and perpendicular to the directrix.

Focus: } **Directrix:** }
- The focus is a point (on the axis of symmetry) inside the parabola.
- The directrix is a line outside the parabola.
- Both the focus and directrix are a distance "p" from the vertex.

Latus Rectum: A chord of the parabola through the focus and perpendicular to the axis of symmetry = $|4p|$ in length.

Roots: The roots of the equation $f(x)$:

(Zeros) **(Solutions)**
- Are the values of x which make the equation equal to zero.
- Are where the graph of the equation crosses the x-axis.

Format		Comments
$f(x) = y = ax^2 + bx + c$ $f(y) = x = ay^2 + by + c$	$a > 0$ $a < 0$ $a > 0$ $a < 0$	—
$(x - h)^2 = 4p(y - k)$ $(y - k)^2 = 4p(x - h)$	$p > 0$ $p < 0$ $p > 0$ $p < 0$	(h, k): the coordinates of the vertex $|p|$: the distance from the vertex to both the focus and the directrix
$x^2 + Dx + Ey + F = 0$ $y^2 + Dx + Ey + F = 0$	—	General form
$x^2 = 4py$ $y^2 = 4px$	$(0,0)$ $p > 0$ $(0,0)$ $p < 0$ $(0,0)$ $p > 0$ $(0,0)$ $p < 0$	Special case when the vertex is at the origin $(0, 0)$ $|p|$: the distance from the vertex (origin) to both the focus and directrix

Equations/Properties

Variable	*x* is the independent variable		*y* is the independent variable	
Equation	$f(x) = y$ $= ax^2 + bx + c$	$(x-h)^2 = 4p(y-k)$	$f(y) = x$ $= ay^2 + by + c$	$(y-k)^2 = 4p(x-h)$
Graph	$a > 0$ or $p > 0$	$a < 0$ or $p < 0$	$a > 0$ or $p > 0$	$a < 0$ or $p < 0$
Axis of Symmetry	$x = \dfrac{-b}{2a}$	$x = h$	$y = \dfrac{-b}{2a}$	$y = k$
Vertex (Turning Point) (Min/Max Point)	$\left(\dfrac{-b}{2a},\ f\left(\dfrac{-b}{2a}\right) \right)$	$(h,\ k)$	$\left(f\left(\dfrac{-b}{2a}\right),\ \dfrac{-b}{2a} \right)$	$(h,\ k)$
Focus	—	$p > 0 \qquad p < 0$ $F: (h,\ k+p)$	—	$p > 0 \qquad p < 0$ $F: (h+p,\ k)$
Directrix	—	$y = k - p$	—	$x = h - p$
Length of Latus Rectum	—	$|4p|$	—	$|4p|$
Eccentricity (e)	Definition: $\dfrac{\text{the distance focus to vertex}}{\text{the distance vertex to directrix}} \Rightarrow$ the eccentricity of a parabola $= \dfrac{p}{p} = 1$			
Roots, Solutions, Zeros	• The values of *x* which make $f(x) = 0$ *or* • The *x*-intercepts	—	• The values of *y* which make $f(y) = 0$ *or* • The *y*-intercepts	—

Roots/Solutions/Zeros

The words **roots, solutions** and **zeros** are **used interchangeably.** There are *three* basic techniques to find the roots of a quadratic equation:

- Factoring
- Completing the square
- Quadratic formula

Without actually finding the roots, several **characteristics** of the roots can be determined:

- **Sum** of the roots
- **Product** of the roots
- **Nature** of the roots

> **Characteristics of the roots**
> (r_1, r_2)

Write the equation as:

$f(x) = y = ax^2 + bx + c$ or

$f(y) = x = ay^2 + by + c$

a is the **coefficient of the x^2 or y^2** (including the $+$ or $-$)

b is the **coefficient of the x or y** (including the $+$ or $-$)

c is the **numerical term without a variable** (include the $+$ or $-$)

Sum of the roots $(r_1 + r_2)$	What the roots add to	$-\dfrac{b}{a}$
Product of the roots $(r_1 \cdot r_2)$	What the roots multiply to	$\dfrac{c}{a}$
Nature of the roots (discriminant)	Are the roots • imaginary or real • The same or different (equal or unequal) • Rational or irrational	$b^2 - 4ac$

Example: $y = x^2 - 3x - 1$

$a = 1$ the coefficient of x^2
$b = -3$ the coefficient of x
$c = -1$ the numerical term

$$\text{Sum} = \frac{-b}{a} = \frac{-(-3)}{1} = 3$$

$$\text{Product} = \frac{c}{a} = \frac{-1}{1} = -1$$

$$\text{Discriminant} = b^2 - 4ac$$
$$= (-3)^2 - 4(1)(-1)$$
$$= 9 + 4$$
$$= 13$$

Analysis of the Discriminant: ($b^2_{-}4ac$)

The roots are:	$b^2 - 4ac < 0$ (a negative value)	$b^2 - 4ac = 0$	$b^2 - 4ac > 0$ (a positive value)
• Real or imaginary	Imaginary	Real	Real
• The same (equal) or different (unequal)		Same (Equal)	Different (Unequal)
• Rational or irrational (has a $\sqrt{\ }$)		Rational	Rational—If the value is a perfect square, i.e., 1, 4, 9, 16, 25, 36 ... Irrational—All other values
Typical graphic representation (Roots are where the graph of the equation crosses the x-axis for $f(x)$)	Discriminant < 0 No real roots	Discriminant = 0 Two equal roots	Discriminant > 0 r_1 r_2
Comments on graphic interpretation	Note: The curve does NOT cross the x-axis; the roots are imaginary.	Note: The curve is TANGENT to the x-axis. Since there are two equal (the same) roots, the curve can only touch the one value given by this root.	Note: The curve crosses the x-axis at two different values. The two crossings can be both positive, both negative, one of each or zero.

Solving a Quadratic Equation

Factoring: The easiest and quickest, *if* the equation can be factored

Completing the Square: Useful to obtain $(x - h)^2$ or $(y - k)^2$ terms

Quadratic formula: Always works, may take a while

Factoring

Find the roots of the equation $f(x) = x^2 - 5x + 6$ **by factoring.**

1. To find the roots set $f(x) = 0$.
2. Factor.
3. Set each factor equal to 0.
4. Solve for x.

$$x^2 - 5x + 6 = 0$$
$$(x - 2) \quad (x - 3) = 0$$
$$x - 2 = 0 \quad | \quad x - 3 = 0$$
$$x = 2 \quad | \quad x = 3$$

The roots are $x = 2$, $x = 3$.

Completing the Square

Find the roots of $f(x) = x^2 - 5x + 6$ **by completing the square.**

1. To find the roots set $f(x) = 0$: $x^2 - 5x + 6 = 0$

2. Algebraically move the numerical term to the other side of the equal sign: $x^2 - 5x \quad = -6$

3. Divide the coefficient of the x by 2 and add this term squared to both sides of the equation:

$$x^2 - 5x + \left(\frac{-5}{2}\right)^2 = -6 + \left(\frac{-5}{2}\right)^2$$
$$\left(x + \frac{-5}{2}\right)^2 = \frac{1}{4}$$

☞ Don't forget to carry the sign of the x terms coefficient.

4. a) Express the left side of the equation as $(x + a)^2$, where a is the value from step 3 found by dividing the coefficient by 2.
   b) Express the right side as a single numerical value.

5. Take the square root of each side of the equation.

$$\sqrt{(x - \tfrac{5}{2})^2} = \pm\sqrt{\tfrac{1}{4}}$$
$$x - \tfrac{5}{2} = \pm\tfrac{1}{2}$$

☞ Don't forget: if $x^2 = b$, then $x = \pm\sqrt{b}$

6. Separate the equation into two parts and solve for x:

$$x - \tfrac{5}{2} = +\tfrac{1}{2} \quad | \quad x - \tfrac{5}{2} = -\tfrac{1}{2}$$
$$x = \tfrac{6}{2} = 3 \quad | \quad x = \tfrac{4}{2} = 2$$

The roots are $x = 2$, $x = 3$.

Quadratic Formula

Find the roots of $f(x) = y = x^2 - 5x + 6$ **using the quadratic formula.**

QUADRATIC FORMULA
$f(x) = ax^2 + bx + c = 0$ $\qquad$ $f(y) = ay^2 + by + c = 0$
$x = \dfrac{-b \pm \sqrt{b^2 - 4ac}}{2a}$ $\qquad$ $y = \dfrac{-b \pm \sqrt{b^2 - 4ac}}{2a}$

To find the roots set $f(x) = 0$:

$$x^2 - 5x + 6 = 0$$

Determine the values of a, b and c.

a is the coefficient of the squared term (including sign):
b is the coefficient of the linear term (including sign):
c is the numerical term (including sign):

$$a = 1$$
$$b = -5$$
$$c = 6$$

Substitute into the quadratic formula:

$$x = \frac{-b \pm \sqrt{b^2 - 4ac}}{2a}$$

Solve for both values of x:

$$= \frac{-(-5) \pm \sqrt{(-5)^2 - 4(1)(6)}}{2(1)}$$

$$= \frac{5 \pm \sqrt{25 - 24}}{2} \quad = \frac{5 \pm \sqrt{1}}{2}$$

$$= \frac{5 \pm 1}{2}$$

$$x = \frac{5 + 1}{2} = \frac{6}{2} \quad \bigg| \quad x = \frac{5 - 1}{2} = \frac{4}{2}$$

$$= 3 \qquad\qquad = 2$$

Graphing a Quadratic Equation

The basic technique for graphing a parabola is to **develop a table with seven ordered pairs.** The turning point should be the center value and three points before and after should be used.

Graph the equation $f(x) = y = 2x^2 - 8x + 3$.

Arrange the equation into $y = ax^2 + bx + c$ form:

$$y = 2x^2 - 8x + 3$$

Determine the axis of symmetry:

$$x = \frac{-b}{2a} = \frac{-(-8)}{2(2)} = 2$$

$$x = -\frac{b}{2a}$$

(This is also the x value of the vertex)

Set up a chart with the value from step 2 at the center.

x	$2x^2 - 8x + 3$	y
2		

4. Choose three values before and after the starting value. Use increments of 1.

5. Determine all seven values of y. Graph the seven ordered pairs; label the ctrve. Graph and label the axis of symmetry (as a dotted line). Label the vertex.

☞ Don't forget to label the axes x and y respectively.

x	$2x^2 - 8x + 3$	y
-1	$2(-1)^2 - 8(-1) + 3 =$	13
0	$2(0)^2 - 8(0) + 3 =$	3
1	$2(1)^2 - 8(1) + 3 =$	-3
2	$2(2)^2 - 8(2) + 3 =$	-5
3	$2(3)^2 - 8(3) + 3 =$	-3
4	$2(4)^2 - 8(4) + 3 =$	3
5	$2(5)^2 - 8(5) + 3 =$	13

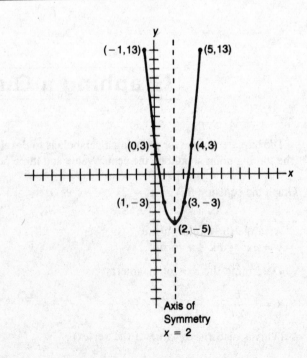

Examples:

1. If $y = x^2 - 2x + 4$, find the sum and product of the roots, the nature of the roots and the solution set.

Solutions:

1. $y = x^2 - 2x + 4 \Rightarrow a = 1, b = -2, c = 4$

$$\text{Sum} = \frac{-b}{a} = \frac{-(-2)}{1} = 2$$

$$\text{Product} = \frac{c}{a} = \frac{4}{1} = 4$$

$$\text{Discriminant} = b^2 - 4ac = (-2)^2 - 4(1)(4)$$
$$= 4 - 16 = -12$$

A negative value indicates the roots are imaginary.

$$x = \frac{-b \pm \sqrt{b^2 - 4ac}}{2a}$$

$$= \frac{-(-2) \pm \sqrt{(-2)^2 - 4(1)(4)}}{2(1)}$$

$$= \frac{2 \pm \sqrt{-12}}{2}$$

$$= \frac{2 \pm 2i\sqrt{3}}{2} = 1 \pm i\sqrt{3} \quad \text{or} \quad 1 \pm \sqrt{3}i$$

Note: $\sqrt{12} = \sqrt{4}\sqrt{3} = 2\sqrt{3}$

2. If $y = x^2 - 2x - 4$, find the coordinates of the vertex and the equation of the axis of symmetry.

2. $y = x^2 - 2x - 4 \Rightarrow a = 1, b = -2, c = -4$

Axis of symmetry:

$$x = \frac{-b}{2a} = \frac{-(-2)}{2(1)} = 1$$

$y = f(x) = x^2 - 2x - 4$
$y = f(1) = 1^2 - 2(1) - 4 = -5$

Vertex: $(1, -5)$

Or change the form:

$$y = x^2 - 2x - 4$$
$$y + 4 = x^2 - 2x$$

Complete the square:

$$y + 4 + (-1)^2 = x^2 - 2x + (-1)^2$$
$$y + 5 = (x - 1)^2$$
$$(x - 1)^2 = y + 5$$
$$(x - h)^2 = 4p(y - k)$$

Axis of symmetry is $x = h$ $\therefore x = 1$
Vertex is (h, k) $\therefore (1, -5)$

3. If $2y + 49 = 16x - y^2$, find the coordinates of the vertex and focus, the equations of the axis of symmetry and the directrix, and also the length of the latus rectum.

3. Rearrange the equation:

$$2y + 49 = 16x - y^2$$
$$y^2 + 2y = 16x - 49$$

Complete the square:

$$y^2 + 2y + \mathbf{(1)}^2 = 16x - 49 + \mathbf{(1)}^2$$
$$(y + 1)^2 = 16x - 48$$
$$(y + 1)^2 = 16(x - 3)$$
$$(y - k)^2 = 4p(x - h)$$

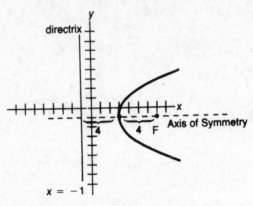

$\therefore k = -1, h = 3$ and $p = 4$
Vertex: $(h, k) \Rightarrow (3, -1)$
Focus: $(h + p, k) \Rightarrow (7, -1)$
Axis of symmetry: $y = k \Rightarrow y = -1$
Directrix: $x = h - p \Rightarrow x = -1$
Latus Rectum: $|4p| \Rightarrow |16| = 16$

4. If the roots of an equation are 3 and -1, find the equation.

4. Method 1:

If the roots are $x = 3$ and $x = -1$, the factors were $(x - 3)(x + 1) = 0$.
Therefore the equation was

$$x^2 - 3x + x - 3 = 0$$
$$x^2 - 2x - 3 = 0$$

Method 2:

$r_1 = 3 \quad r_2 = -1$
$x^2 + bx + c = 0$ (Let $a = 1$)

$$\text{Sum} = \frac{-b}{1} = -b$$

$\text{Sum} = r_1 + r_2 = 3 + (-1) = 2$
$-b = 2 \quad b = -2$

$$\text{Product} = \frac{c}{1} = c$$

$$\text{Product} = r_1 \cdot r_2 = 3(-1) = -3$$

$c = -3$
$x^2 - 2x - 3 = 0$

5. If 6 is a root of the equation $y = 2x^2 - 7x + k$, find the value k.

5. A root is the value of x which makes the equation equal to zero:

$$y = 2(6)^2 - 7(6) + k = 0$$
$$72 - 42 + k = 0$$
$$30 + k = 0$$
$$k = -30$$

Practice Problems 13.4

1. The quadratic equation whose roots are 1 and 2 is

 A) $x^2 + 2x + 3$ B) $x^2 - 2x + 3$
 C) $x^2 + 3x + 2$ D) $x^2 - 3x + 2$
 E) $x^2 - 2x - 3$

2. For which value of c will the roots of the equation $x^2 + 4x + c = 0$ be real and equal?

 A) 1 B) 2 C) 3 D) 4 E) 5

3. Find the focus of the equation $y = x^2 - 4x$.

 A) $(2, -3)$ B) $(2, -4)$ C) $(2, -5)$
 D) $(1, -3)$ E) $(1, -4)$

4. The directrix of $x^2 = 12(y - 4)$ is

 A) $y = 0$ B) $y = 1$ C) $y = 3$
 D) $x = 1$ E) $x = 3$

5. The solution set of $f(x) = x^2 - 4x - 5$ is

 A) $\{2, -3\}$ B) $\{1, 5\}$ C) $\{-1, 5\}$
 D) $\{-5, 1\}$ E) $\{-1, -5\}$

6. If $x = 2$ is a root of $y = x^2 + kx - 2$, find the value of k.

 A) -1 B) 0 C) 1 D) 2 E) 3

7. For which values of k is the graph $f(x) = kx^2 - 4x + k$ tangent to the x-axis?

 A) 1 and 4 B) 2 and -2 C) 2 only
 D) 0 and 5 E) -2 only

8. Find the roots of $2x^2 - 4 = 5x$.

 A) $\left\{\dfrac{-5 \pm \sqrt{57}}{4}\right\}$ B) $\left\{\dfrac{5 \pm \sqrt{57}}{4}\right\}$

 C) $\left\{\dfrac{5 \pm \sqrt{-7}}{4}\right\}$ D) $\left\{\dfrac{-5 \pm \sqrt{-7}}{4}\right\}$ E) $\{\pm 3\}$

Solutions on page 255

13.5 Circles

A circle is represented by a second degree equation in both x and y. When the x^2 and y^2 terms are on the same side of the equation, they have the **same sign** and the **same coefficients**.

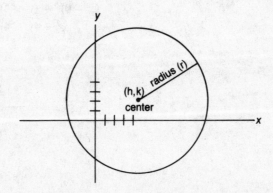

Format		
$(x - h)^2 + (y - k)^2 = r^2$	Standard form	center: (h, k) r = radius
$\dfrac{(x-h)^2}{r^2} + \dfrac{(y-k)^2}{r^2} = 1$		center: (h, k) r = radius
$x^2 + y^2 + Dx + Ey + F = 0$	General form	center: $\left(\dfrac{-D}{2}, \dfrac{-E}{2}\right)$ $r = \tfrac{1}{2}\sqrt{D^2 + E^2 - 4F}$
$x^2 + y^2 = r^2$	—	center: $(0, 0)$ r = radius

Graphing a Circle

The basic technique is to determine the center and radius. From the center mark points to the right, left, up and down of center at a distance equal to the radius.

Graph the equation $x^2 + 2x + y^2 - 4y = 4$.

1. Arrange the equation into $(x - h)^2 + (y - k)^2 = r^2$ form.

$$x^2 + 2x \qquad + y^2 - 4y \qquad = 4$$

Complete the square on x and y.

$$x^2 + 2x + (1)^2 + y^2 - 4y + (-2)^2 = 4 + (1)^2 + (-2)^2$$
$$(x + 1)^2 + (y - 2)^2 = 9$$
Center: $(-1, 2)$ $r = 3$

2. Determine the center and radius.

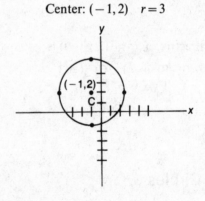

3. Graph the center; locate the four convenient points at $r = 3$.
4. Connect the graphed points.

Examples:

1. Write the equation of the circle with center: $(0,4)$ and radius $= \sqrt{5}$.

Solutions:

1. $(x - h)^2 + (y - k)^2 = r^2$
$$(x - 0)^2 + (y - 4)^2 = (\sqrt{5})^2$$
$$x^2 + (y - 4)^2 = 5$$

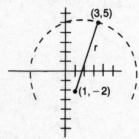

2. Write the equation of the circle with center: $(1, -2)$ and passing through $(3, 5)$.

2. **Method 1**

radius = distance $(1, -2)$ to $(3, 5)$

$$r = d = \sqrt{(5 - (-2))^2 + (3 - 1)^2}$$
$$= \sqrt{7^2 + 2^2}$$
$$= \sqrt{53}$$
$$(x - 1)^2 + (y + 2)^2 = 53$$

Method 2:

$$(x - 1)^2 + (y + 2)^2 = r^2$$

Substitute the values of x and y into the equation:

$$(3 - 1)^2 + (5 + 2)^2 = r^2$$
$$4 + 49 = r^2$$
$$r^2 = 53$$
$$\therefore (x - 1)^2 + (y + 2)^2 = 53$$

3. Find the coordinates of the center and the length of the radius of $x^2 + y^2 - 2x - 2y + 1 = 0$.

3. Rearrange and complete the square:

$$x^2 - 2x + (-1)^2 + y^2 - 2y + (-1)^2$$
$$= -1 + (-1)^2 + (-1)^2$$
$$(x - 1)^2 + (y - 1)^2 = 1$$
$$\text{Center: } (1, 1) \quad r = 1$$

Practice Problems 13.5

1. Find the equation of a circle with *Center*: $(-1, 1)$ and $r = 3$.

   A) $(x - 1)^2 + (y + 1)^2 = 3$
   B) $(x + 1)^2 + (y - 1)^2 = 3$
   C) $(x - 1)^2 + (y - 1)^2 = 9$
   D) $(x + 1)^2 + (y + 1)^2 = 9$
   E) $(x + 1)^2 + (y - 1)^2 = 9$

2. The radius of the circle $x^2 + y^2 = 8$ is

   A) 2   B) $2\sqrt{2}$ C) $4\sqrt{2}$ D) 4   E) 8

3. Find the center of $x^2 + y^2 - 6x + 4y + 10 = 0$.

   A) $(3, -2)$ B) $(2, -3)$ C) $(2, 3)$
   D) $(-2, 3)$ E) $(-3, 2)$

4. Write the equation of the circle with center: $(-1, -1)$ and passing through the point $(4, -5)$.

   A) $x^2 + y^2 = 41$
   B) $(x - 1)^2 + (y - 1)^2 = \sqrt{41}$
   C) $(x + 1)^2 + (y + 1)^2 = \sqrt{41}$
   D) $(x - 1)^2 + (y - 1)^2 = 41$
   E) $(x + 1)^2 + (y + 1)^2 = 41$

Solutions on page 255

13.6 Ellipses

An ellipse is represented by a second degree equation in both x^2 and y^2. When the x^2 and y^2 terms are put on the same side of the equation, they have the **same sign** and **different coefficients**.

Center: Has coordinates (h, k). Is at the midpoint of both the major vertex and minor vertex.

Major Axis: The longer line segment between vertices and passing through the center of the ellipse.

Minor Axis: The shorter line segment between vertices and passing through the center. The perpendicular bisector of the major axis.

Major Vertices: The endpoints of the major axis. A distance "a" from the center.

Minor Vertices: The endpoints of the minor axis. A distance "b" from the center.

Focus: A distance "c" from the center.

$$b^2 + c^2 = a^2$$

Latus Rectum: A chord of the ellipse through the focus and perpendicular to the major axis.

$$\text{Length} = \left| \frac{2b^2}{a} \right|$$

Directrix: A line outside the ellipse perpendicular to the major axis at a distance $\frac{a^2}{c}$ or $\frac{a}{e}$ from the center. Their equations are:

$$y = k \pm \frac{a}{e}$$

$$x = h \pm \frac{a}{e}$$

Eccentricity (e): A measurement defined as $\frac{c}{a}$.

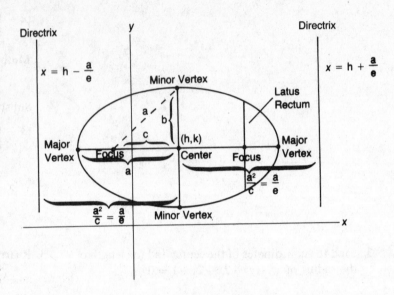

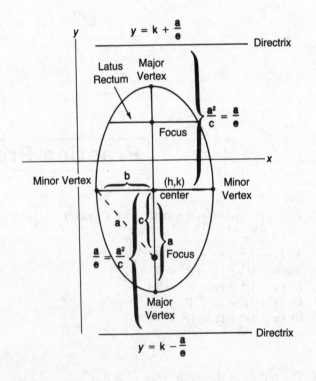

☞ The focus, directrix and latus rectum are associated with the major axis.

The eccentricity (e) for an ellipse is less than 1: $e < 1$.

	center: (h, k)
$\dfrac{(x - h)^2}{a^2} + \dfrac{(y - k)^2}{b^2} = 1$ $\left.\rule{0pt}{48pt}\right\}$ Standard form $\dfrac{(y - k)^2}{a^2} + \dfrac{(x - h)^2}{b^2} = 1$	"a" distance from center to major vertex "b" distance from center to minor vertex $a^2 > b^2$
$Ax^2 + Cy^2 + Dx + Ey + F = 0$ (A and C have the same sign)	General form
$\dfrac{x^2}{a^2} + \dfrac{y^2}{b^2} = 1$ $\dfrac{y^2}{a^2} + \dfrac{x^2}{b^2} = 1$	Special case: Center is at origin $(0, 0)$

☞ a^2 always represents the larger of the two denominator values and is always associated with the major axis.

Graphing an Ellipse

The usual approach to sketching an ellipse is to determine its center, and the coordinates of its major and minor vertices.

Sketch a graph of the ellipse given by the equation $4x^2 + 9y^2 + 16x - 18y - 11 = 0$.

1. Rearrange the equation into a more useful form.

$$4x^2 + 9y^2 + 16x - 18y = 11$$

Associate the x and y terms separately:

$$4x^2 + 16x \quad + 9y^2 - 18y \quad = 11$$

Complete the square.

When completing the square the x^2 or y^2 terms should have a coefficient of 1.

$$4(x^2 + 4x + (+2)^2) + 9(y^2 + 2y + (+1)^2)$$
$$= 11 + 4(+2)^2 + 9(+1)^2$$

 The factored out numbers, in this case a 4 and 9, multiply the $(+2)^2$ and $(+1)^2$, respectively, adding $4(+2)^2$ and $9(+1)^2$ to the other side of the equation.

$$4(x + 2)^2 + 9(y + 1)^2 = 11 + 16 + 9$$
$$= 36$$

$$\frac{4(x + 2)^2}{36} + \frac{9(y + 1)^2}{36} = \frac{36}{36}$$

$$\frac{(x + 2)^2}{9} + \frac{(y + 1)^2}{4} = 1$$

2. Determine the center and vertices.

Center: $(-2, -1)$

Major vertices ± 3 from center
Minor vertices ± 2 from center

The major axis is parallel to the x axis.

3. Graph the information from step 2 and sketch the ellipse.

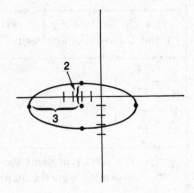

Examples:

1. Find the equation of an ellipse having foci at $(3, 8)$ and $(3, 2)$, and whose eccentricity is $\frac{3}{5}$.

Solutions:

1. Draw a sketch to assist.

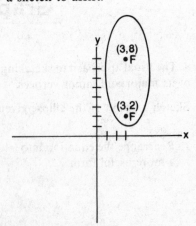

Since the foci are on the major axis the form is

$$\frac{(y-k)^2}{a^2} + \frac{(x-h)^2}{b^2} = 1$$

Since the foci are equidistant from the center, the center is the midpoint between the foci.

$$\text{center} = MP = \left(\frac{3+3}{2}, \frac{8+2}{2}\right) = (3, 5)$$

The distance "c" center to focus is 3.

$$c = 3$$

The eccentricity $e = \frac{3}{5} = \frac{c}{a}$

Substituting $c = 3 \Rightarrow a = 5$

$$b^2 + c^2 = a^2$$
$$b^2 \quad = 5^2 - 3^2 = 16$$

Center: (3, 5) $a = 5$, $b = 4$

$$\frac{(y-5)^2}{25} + \frac{(x-3)^2}{16} = 1$$

2. Find the coordinates of the center, foci, major vertices, minor vertices and the length of the latus rectum for the ellipse $2x^2 + 3y^2 - 4x - 18y + 23 = 0$.

2. Rearrange the equation.
First separate the variables and complete the square.

 Be sure the x^2 and y^2 terms have a coefficient of 1 when completing the square

$$2x^2 - 4x + 3y^2 - 18x = -23$$
$$2(x^2 - 2x + (-1)^2) + 3(y^2 - 6x + (-3)^2)$$
$$= -23 + 2(-1)^2 + 3(-3)^2$$
$$2(x-1)^2 + 3(y-3)^2 = -23 + 2 + 27$$
$$= 6$$
$$\frac{2(x-1)^2}{6} + \frac{3(y-3)^2}{6} = \frac{6}{6}$$
$$\frac{(x-1)^2}{3} + \frac{(y-3)^2}{2} = 1$$

● Center: (1, 3)

$a = \sqrt{3}$
$b = \sqrt{2}$

● Major axis is parallel to the x axis
$$b^2 + c^2 = a^2$$
$$(\sqrt{2})^2 + c^2 = (\sqrt{3})^2$$
$$2 + c^2 = 3$$
$$c = 1$$

- Foci: (2, 3)
  (0, 3)

- Major vertices: $(1 + \sqrt{3}, 3)$
  $(1 - \sqrt{3}, 3)$

- Minor vertices: $(1, 3 + \sqrt{2})$
  $(1, 3 - \sqrt{2})$

- Length of latus rectum:

$$\left|\frac{2b^2}{a}\right| = \frac{2(2)}{\sqrt{3}} = \frac{4\sqrt{3}}{3}$$

Practice Problems 13.6

1. Find the equation of the ellipse with $V(0, \pm 6)$ and $F(0, \pm 2)$.

   A) $\dfrac{x^2}{36} + \dfrac{y^2}{32} = 1$ B) $\dfrac{y^2}{36} + \dfrac{x^2}{32} = 1$

   C) $\dfrac{y^2}{36} + \dfrac{x^2}{4} = 1$ D) $\dfrac{x^2}{36} + \dfrac{y^2}{4} = 1$

   E) $\dfrac{y^2}{32} + \dfrac{x^2}{4} = 1$

2. Find the distance from the center to the major vertex for the ellipse with equation

$$\frac{(x-1)^2}{4} + \frac{(y-2)^2}{3} = 3$$

   A) $\sqrt{3}$ B) 2   C) $\dfrac{2\sqrt{3}}{3}$ D) $\dfrac{4}{3}$ E) $\sqrt{12}$

3. If $4x^2 + 9y^2 = 36$, find the length of the latus rectum.

   A) $\dfrac{4}{3}$ B) $\dfrac{10}{3}$ C) $\dfrac{8}{3}$ D) $\dfrac{81}{2}$ E) 9

4. Find the distance from the center to a focus for the ellipse

$$\frac{25x^2}{9} + 4y^2 = 1$$

   A) $\dfrac{\sqrt{11}}{10}$ B) $\dfrac{\sqrt{11}}{6}$ C) $\dfrac{5}{2}$ D) $\dfrac{4}{3}$ E) $\dfrac{7}{6}$

Solutions on page 256

13.7 Hyperbola

A hyperbola is represented by a second degree equation in both x^2 and y^2. When the x^2 and y^2 terms are on the same side of the equation, the two terms have **different signs.**

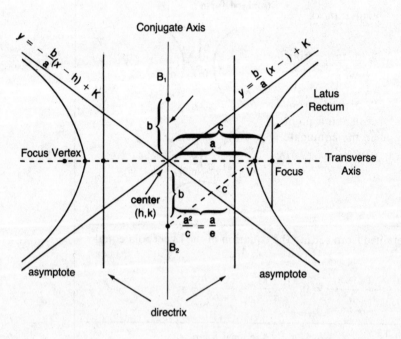

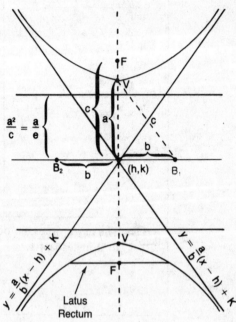

Center:	Has the coordinates (h, k). Is at the midpoint of the line segment connecting the two vertices.
Vertices:	The turning points. A distance **'a'** from the center.
Foci:	A distance **'c'** from the center.
Transverse Axis:	The line through the foci, vertices and center.
Conjugate Axis:	The line through the center perpendicular to the transverse axis. It extends between two points (B_1, B_2) each a distance **'b'** from the center.

☞ The hyperbola does not cross the conjugate axis but points B_1 and B_2 are useful in its construction.

Latus Rectum:	A chord of the hyperbola perpendicular to the transverse axis, passing through the focus. Its length is $\left\lvert \dfrac{2b^2}{a} \right\rvert$.
Directrices:	Two lines, each a distance $\dfrac{a^2}{c}$ or $\dfrac{a}{e}$ from the center.
Asymptotes:	Two lines which the hyperbola branches approach but do not touch.

Their equation are $\begin{cases} y = \pm b/a(x - h) + k & \text{for} \quad \rightarrow\leftarrow \quad \text{hyperbolas} \\ y = \pm a/b(x - h) + k & \text{for} \quad \updownarrow \quad \text{hyperbolas} \end{cases}$

$$\boxed{b^2 + a^2 = c^2}$$

Eccentricity: (e)   A measurement defined as $\dfrac{c}{a}$.

 The eccentricity (e) of a hyperbola is greater than $e > 1$.

$\dfrac{(x-h)^2}{a^2} - \dfrac{(y-k)^2}{b^2} = 1$	**Standard form** center: (h, k) 'a' distance from center to vertex
$\dfrac{(y-k)^2}{a^2} - \dfrac{(x-h)^2}{b^2} = 1$	'b' distance from center to a point on the conjugate axis 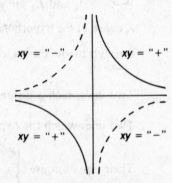

 The asymptotes are obtained from setting the equation of the hyperbola equal to 0 instead of 1:

$$\frac{(x-h)^2}{a^2} - \frac{(y-k)^2}{b^2} = 0, \quad \frac{(y-k)^2}{a^2} - \frac{(x-h)^2}{b^2} = 0$$

$Ax^2 + Cy^2 + Dx + Ey + F = 0$ (A and C have different signs)	General form
$\dfrac{x^2}{a^2} - \dfrac{y^2}{b^2} = 1$	Special case: center: $(0, 0)$ Asymptotes are: $y = \pm\dfrac{a}{b}x$
$\dfrac{y^2}{a^2} - \dfrac{x^2}{b^2} = 1$	Asymptotes are: $y = \pm\dfrac{b}{a}x$

☞ a^2 is always associated with the positive term.

$xy = a$	The hyperbola whose asymptotes are the x and y axis

$xy = "-"$ $xy = "+"$

$xy = "+"$ $xy = "-"$

$a > 0$ the curve is in the I and III quadrants
$a < 0$ The curve is in the II and IV quadrants

Graphing a Hyperbola

The usual approach to sketching a hyperbola is to determine its center, vertices and the points B_1 and B_2 on the conjugate axis. Then the asymptotes are constructed by completing a rectangle containing the vertices, B_1 and B_2. The diagonals of the rectangle form the asymptotes.

Sketch a graph of the hyperbola given by the equation $4x^2 - 9y^2 + 16x - 18y - 29 = 0$

1. Rearrange the equation into a more useful form.

$$4x^2 - 9y^2 + 16x - 18y = 29$$

Associate the x and y terms separately and complete the square.

$$4x^2 + 16x - 9y^2 - 18y = 29$$
$$4(x^2 + 4x + (2)^2) - 9(y^2 + 2y + (1)^2)$$
$$= 29 + 4(2)^2 - 9(1)^2$$
$$4(x + 2)^2 - 9(y + 1)^2 = 29 + 16 - 9$$
$$= 36$$
$$4(x + 2)^2 - 9(y + 1)^2 = 36$$

divide by 36 to make the equation equal to 1

$$\frac{(x + 2)^2}{9} - \frac{(y + 1)^2}{4} = 1$$

2. Determine the center, vertices, 'a' and 'b'.

Center: $(-2, -1)$

Vertices are a distance 3 from the center.

$$a = 3$$

The points on the conjugate axis are a distance 2 from the center.

$$b = 2$$

$$B_1(-2, 1) \qquad B_2(-2, -3)$$

3. Graph the center, vertices, B_1 and B_2, complete the rectangle and draw the asymptotes.

4. Sketch the hyperbola.

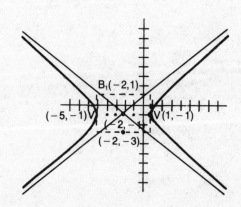

Examples:

1. Find the center, foci, vertices and eccentricity of the hyperbola $4x^2 - 9y^2 - 8x + 36y = 68$.

Solutions:

1. Put in standard form by completing the square.

$(4x^2 - 8x) \qquad -(9y^2 + 36y) \quad = 68$
$4(x^2 - 2x + (-1)^2) - 9(y^2 - 4y + (-2)^2)$
$$= 68 + 4(-1)^2 - 9(-2)^2$$

$4(x-1)^2 - 9(y-2)^2 = 36$

divide each term by 36 to make the equation equal to 1.

$$\frac{(x-1)^2}{9} - \frac{(y-2)^2}{4} = 1$$

center: $(1, 2)$ $\qquad a = 3 \qquad b = 2$

$$c^2 = a^2 + b^2$$
$$= 9 + 4$$
$$= 13$$
$$c = \sqrt{13}$$

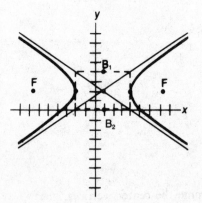

center: $(1, 2)$
Vertices: $(-2, 2)$, $(4, 2)$
Foci: $(1 - \sqrt{13}, 2)$, $(1 + \sqrt{13}, 2)$

Eccentricity: $e = \dfrac{c}{a} = \dfrac{\sqrt{13}}{3}$

2. Write the equation of the hyperbola with $C: (0, 0)$, one vertex at $(0, 6)$ and focus at $(0, 10)$.

2.

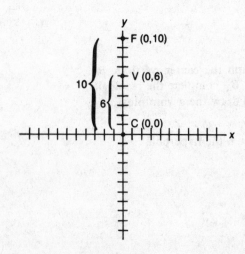

The format is $\dfrac{y^2}{a^2} - \dfrac{x^2}{b^2} = 1$.

$$a = 6 \qquad c = 10$$

$$\therefore b^2 + a^2 = c^2$$
$$b^2 = 10^2 - 6^2$$
$$= 64$$
$$b = 8$$

$$\frac{y^2}{36} - \frac{x^2}{64} = 1$$

3. Find the equation of the asymptotes for the hyperbola $4(y - 4)^2 - 9(x - 2)^2 = 36$.

3. $4(y-4)^2 - 9(x-2)^2 = 36$

divide each term by 36 to make the equation equal to 1

$$\frac{(y - 4)^2}{9} - \frac{(x - 2)^2}{4} = 1$$

center: $(2, 4)$ $\quad a = 3, b = 2$

$$y = \pm\tfrac{3}{2}(x - 2) + 4$$

Practice Problems 13.7

1. Find the length of the latus rectum for the hyperbola

$$\frac{x^2}{25} - \frac{y^2}{9} = 1$$

A) $\frac{50}{9}$ B) $\frac{18}{5}$ C) 3 D) $\frac{50}{3}$ E) $\sqrt{34}$

2. Find the coordinates of one of the foci of

$$\frac{(x - 1)^2}{2} - \frac{(y + 11)^2}{2} = 1$$

A) $(1, -11)$ B) $(-1, -11)$ C) $(2, -11)$
D) $(1, 13)$ E) $(1, 9)$

3. Find the equation of one of the asymptotes for the hyperbola

$$16x^2 + 96x - 9y^2 + 36y = 36$$

A) $3y - 4x - 18 = 0$ B) $4y - 3x - 17 = 0$
C) $96x + 36y = 36$ D) $9y - 16x - 62 = 0$
E) $9y + 16x + 30 = 0$

4. The graph of $xy = -3$ is in which quadrants?

A) I and II only B) I and III only
C) II and IV only D) I and IV only
E) II only

Solutions on page 257

13.8 Summary of Characteristics

Type of Relation	Characteristics of Equations	Examples	Usual Format	Graphic Example	Misc.
Line	All linear terms	$x = 4, \; y = 7$ $3x + 2y = 9$ $y = 5x + 3$	$y = mx + b$		
Parabola	Only one variable is squared, the other is linear	$y^2 = 8, \; y^2 + x = 7y$ $y = 3x^2 - 9x + 2$	$(y - k) = 4p(x - h)^2$ and $(x - h) = 4p(y - k)^2$ or $y = ax^2 + bx + c$ and $x = ay^2 + by + c$		$e = 1$

With x^2 and y^2 on the same side of the equation

Circle	Both variables squared, same signs, same coefficients	$x^2 + y^2 = 9$ $3x^2 + 3y^2 = 9$ $(x - 2)^2 + (y + 1)^2 = 9$	$(x - h)^2 + (y - k)^2 = r^2$		
Ellipse	Both variables squared, same signs, different coefficients	$x^2 + 2y^2 = 9$ $3x^2 + 4y^2 = 9$ $\dfrac{x^2}{2} + \dfrac{y^2}{4} = 9$ $\dfrac{(x-1)^2}{2} + \dfrac{(y+2)^2}{3} = 1$	$\dfrac{(x - h)^2}{a^2} + \dfrac{(y - k)^2}{b^2} = 1$ $\dfrac{(y - k)^2}{a^2} + \dfrac{(x - h)^2}{b^2} = 1$		$e < 1$
Hyperbola	Both variables squared, different signs (coefficients can be the same or different)	$x^2 - 2y^2 = 9$ $\dfrac{y^2}{3} - \dfrac{x^2}{4} = 1$ $\dfrac{(x-1)^2}{4} - \dfrac{(y+2)^2}{3} = 1$	$\dfrac{(x - h)^2}{a^2} - \dfrac{(y - k)^2}{b^2} = 1$ $\dfrac{(y - k)^2}{a^2} - \dfrac{(x - h)^2}{b^2} = 1$		$e > 1$
	$xy = a$	$xy = 6, \quad x = \dfrac{2}{y}$	$xy = a$		

Practice Problems 13.8

1. The equation

$$2x^2 + 4y^2 + 2x + 4y + 2 = 0$$

is a(n)

A) circle   B) parabola   C) ellipse
D) hyperbola   E) line

2. The equation

$$2x^2 + 2y^2 + 2x + 2y + 2 = 0$$

is a(n)

A) circle   B) parabola   C) ellipse
D) hyperbola   E) line

3. The equation

$$2x^2 - 4y^2 + 2x + 4y + 2 = 0$$

is a(n)

A) circle   B) parabola   C) ellipse
D) hyperbola   E) line

4. The equation

$$2x^2 - 4y + 2x + 2 = 0$$

is a(n)

A) circle   B) parabola   C) ellipse
D) hyperbola   E) line

Solutions on page 257

14. Special Products, Factoring and Applications of Factoring

14.1 Special Products

These are frequently used algebraic multiplications. They serve as a model for more complex products of the same form. Because of their frequency of occurrence math students may find it useful to commit these to memory. These products are also used in reverse to convert from a polynomial into factors.

$$(x + y)^2 = (x + y)(x + y) = x^2 + 2xy + y^2$$
$$(x - y)^2 = (x - y)(x - y) = x^2 - 2xy + y^2$$
$$(x + y)(x - y) = x^2 - y^2$$

$$(x + y)(x^2 - xy + y^2) = x^3 + y^3$$
$$(x - y)(x^2 + xy + y^2) = x^3 - y^3$$

  $(a+b)^2 \neq a^2 + b^2$ unless $ab = 0$
$(a-b)^2 \neq a^2 + b^2$ unless $ab = 0$

Examples:

1. Multiply $(3a + 2b)^2$.

2. Multiply $(7n - 8d)(7n + 8d)$.

Solutions:

1. Using the model
   $(x + y)^2 = (x + y)(x + y) = x^2 + 2xy + y^2$
   where $x = 3a$ and $y = 2b$
   $(3a + 2b)^2 = (3a)^2 + 2(3a)(2b) + (2b)^2$
   $ = 9a^2 + 12ab + 4b^2$

2. Using the model
   $(x - y)(x + y) = x^2 - y^2$
   where $x = 7n$ and $y = 8d$
   $(7n - 8d)(7n + 8d) = (7n)^2 - (8d)^2$
   $ = 49n^2 - 64d^2$

3. Multiply $(7l + 3)(49l^2 - 21l + 9)$.

3. Using the model

$$(x+y)(x^2-xy+y)^2 = x^3+y^3$$

where $x = 7l$ and $y = 3$

$$(7l+3)(49l^2-21l+9) = (7l)^3+(3)^3$$
$$= 343l^3+27$$

Practice Problems 14.1

1. Multiply $(3f - 6g^2)^2$.

   A) $9f^2 - 36fg^2 + 36g^4$ B) $9f^2 - 18fg^2 + 36g^4$
   C) $9f - 18fg + 36g^2$
   D) $9f^2 + 36g^4$ E) $9f^2 - 36g^4$

2. Multiply $(5m^3 - 3)(5m^3 + 3)$

   A) $25m^6 - 9$ B) $25m^9 - 9$
   C) $25m^6 - 6$ D) $25m^9 + 6$
   E) $10m^6 - 9$

3. Multiply $(5 - 3x)(25 + 15x + 9x^2)$.

   A) $27x^3 - 125$ B) $27x^2 + 45x - 30$
   C) $27x^3 - 12x + 125$ D) $125 - 27x^3$
   E) $9x^2 + 12x + 30$

4. Multiply $(7\sqrt{a} + 3\sqrt{b})(7\sqrt{a} - 3\sqrt{b})$.

   A) 58 B) $49a - 9b$ C) $49a^2 - 9b^2$
   D) $49a^2 + 9b^2$
   E) $49a + 9b$

Solutions on page 258

14.2 Factoring/Factoring by Greatest Common Factor

Factors are numbers, terms or a series of terms which divide without remainder into the original expression. Expressions which can be factored may be stated as a product.

There are three basic types of factoring:

- **Greatest Common Factor** (GCF)
- **Special Products** $\Rightarrow x^2 + 2xy + y^2$
  $$x^2 - 2xy + y^2$$
  $$x^2 - y^2$$
  $$x^3 + y^3$$
  $$x^3 - y^3$$
- **Trinomial** $(ax^2 + bx + c)$

The basic starting point is to attempt factoring in the order provided above: GCF first, Special Products second and Trinomial last.

Factoring by **Greatest Common** *Factor:*

- Find the greatest common monomial factor of each term.
- Divide the GCF into each term and obtain a quotient.
- Express the original expression as the product of the GCF and quotient obtained (Original Expression) = (GCF)(Quotient).

Examples:

1. Find the GCF of $18x^3y^2$ and $27x^5y$.

2. Factor $18x^3y^2 + 27x^5y$.

3. Factor $3a^2b^3 - 6a^3b^2 + 9a^4$.

Solutions:

1. 9 is the largest common numerical factor.
   x^3 is the highest common power of x.
   y is the highest common power of y.
   The GCF $= 9x^3y$.

2. 1. The GCF $= 9x^3y$.

   2. $\dfrac{18x^3y^2 + 27x^5y}{9x^3y} = \dfrac{18x^3y^2}{9x^3y} + \dfrac{27x^5y}{9x^3y}$

   $= 2y + 3x^2$

   3. $18x^3y^2 + 27x^5y = 9x^3y(2y + 3x^2)$

3. 3 is the greatest common numerical factor.
   a^2 is the highest common power of x.
   There is no common factor of b.

   1. The GCF $= 3a^2$.

   2. $\dfrac{3a^2b^3 - 6a^3b^2 + 9a^4}{3a^2} = b^3 - 2ab^2 + 3a^2$

   3. $3a^2 - 6a^3b^2 + 9a^4 = 3a^2(b^3 - 2ab^2 + 3a^2)$

Practice Problems 14.2

1. Factor $9x^2 - 9x^2y$.

   A) $9(x^2 - y)$ B) $9x^2(1 - y)$
   C) $9y(1 - x^2)$ D) $9x^2(y)$
   E) $9x^2(y - 1)$

2. Factor $3m^2n^2 + 3m^3n^3$.

   A) 3 B) $3m^2n^2$ C) $3m^2n^2(1 + mn)$
   D) $3m^2n^2(mn)$ E) $3mn(1 + mn)$

3. Factor $4a^2 + 20a + 5a^3b$.

   A) $a(4a + 20 + 5a^2b)$
   B) $4a(a + 5 - a^2b)$ C) $5a(a + 4 - a^2b)$
   D) $a(4a^2 + 20 + ab)$ E) $ab(4a + 1)$

4. Factor $3n^2l^2 + 6nl + 9n^3l^3$.

   A) $nl(nl + 3 + 3n^2l^2)$
   B) $n^2l^2(3 + 2nl + 3n^2l^2)$
   C) $3(7nl + 3n^3l^3)$
   D) $3nl(nl + 2 + 3n^2l^2)$
   E) $3n^3l^3(3nl + 2n^2l^2 + 1)$

Solutions on page 258

14.3 Factoring by Special Products

- Match the pattern of the polynomial to be factored to the formats of the special products.

Special Product	Factors	Name of Special Product
$x^2 + 2xy + y^2$	$= (x+y)(x+y) = (x+y)^2$	Perfect square
$x^2 - 2xy + y^2$	$= (x-y)(x-y) = (x-y)^2$	trinomials
$x^2 - y^2$	$= (x+y)(x-y)$	Difference of two squares
$x^3 + y^3$	$= (x+y)(x^2 - xy + y^2)$	Sum of two cubes
$x^3 - y^3$	$= (x-y)(x^2 + xy + y^2)$	Difference of two cubes

● If there is a match, assign appropriate values to the x and y and use the chart above.

 Students will find memorizing the special products and their factors useful.

Examples:

1. Factor $4a^2 - 1$.

2. Factor $8a^3 - 1$.

3. Factor $4a^2 + 4ab + b^2$.

Solutions:

1. Using the format

   $x^2 - y^2$

   where $x = 2a$ and $y = 1$

   $x^2 - y^2 = (x+y)(x-y)$
   $4a^2 - 1 = (2a+1)(2a-1)$

2. Using the format

   $x^3 - y^3$

   where $x = 2a$ and $y = 1$

   $x^3 - y^3 = (x-y)(x^2 + xy + y^2)$
   $8a^3 - 1 = (2a-1)(4a^2 + 2a + 1)$

3. Using the format

   $x^2 + 2xy + y^2$

   where $x = 2a$ and $y = b$

   $y = b$

   $x^2 + 2xy + y^2 = (x+y)(x+y)$
   $4a^2 + 4ab + b^2 = (2a+b)(2a+b)$

Practice Problems 14.3

1. Factor $9a^2 - 16$.

   A) $(9a)(4)$ B) $(9a - 16)(9a + 16)$
   C) $(3a + 4)(3a + 4)$ D) $(3a - 4)(3a - 4)$
   E) $(3a + 4)(3a - 4)$

2. Factor $16 - 9a^2$.

   A) $(3a + 4)(3a - 4)$ B) $(4 + 3a)(4 - 3a)$
   C) $(3a - 4)(3a - 4)$ D) $(4 - 3a)(4 - 3a)$
   E) $(4 + 3a)(4 + 3a)$

3. Factor $a^2 + 10a + 25$.

A) $(a + 1)(a + 25)$ B) $(a + 4)(a + 5)$
C) $(a + 5)(a - 5)$ D) $(a + 5)(a + 5)$
E) $(a - 5)(a - 5)$

4. Factor $27a^3 - 8$.

A) $(3a + 2)(3a - 2)$ B) $(3a + 2)(9a^2 - 6a + 4)$
C) $(3a - 2)(9a^2 - 6a + 4)$
D) $(3a - 2)(9a^2 + 6a + 4)$
E) $(3a - 2)(9a^2 + 6a + 2)$

5. Factor $27a^3 + 8$.

A) $(3a + 2)(3a - 2)$
B) $(3a + 2)(9a^2 - 6a + 4)$
C) $(3a - 2)(9a^2 - 6a + 4)$
D) $(3a - 2)(9a^2 + 6a + 4)$
E) $(3a - 2)(9a^2 + 6a + 12)$

6. Factor $9a^2 - 12a + 4$.

A) $(3a + 2)(3a + 2)$ B) $(3a + 2)(3a - 2)$
C) $(3a - 2)(3a - 2)$ D) $(3a - 2)(3a - 10)$
E) $(9a + 1)(a + 4)$

Solutions on page 258

14.4 Factoring Trinomials and Factoring Completely

Factoring **trinomials** of the form $ax^2 + bx + c$ is a trial and error process. The approach varies slightly depending on the value of a.

 Always see if a greatest common monomial factor, different from 1, exists. If so, factor out the GCF before attempting trinomial factoring.

There are four types to be considered:

- $ax^2 + bx + c$ when **a = 1** $\Rightarrow$ $x^2 + bx + c$
- $ax^2 + bx + c$ when **a is a prime number greater than 1** $\Rightarrow$ $2x^2 + bx + c$
  $3x^2 + bx + c$
  $5x^2 + bx + c$
  $7x^2 + bx + c$
  $11x^2 + bx + c$
- $ax^2 + bx + c$ when **a is a non-prime number** $\Rightarrow$ $4x^2 + bx + c$
  $6x^2 + bx + c$
  $8x^2 + bx + c$
  $9x^2 + bx + c$
- $ax^2 + bx + c$ **non-factorable**

Factoring $ax^2 + bx + c$ when $a = 1$

1. Write the factors as $(x \quad)(x \quad)$.
2. Determine all the corresponding pairs of factors of c. Note: If $c > 0$ use pairs of positive factors and pairs of negative factors.
   If $c < 0$ use pairs of factors with different signs.
3a. Insert each pair of factors into the parentheses on a trial and error basis. Do one pair at a time.
3b. Insert the first pair in order. Put the first number of the pair into the first set of parentheses and the

Factor $x^2 + 6x + 5$.

1. $(x \quad)(x \quad)$
2. $+5, +1$
   $-5, -1$

3. $(x + 5)(x + 1)$
   $(+5)(x) = +5x$
   $x(+1) = +1x$

$+5x + 1x = 6x \checkmark$

second number of the pair into the second set of parentheses.

3c. Test for the middle term by multiplying the inner and outer terms (including signs). Add the results.

3d. Does the sum equal the middle term of the trinomial? If yes—these are the factors. If no—continue the process with the next pair of factors.

The factors are

$$(x + 5)(x + 1)$$

Examples:

1. Factor $x^2 + 2x - 8$.

Solutions:

1. $c = -8$, middle term $= +2x$

   1. $(x\quad)(x\quad)$ the trinomial has $a = 1$.

   2. $-8, +1$ Identify the pairs of factors of $c = -8$.
     $+8, -1$
     $-4, +2$
     $+4, -2$

   3. Insert each pair and test.

$$(x - 8)(x + 1) \quad -8x + 1x = -7x \neq 2x$$

$$(x + 8)(x - 1) \quad 8x + (-1x) = 7x \neq 2x$$

$$(x - 4)(x + 2) \quad -4x + 2x = -2x \neq 2x$$

$$(x + 4)(x - 2) \quad +4x + (-2x) = 2x = 2x \checkmark$$

Factors are $(x + 4)(x - 2)$.

2. Factor $x^2 + 9x + 20$.

2. $c = 20$, middle term $= +9x$

   1. $(x\quad)(x\quad)$ The trinomial has $a = 1$.

   2. $+4, +5$ Identify the pairs of factors of $c = 20$.
     $-4, -5$
     $+10, +2$
     $-10, -2$
     $+20, +1$
     $-20, -1$

   3. Insert each pair and test.

$$(x + 4)(x + 5) \quad 4x + 5x = 9x = 9x \checkmark$$

Factors are $(x + 4)(x + 5)$.

Factoring $ax^2 + b + c$ when a is a prime number greater than one

1. Write the factors as $(ax\quad)(x\quad)$.
2. Determine all the corresponding pairs of factors of c. Note: If $c>0$ use pairs of positive factors and pairs of negative factors.
   If $c<0$ use pairs of factors with different signs.
3a. Insert each pair of factors into the parentheses on a trial and error basis. Do one pair at a time.
3b. Insert the first pair in order. Put the first number of the pair into the first set of parentheses and the second number of the pair into the second set of parentheses.
3c. Test for the middle term by multiplying the inner and outer terms (including signs). Add the results.
3d. Does the sum equal the middle term of the trinomial? If yes—these are the factors.
   If no—reverse the pair of numbers. Put the first number of the pair in the second set of parentheses and the second number of pair in the first set of parentheses.
3e. Test for the middle term by multiplying the inner and outer terms (including signs). Add the results.
3f. Does the sum equal the middle term of the trinomial? If yes—these are the factors.
   If no—continue the process with the next pair of factors.

Factor $3x^2 - 5x - 12$.

1. $(3x\quad)(x\quad)$
2. $-12, +1$
   $+12, -1$
   $-4, +3$
   $+4, -3$
   $-6, +2$
   $+6, -2$

3. $(3x - 12)(x + 1)$
   $(-12x) + (+3x) = -9x \neq -5x$

   $(3x + 1)(x - 12)$
   $(+1x) + (-36x) = -35x \neq -5x$

   $(3x + 12)(x - 1)$
   $12x + (-3x) = 9x \neq -5x$

   $(3x - 1)(x + 12)$
   $-x + 36x = 35x \neq -5x$

   $(3x - 4)(x + 3)$
   $-4x + 9x = 5x \neq -5x$

   $(3x + 3)(x - 4)$
   $3x(-12x) = -9x \neq -5x$

   $(3x + 4)(x - 3)$
   $4x + (-9x) = -5x = -5x \checkmark$

   Factors are $(3x + 4)(x - 3)$.

Examples:

3. Factor $5x^2 - 24x - 5$.

4. Factor $11x^2 - 35x + 6$.

Solutions:

3. $c = -5$, middle term $= -24x$

1. $(5x\quad)(x\quad)$ 5 is a prime number.

2. $-5, +1$ Identify the pairs of factors of $c = 5$.

3. Insert each pair and test. Also, be sure to reverse each pair.

$$(5x\overset{-5x}{\underset{+5x}{- 5)(x + 1)}}$$

$$-5x + 5x = 0 \neq -24x$$

$$(5x\overset{x}{\underset{-25x}{+ 1)(x - 5)}}$$

$$x + (-25x) = -24x = -24x$$

The factors are $(5x + 1)(x - 5)$.

4. $c = 6$, middle term $= -35x$

1. $(11x\quad)(x\quad)$ 11 is a prime number.

2. $+6, +1$ Identify the pairs of factors of $c = +6$.
$-6, -1$
$+3, +2$
$-3, -2$

3. Insert each pair of factors and test. Be sure to reverse each pair.

$$(11x\overset{6x}{\underset{11x}{+ 6)(x + 1)}}$$
$$6x + 11x = 17x \neq -35x$$

$$(11x\overset{x}{\underset{66x}{+ 1)(x + 6)}}$$
$$x + 66x = 67x \neq -35x$$

$$(11x\overset{-6x}{\underset{-11x}{- 6)(x - 1)}}$$
$$-6x + (-11x) = -17x \neq -35x$$

$$(11x\overset{-x}{\underset{-66x}{- 1)(x - 6)}}$$
$$-x + (-66x) = -67x \neq -35x$$

$$(11x\overset{3x}{\underset{22x}{+ 3)(x + 2)}}$$
$$3x + 22x = 25x \neq -35x$$

$$(11x + \overset{2x}{\underset{33x}{2)(x + 3)}}$$

$$2x + 33x = 35x \neq -35x$$

$$(11x - \overset{-3x}{\underset{-22x}{3)(x - 2)}}$$

$$-3x + (-22x) = -25x \neq -35x$$

$$(11x - \overset{-2x}{\underset{-33x}{2)(x - 3)}}$$

$$-2x + (-33x) = -35x = -35x \checkmark$$

Factors are $(11x - 2)(x - 3)$.

Factoring $ax^2 + bx + c$ when a is a non-prime number

1. List all the positive sets of factors for a. If for some reason a is negative, factor out a GCF of -1 before starting.

2. Write the factors as $(a'x \quad)(a''x \quad)$, where a' and a'' represent the first set of positive factors of a.

3. Determine all the corresponding pairs of factors of c. Note: If $c > 0$ use pairs of positive factors and pairs of negative factors.
   If $c < 0$ use pairs of factors with different signs.

4a. Insert each pair of factors **of c** into the parentheses on a trial and error basis. Do one pair at a time.

4b. Insert the first pair in order. Put the first number of the pair into the first set of parentheses and the second number of the pair into the second set of parentheses.

4c. Test for the middle term by multiplying the inner and outer terms (including signs). Add the results.

4d. Does the sum equal the middle term of the trinomial? If yes—these are the factors.
    If no—reverse the pair of numbers. Put the first number of the pair in the second set of parentheses and the second number of the pair in the first set of parentheses

4e. Test for the middle term by multiplying the inner and outer terms (including signs). Add the results.

4f. Does the sum equal the middle term of the trinomial? If yes—these are the factors.
    If no— continue the process with the next pair of factors **of c**.

5. If none of the factors or their reverses work, change the original a' and a'' to the next possible set of positive factors of a and repeat step 4a.

Factor $4x^2 + x - 3$.

1. 4, 1
   2, 2

2. $(4x \quad)(1x \quad)$

3. $-3, +1$ middle term $= +1x$
   $+3, -1$

4. $(4x - \overset{-3x}{\underset{4x}{3)(1x + 1)}}$

   $-3x + 4x = x = x \checkmark$

   The factors are $(4x - 3)(x + 1)$.

Examples:

5. Factor $6x^2 + 7x - 5$.

Solutions:

5. $a = 6$, $c = -5$, middle term $= 7x$

1. List all the positive pairs of factors of a
   6, 1
   3, 2

2. Use the first positive pair.
   $(6x \quad)(1x \quad)$

3. Pairs of factors of c
   $-5, 1$
   $5, -1$

4. $(6x \underset{+6x}{\overset{-5x}{\underline{- 5)(1x} + 1)}}$ Insert and test.

   $-5x + 6x = x \neq 7x$ Be sure to reverse each pair.

$(6x \underset{-30x}{\overset{x}{\underline{+ 1)(1x} - 5)}}$

$x + (-30x) = -29x \neq 7x$

$(6x \underset{-6x}{\overset{5x}{\underline{+ 5)(1x} - 1)}}$

$5x + (-6x) = -x \neq 7x$

$(6x \underset{+30x}{\overset{-x}{\underline{- 1)(1x} + 5)}}$

$-x + (30x) = 29x \neq 7x$

---

5. Use the next set of factors of a and repeat step 4a.

$(3x \underset{+3x}{\overset{-10x}{\underline{- 5)(2x} + 1)}}$

$-10x + 3x = -7x \neq 7x$

$(3x \underset{-15x}{\overset{2x}{\underline{+ 1)(2x} - 5)}}$

$2x + (-15x) = -13x \neq 7x$

$(3x \underset{-3x}{\overset{10x}{\underline{+ 5)(2x} - 1)}}$

$10x + (-3x) = 7x = 7x \checkmark$

Factors are: $(3x + 5)(2x - 1)$

6. Factor $18x^2 - 6x - 24$.

6. Note: There is a greatest common monomial factor of 6. First factor out the GCF of 6.

$$18x^2 - 6x - 24 = 6(3x^2 - x - 4)$$

The trinomial $3x^2 - x - 4$ now has a prime

number as the coefficient of the x^2.

1. $c = -4$, middle term $= -1x$

   $(3x \quad)(x \quad)$

2. $+4, -1$

   $-4, +1$

   $+2, -2$

   Note: $-2, +2$ is the same as $+2, -2$. $\therefore$ Testing both pairs of numbers is not necessary.

3. $(3x + 4)(x - 1)$

   $4x + (-3x) = 1x \neq -1x$

   _____

   $(3x - 1)(x + 4)$

   $-x + (12x) = 11x \neq -1x$

   _____

   $(3x - 4)(x + 1)$

   $-4x + 3x = -1x = -1x \checkmark$

Factors of $3x^2 - x - 4 = (3x - 4)(x + 1)$

Factors of

$$18x^2 - 6x - 24 = 6(3x^2 - x - 4)$$
$$= 6(3x - 4)(x + 1)$$

> **When $ax^2 + bx + c$ is unfactorable**

Trinomials of the form $ax^2 + bx + c$ are not always factorable. **If none of the procedures work leave the trinomial is $ax^2 + bx + c$ form.**

7. Can $x^2 + 2x - 7$ be factored?

7. $c = -7$, middle term $= +2x$

1. $(x \quad)(x \quad)$ The trinomial has $a = 1$.

2. $+7, -1$ Identify the factor pairs of c.

   $-7, +1$

3. Insert and test.

   $(x + 7)(x - 1)$

   $7x + (-1x) = 6x \neq 2x$

   _____

   $(x - 7)(x + 1)$

   $-7x + 1x = -6x \neq 2x$

   _____

None of the possibilities work.

$x^2 + 2x - 7$ is not factorable.

Combined Factoring or Factor Completely

Some expressions can be factored into more than two factors. To explore this possibility:

> ● Factor out any Greatest Common Factor (GCF).
> ● Factor the resulting polynomial, if possible.
> ● Factor any resulting polynomial factors, if possible.
> ● Express the original polynomial as a product of its factors.

Examples:

8. Factor $3x^4 - 48$.

Solutions:

8. $3x^4 - 48$ There is a GCF of 3
$= 3(x^4 - 16)$

$x^4 - 16$ is the difference of two squares
$x^2 - y^2 = (x + y)(x - y)$
$\therefore x^4 - 16 = (x^2 + 4)(x^2 - 4)$

$3(x^4 - 16)$
$= 3(x^2 + 4)(x^2 - 4)$

$x^2 - 4$ is the difference of two squares
$\therefore x^2 - 4 = (x + 2)(x - 2)$

$3(x^2 + 4)(x^2 - 4)$
$= 3(x^2 + 4)(x + 2)(x - 2)$

Practice Problems 14.4

1. Factor $x^2 - 7x + 10$.

   A) $(x - 10)(x - 1)$ B) $(x + 10)(x + 1)$
   C) $(x - 5)(x - 2)$ D) $(x + 2)(x + 5)$
   E) $(x - 2)(x + 5)$

2. Factor completely $3x^2 - 27$.

   A) $3(x^2 - 9)$ B) $3(x^2 + 3)(x^2 - 3)$
   C) $3(x - 3)(x - 3)$ D) $3(x + 3)(x - 3)$
   E) $3x(x - 9)$

3. Factor $3x^2 - 4x + 4$.

   A) $(3x + 2)(x - 2)$ B) $(3x - 2)(x + 2)$
   C) $(3x - 2)(3x + 2)$ D) $(3x + 2)(x + 2)$
   E) $(3x - 2)(x - 2)$

4. Factor $15y^2 - 5y$.

   A) $(5y - 1)(3y + 5)$ B) $(5y)(3y)$
   C) $(5y)(3y - 1)$ D) $(5y - y)(3y + 5)$
   E) $(3y - 1)(5y + y)$

5. Factor completely $2x^3 - 16$.

   A) $2(x^3 - 8)$ B) $2(x + 4)(x^2 - 4)$
   C) $2(x^2 + 4)(x - 4)$ D) $2(x^2 + 2x + 4)$
   E) $2(x - 2)(x^2 + 2x + 4)$

6. Factor completely $4x^2 + 12x - 16$.

   A) $(4x - 4)(x + 4)$ B) $4(x + 4)(x - 1)$
   C) $4(x - 4)(x + 1)$ D) $(4x - 16)(x + 1)$
   E) $(x + 4)(x - 4)$

7. Factor completely $8x^8 - 8$.

   A) $8(x^8 - 1)$ B) $8(x^4 - 8)(x^4 + 1)$
   C) $8(x^6 - 8)(x^2 + 1)$
   D) $8(x^4 + 1)(x^2 + 1)(x + 1)(x - 1)$
   E) $8(x^4 - 1)(x^2 - 1)(x + 1)(x - 1)$

8. Factor $4x^2 - 8x + 3$.

   A) $(2x - 3)(2x - 1)$ B) $(4x - 3)(x - 1)$
   C) $(2x - 1)(x - 3)$ D) $(4x - 1)(x - 3)$
   E) $4x(x - 8 + 3x^2)$

9. Factor $7x^2 - 13x - 2$.

   A) $(7x - 2)(x + 1)$ B) $(7x + 1)(x - 2)$
   C) $(7x^2)(13x - 2)$
   D) $(7x - 6)(x - 7)$ E) $(7x + 1)(x + 2)$

10. Factor $x^2 - 5x + 6$.

    A) $(x + 6)(x - 1)$ B) $(x - 6)(x + 1)$
    C) $(x + 3)(x - 2)$ D) $(x - 3)(x - 2)$
    E) $(x - 3)(x + 2)$

11. Factor $2x^2 + x - 3$.

    A) $(2x - 3)(x + 1)$ B) $(2x + 1)(x - 3)$
    C) $(2x + 3)(x - 1)$ D) $(2x - 1)(x + 3)$
    E) $2(x - 3)(x + 1)$

12. Factor completely $4x^3 - 49x$.

    A) $4(x^3 - 12x)$ B) $4x(x^2 - 12)$
    C) $(2x - 7)(2x^2 + 7)$ D) $x(4x^2 - 49)$
    E) $x(2x + 7)(2x - 7)$

Solutions on page 259

14.5 Applications of Factoring I (Multiplying and Dividing Polynomial Fractions)

- Express the problem as a series of fraction multiplications.

$$\frac{a}{b} \cdot \frac{c}{d} \cdot \frac{e}{f} \cdots$$

- If there is division, invert the fraction after the division sign and multiply it by the first fraction.

$$\left. \begin{array}{c} \dfrac{a}{b} \div \dfrac{x}{y} \\[2em] \dfrac{\dfrac{a}{b}}{\dfrac{x}{y}} \end{array} \right\} \Rightarrow \frac{a}{b} \cdot \frac{y}{x}$$

- Factor all the polynomials which can be factored.
- Reduce like factors and opposite factors to $\frac{1}{1}$ and $\frac{-1}{1}$, respectively.

Like Factors: Every term in each factor has a correspondence with the same sign.
Opposite Factors: Every term in each factor has a correspondence with the opposite sign.

$$\frac{x+1}{x+1} \Rightarrow \frac{x \text{ is "+", } 1 \text{ is "+"}}{x \text{ is "+", } 1 \text{ is "+"}} = \frac{1}{1} = 1 \qquad \frac{x-1}{1-x} \Rightarrow \frac{x \text{ is "+", } 1 \text{ is "−"}}{x \text{ is "−", } 1 \text{ is "+"}} = \frac{-1}{1} = -1$$

$$\frac{-1-x}{1-x} \Rightarrow \frac{1 \text{ is "−", } x \text{ is "−"}}{1 \text{ is "+", } x \text{ is "+"}} = \frac{-1}{1} = -1 \qquad \frac{x-1}{x-1} \Rightarrow \frac{x \text{ is "+", } 1 \text{ is "−"}}{x \text{ is "+", } 1 \text{ is "−"}} = \frac{1}{1} = 1$$

$$\left.\frac{x+1}{x-1} \Rightarrow \frac{x \text{ is "+", } 1 \text{ is "+"}}{x \text{ is "+", } 1 \text{ is "−"}}\right\} \text{ cannot be reduced}$$

$$\frac{-x+1}{1-x} \Rightarrow \frac{x \text{ is "−", } 1 \text{ is "+"}}{x \text{ is "−", } 1 \text{ is "+"}} = \frac{1}{1} = 1$$

 $$\frac{x^2+2}{2} \neq x^2$$

$$\neq x^2 + 1$$

 $$\frac{3x^2 + 2x}{3x^2} \neq 2x$$

$$\neq 1 + 2x$$

● Multiply and simplify the remaining factors.

$$\frac{a}{b} \cdot \frac{c}{d} \cdot \frac{e}{f} = \frac{ace}{bdf}$$

Examples:

1. Multiply $\dfrac{x^2+x}{x^2-6x-7} \cdot \dfrac{3x-21}{x^2-x}$.

2. Multiply $\dfrac{x^2-16}{3-3x} \cdot \dfrac{3x^2-3}{x^2-3x-4}$.

3. Divide $\dfrac{a^3+6a^2}{a^2+12a+36} \div \dfrac{6a-36}{a^2-36}$.

Solutions:

1. $$\frac{\overset{(1)}{\cancel{x}}(\overset{(1)}{\cancel{x+1}})}{(\cancel{x+1})(\cancel{x-7})} \cdot \frac{3(\overset{(1)}{\cancel{x-7}})}{\cancel{x}(x-1)} = \frac{3}{x-1}$$

2. $$\frac{(x+4)(\overset{(1)}{\cancel{x-4}})}{\cancel{3}(1-x)} \cdot \frac{\cancel{3}(\overset{(1)}{\cancel{x+1}})(\overset{(-1)}{\cancel{x-1}})}{(\cancel{x+1})(\cancel{x-4})} = -(x+4)$$

Note: $\dfrac{x-1}{1-x} \Rightarrow \dfrac{x \text{ is "+", } 1 \text{ is "−"}}{x \text{ is "−", } 1 \text{ is "+"}} = \dfrac{-1}{1}$.

3. Division by $\dfrac{6a-36}{a^2-36}$ should be changed to multiplication by $\dfrac{a^2-36}{6a-36}$

$$\frac{a^3+6a^2}{a^2+12a+36} \cdot \frac{a^2-36}{6a-36}$$

$$= \frac{a^2(\overset{(1)}{\cancel{a+6}})}{(\cancel{a+6})(\cancel{a+6})} \cdot \frac{(\overset{(1)}{\cancel{a+6}})(\overset{(1)}{\cancel{a-6}})}{6(\cancel{a-6})}$$

$$= \frac{a^2}{6}$$

4. Simplify $\dfrac{x^2 - 6x - 16}{3x^2 + 30x + 48} \div \dfrac{2x^2 - 128}{x^2 + 16x + 64}$.

4. $\dfrac{x^2 - 6x - 16}{3x^2 + 30x + 48} \div \dfrac{2x^2 - 128}{x^2 + 16x + 64}$

Division should first be changed to multiplication

$$= \dfrac{x^2 - 6x - 16}{3x^2 + 30x + 48} \cdot \dfrac{x^2 + 16x + 64}{2x^2 - 128}$$

$$= \dfrac{\overset{(1)}{(x \cancel{+} 2)}\overset{(1)}{(x \cancel{+} 8)}}{3(x \cancel{+} 2)(x \cancel{+} 8)} \cdot \dfrac{\overset{(1)}{(x \cancel{+} 8)}\overset{(1)}{(x \cancel{+} 8)}}{2(x \cancel{+} 8)(x \cancel{+} 8)} = \dfrac{1}{6}$$

Practice Problems 14.5

1. Simplify $\dfrac{x+3}{2x+4} \cdot \dfrac{x^2-x-6}{2x+6}$.

    A) $\dfrac{4}{x-3}$ B) $\dfrac{x-3}{4}$ C) $\dfrac{1}{x-3}$

    D) $x-3$ E) $\dfrac{1}{4}$

3. Simplify $\dfrac{x^2 + 3x + 2}{x^2 - x - 2} \cdot \dfrac{x^2 + x - 6}{x^2 - x - 12}$.

    A) $\dfrac{x+2}{x-4}$ B) $\dfrac{x-4}{x+2}$ C) $-\dfrac{1}{2}$

    D) $\dfrac{1}{2}$ E) $\dfrac{(x+1)(x+2)}{(x+3)(x-4)}$

2. Simplify $\dfrac{a^2 + 8a + 15}{a^2 + 5a} \div \dfrac{a^2 - 9}{a^2 + 2a - 3}$.

    A) $\dfrac{a-1}{a}$ B) $\dfrac{a(a-3)}{(a-1)(a+3)}$ C) $\dfrac{a}{a-1}$

    D) $\dfrac{a-1}{a+1}$ E) $\dfrac{(a+3)(a-1)}{a(a-3)}$

4. Simplify $\dfrac{a+2}{2a} \cdot \dfrac{2a-4}{2+a} \cdot \dfrac{a^2-4}{12-6a} \div \dfrac{2+a}{12a^2}$.

    A) $-2(a-2)$ B) $2a(a-2)$
    C) $-2a(a-2)$ D) $(a+2)(a-2)$
    E) $\dfrac{-2}{2+a}$

Solutions on page 261

14.6 Applications of Factoring II (Solving Equations)

> To solve an equation by factoring:
>
> - Rearrange the equation into $f(x) = 0$ form, $ax^2 + bx + c = 0$ if it's a quadratic.
> - Factor the equation completely (if possible).
> - Set each factor separately equal to zero and solve.
>
> This procedure is predicated on the property: $lmn = 0$ if and only if $l = 0$ or $m = 0$ or $n = 0$.

Examples:

1. Solve $(x - 2)(x + 3) = 0$.

2. Solve $(x - 2)(x + 3) = 6$.

3. Find the solution set of the equation
$x^3 - 2x + x = 0$.

Solutions:

1. $\quad (x - 2)(x + 3) = 0$

$x - 2 = 0 \mid x + 3 = 0$
$\quad x = 2 \mid \quad x = -3$

$x = 2$ or $x = -3$

2. Since the equation $\neq 0$, it must be rearranged.

$$(x - 2)(x + 3) = 6$$
$$x^2 + x - 6 = 6$$
$$x^2 + x - 12 = 0$$
$$(x - 3)(x + 4) = 0$$

$x - 3 = 0 \mid x + 4 = 0$
$\quad x = 3 \mid \quad x = -4$

$x = 3$ or $x = -4$

3. $\quad\quad\quad x^3 - 2x^2 + x = 0$

$\quad\quad x(x^2 - 2x + 1) = 0 \;\; \text{GCF: } x$

$\quad\quad x(x - 1)(x - 1) = 0$

$x = 0 \mid x - 1 = 0 \mid x - 1 = 0$
$\quad\quad\quad\quad x = 1 \mid \quad\quad x = 1$

$x = 0$ or $x = 1$ (double root)

$\{0, 1\}$

Practice Problems 14.6

1. Solve for x: $x^2 - 4 = 0$.

A) $x = 4$ or $x = 1$ B) $x = 2$ only
C) $x = 2$ or $x = -2$ D) $x = -2$ only
E) $x = -4$ or $x = 1$

2. Solve for x: $x^2 + 9 = 10x$.

A) $x = 1$ or $x = 9$ B) $x = -1$ or $x = -9$
C) $x = 1$ or $x = -9$ D) $x = -1$ or $x = 9$
E) $x = 1$ only

3. Find the solution set of the equation

$2n^3 - 2n = 0$

A) $\{0\}$ B) $\{1\}$ C) $\{-1\}$
D) $\{1, -1\}$ E) $\{0, 1, -1\}$

4. Find the solution set for the equation

$8x^2 + 10x - 3 = 0$

A) $\{0, 3\}$ B) $\left\{\dfrac{1}{4}, \dfrac{-3}{2}\right\}$ C) $\left\{\dfrac{-3}{8}, \dfrac{-1}{4}\right\}$

D) $\left\{\dfrac{-1}{4}, \dfrac{3}{2}\right\}$ E) $\left\{\dfrac{3}{8}, \dfrac{1}{4}\right\}$

Solutions on page 261

15. Solving Equations

This section reviews a multiplicity of solution techniques as applied to equations or systems of equations. Included are solution techniques for:

Equalities

Type of Equation or System	Solution Technique		
	Algebra	Graphic	Determinants
15.1 Linear	√		
15.2 Proportions	√		
15.3 Literal	√		
15.4 Linear-Linear	√	√	√
15.5 Quadratic	√	√	
15.6 Linear-Quadratic	√	√	
15.7 Quadratic-Quadratic	√	√	
15.8 Radical	√		
15.9 Absolute Value	√		
15.10 3 Simultaneous Linear	√		√

Inequalities

		Algebra	Graphic
15.11 Linear	1 Variable	√	Graphic Solution Representation
	2 Variables		√
15.12 Linear-Linear			√
15.13 Quadratic	1 Variable		Graphic Solution Representation
	2 Variables		√
15.14 Quadratic-Linear Quadratic-Quadratic			√

15.1 Solving Linear Equations

The technique for solving a linear equation with one variable is outlined below.

Step	
1	Remove fractions or decimals by multiplication. For fractions the multiplier is the lowest common denominator (LCD). For decimals it is the smallest power of 10 which clears all the decimal places. $2\left(\dfrac{x}{2} - 3 = x\right) \Rightarrow x - 6 = 2x \quad 100(.3x + 1 = .01) = 30x + 100 = 1$
2	Remove all the parentheses by using the distributive law of multiplication. Don't forget to distribute minus signs. $7 - (x + 3) = 7 - x - 3$
3	Collect all terms containing the unknown (for which you are solving) on the same side of the equal sign. All terms *not* containing the unknown should be brought to the *other* side of the equation. Usually the terms containing the unknown are brought to the left side of the equation. Remember, whenever a term crosses from one side of the equal sign to the other it changes sign. $2x - 3 = x - 2 \quad \Rightarrow \quad 2x - x = -2 + 3$
4	Determine the coefficient of the unknown either by combining similar terms or factoring when terms cannot be combined. Similarly combine all common terms on the other side of the equation. $3x + 2x = 12 + 7 \Rightarrow 5x = 19$ $ax + bx = 12 - 1 \Rightarrow x(a + b) = 11$
5	Divide both sides of the equation by the coefficient of the unknown. $\dfrac{5x}{5} = \dfrac{19}{5} \quad \Bigg\vert \quad \dfrac{x(a + b)}{a + b} = \dfrac{11}{a + b}$ $x = \dfrac{19}{5} \quad \Bigg\vert \quad x = \dfrac{11}{a + b}$

 An equation is balanced to the left and right of the equal sign. You can apply any legitimate mathematical operation to either side of the equation provided:

1. it is applied to the side as a whole.
2. it is applied to the other side as a whole, to maintain the balance.

Examples:

1. Solve $3x - 15 = x + 1$ for x.

Solutions:

1. Steps 1 and 2 do not apply.

$$3x - 15 = x + 1$$

Step 3: Collect the variable terms on one side of the equation:

$$3x - x = 1 + 15$$

Step 4: Combine common terms:

$$2x = 16$$

Step 5: Divide by the coefficient of the variable:

$$x = 8$$

2. Solve $\dfrac{2x}{5} + 9 = \dfrac{x}{3} + 10$.

2. Step 1: Clear all fractions:

$$15\left(\frac{2x}{5} + 9 = \frac{x}{3} + 10\right) \Rightarrow 6x + 135 = 5x + 150$$

Step 2 does not apply.

Steps 3, 4, 5:
$$6x + 135 = 5x + 150$$
$$6x - 5x = 150 - 135$$
$$x = 15$$

3. Solve $2(x + 1) - 10 = 7 - (x + 3)$.

3. Step 1 does not apply.

$$2(x + 1) - 10 = 7 - (x + 3)$$

Step 2: Remove parentheses:

$$2x + 2 - 10 = 7 - x - 3$$

Steps 3, 4, 5:
$$2x - 8 = 4 - x$$
$$2x + x = 4 + 8$$
$$\frac{3x}{3} = \frac{12}{3}$$
$$x = 4$$

4. Solve $.4x + .16 = 1.36$.

4. $100(.4x + .16 = 1.36)$
$$40x + 16 = 136$$
$$40x = 120$$
$$x = 3$$

Practice Problems 15.1

1. Solve for x: $3x + 5 = 5x - 3$.

   A) 4   B) 1   C) $\frac{1}{8}$ D) 0 E) -1

2. Solve for d: $\frac{d}{2} + 6 = 4$.

   A) 4 B) -4 C) 2 D) -2 E) -1

3. Solve for y: $x - .2 = 1.8$.

   A) .2 B) 1.6 C) 2 D) 16 E) 20

4. Solve for b: $4(b - 1) - 3 = 17$.

   A) -4 B) 2.5 C) 4 D) 6 E) $\frac{21}{4}$

5. Solve for n: $\frac{2n}{3} - \frac{5n}{12} = \frac{5}{4}$.

   A) $-\frac{5}{3}$ B) $\frac{5}{3}$ C) $\frac{40}{9}$ D) $\frac{1}{2}$ E) 5

6. Solve for a: $5 + 3(a + 2) = 14$.

   A) 1   B) $\frac{3}{2}$ C) $-\frac{1}{4}$ D) $-\frac{1}{2}$ E) -1

Solutions on page 261

15.2 Solving Proportions

Proportions are ratios that are equal. There are two common formats:

$$a\!:\!\underbrace{\overset{\text{means}}{b = c}}_{\text{extremes}}\!:\!d \quad \text{or} \quad \frac{a}{b} = \frac{c}{d}$$

The symbol ":" is read "is to" and the symbol "=" is read "as". Therefore $2 : 3 = x : 4$ is read "Two *is to* three *as* x *is to* four".
The positions a and d are called the extremes.
The positions b and c are called the means.

The procedure for solving an equation in the format of a proportion is quite simple.

Step

1   Convert the proportion to $\dfrac{a}{b} = \dfrac{c}{d}$ format.

$$2\!:\!3 = x\!:\!7 \Rightarrow \frac{2}{3} = \frac{x}{7}$$

2   Cross multiply and solve as a linear equation.

$$\frac{2}{3} = \frac{x}{7} \Rightarrow 3x = 14$$
$$x = \tfrac{14}{3}$$

 A formal statement of the procedure used in step 2 is:
"The product of the means equals the product of the extremes."

Examples:

1. Solve for n: $\dfrac{n}{6} = \dfrac{11}{2}$.

2. Solve for x: $\dfrac{x+8}{5} = \dfrac{x+2}{3}$.

Solutions:

1. Cross multiply:

$$\frac{n}{6} = \frac{11}{2}$$
$$2n = 66$$
$$n = 33$$

2. Crossmultiply:

$$\frac{x+8}{5} = \frac{x+2}{3}$$
$$3(x+8) = 5(x+2)$$
$$3x + 24 = 5x + 10$$
$$2x = 14$$
$$x = 7$$

Practice Problems 15.2

1. Solve for a: $\dfrac{3a+1}{4} = \dfrac{5}{2}$.

   A) 4 B) 3 C) 2 D) $\dfrac{19}{6}$ E) 1

2. Solve for m: $\dfrac{m}{4} = \dfrac{5}{2}$.

   A) .1 B) 1 C) 10 D) 100 E) 0

3. Solve for x:

$$\frac{x+1}{8} = \frac{2x+1}{4}$$

   A) 3 B) -3 C) $\dfrac{1}{3}$ D) $-\dfrac{1}{3}$ E) 2

4. Solve for l:

$$\frac{3l-1}{5l+2} = \frac{2}{7}$$

   A) 1 B) $-\dfrac{3}{2}$ C) $\dfrac{3}{11}$ D) $\dfrac{-3}{11}$ E) $\dfrac{7}{11}$

Solutions on page 262

15.3 Literal Equations

Literal equations are equations in which many or all of the constants are expressed in terms of letters rather than numbers. It is paramount to keep track of which letter you are solving for.

> The procedure for solving linear literal equations is identical to that for linear equations or proportions as applicable.

Examples:

1. Solve for x: $ax + b = c$.

2. Solve for x: $\dfrac{1}{x} + a = 2b$.

3. Solve for n: $\dfrac{cn + 1}{x} = \dfrac{2y}{3}$.

Solutions:

1. Collect all the terms containing the unknown on one side of the equation. Divide by the coefficient of the unknown:

$$ax + b = c$$

$$\frac{ax}{a} = \frac{c - b}{a}$$

$$x = \frac{c - b}{a}$$

2. Clear the fraction:

$$x\left(\frac{1}{x} + a = 2b\right) \Rightarrow 1 + ax = 2bx$$

Bring all x terms to one side of the equation:

$$1 + ax = 2bx$$

$$ax - 2bx = -1$$

Factor out a single x:

$$x(a - 2b) = -1$$

Divide by the coefficient of the x:

$$\frac{x(a - 2b)}{(a - 2b)} = \frac{-1}{a - 2b}$$

$$x = \frac{-1}{a - 2b} \quad \text{or} \quad \frac{1}{2b - a}$$

☞ If $x = \dfrac{-1}{a - 2b}$, multiplying the top and bottom of the fraction by -1 yields

$$x = \frac{-1}{(a - 2b)}\frac{(-1)}{(-1)} = \frac{1}{2b - a}$$

3. $\dfrac{cn + 1}{x} = \dfrac{2y}{3}$

Cross multiply:

$$= 3(cn + 1) = 2xy$$

Bring all n terms to one side of the equation:

$$3cn + 3 = 2xy$$

$$3cn \quad = 2xy - 3$$

Divide by the coefficient of the unknown:

$$\frac{3cn}{3c} = \frac{2xy - 3}{3c}$$

$$n = \frac{2xy - 3}{3c}$$

Practice Problems 15.3

1. Solve for a: $\dfrac{cx + 1}{a} = \dfrac{b}{3}$

   A) $\dfrac{ab - 1}{3}$ B) $\dfrac{3cx + 1}{b}$ C) $\dfrac{3b}{cx + 1}$

   D) $\dfrac{3(cx + 1)}{b}$ E) $3c$

3. Solve for t: $a = \dfrac{t + b}{t - c}$

   A) $\dfrac{b}{c}$ B) $\dfrac{ac}{b}$ C) $\dfrac{b + ac}{a - 1}$

   D) $\dfrac{c + b}{a - 1}$ E) $\dfrac{ac + b}{a + 1}$

2. Solve for x: $3(x + 2) = a(b + x)$

   A) $\dfrac{ab - 6}{3 - a}$ B) $\dfrac{ab + 2}{2}$ C) $\dfrac{ab - 2}{2}$

   D) $\dfrac{ab + 6}{a - 3}$ E) $\dfrac{3 - a}{ab - 6}$

4. Solve for n: $\dfrac{1}{a} = \dfrac{1}{n} + \dfrac{1}{b}$

   A) $a + b$ B) $\dfrac{b - a}{ab}$ C) $\dfrac{ab}{b - a}$

   D) 1 E) ab

Solutions on page 262

15.4 Linear-Linear Systems

The solution set to a linear-linear system is the set of values of the variables which satisfy both equations simultaneously. These are also called simultaneous equations. Graphically the solution set is where the two graphs intersect. There are three types of linear systems:

- **Consistent:** There is one solution

- **Inconsistent:** There are no solutions. Parallel lines are an example.

- **Dependent:** Infinite solutions. The two lines are coincident. They overlap.

There are two algebraic solution techniques:

- **Elimination**
- **Substitution**

Both are discussed and demonstrated here. This is followed by the graphical approach and lastly a solution by determinants is presented.

> • **Elimination:** Multiply one or both of the equations by suitable constants in order to make the coefficients of one of the variables the same—but opposite in sign. Remember that multiplying *all* terms in an equation by the same constant does not upset the balance of the equation. One of the unknowns can then be eliminated by adding the two equations. When working with simultaneous equations, always be sure to have the terms containing the variables on one side of the equation and the remaining terms on the other side.
>
> Once one of the variables is eliminated solve the linear equation. Using the value for the found variable, substitute this value back into one of the original equations to find the value of the other variable.

Example 1: Solve: $3x + y = 4$
$$x - 2y = 6$$

Plan: Eliminate y
$$3x + y = 4$$
$$x - 2y = 6$$

Multiply the first equation by 2:
$$2(3x + y = 4) \Rightarrow 6x + 2y = 8$$

Add the two equations:
$$6x + 2y = 8$$
$$\underline{x - 2y = 6}$$
$$7x \quad = 14$$
$$x \quad = 2$$

Substitute $x = 2$ into one of the original equations:
$$3x + y = 4$$
$$3(2) + y = 4$$
$$6 + y = 4$$
$$y = -2$$

Solution: $x = 2$, $y = -2$ or written as an ordered pair $(2, -2)$

Example 2: Solve: $2x - 3y = 11$
$$3x - 4y = -9$$

Plan: Eliminate x
$$2x - 3y = 11$$
$$3x + 4y = -9$$

Multiply the first equation by -3 and the second by 2:
$$-3(2x - 3y = 11)$$
$$2(3x + 4y = -9)$$

Add the equations:
$$-6x + 9y = -33$$
$$\underline{6x + 8y = -18}$$
$$17y = -51$$
$$y = -3$$

Substitute $y = -3$ into one of the original equations:
$$2x - 3y = 11$$
$$2x - 3(-3) = 11$$
$$2x + 9 = 11$$
$$2x = 2$$
$$x = 1$$

Solution:
$$x = 1, y = -3$$

or written as an ordered pair:
$$(1, -3)$$

stitution: Solve the easiest of the two equations for the value of either variable. stitute this value into the second equation as appropriate. Solve the linear ation. Using the value for the found variable substitute this value back into of the original equations to find the value of the other variable.

Example 3: Solve: $3x + y = 4$
$$x - 2y = 6$$

Plan: Solve the
second equation for x

$$x - 2y = 6$$
$$x = 6 + 2y$$

Substitute $x = 6 + 2y$ into the first equation at each appearance of x:

$$3x + y = 4$$
$$3(6 + 2y) + y = 4$$
$$18 + 6y + y = 4$$
$$7y = -14$$
$$y = -2$$

Substitute $y = -2$ into one of the original equations:

$$x - 2y = 6$$
$$x - 2(-2) = 6$$
$$x + 4 = 6$$
$$x = 2$$

Solution:

$$x = 2, y = -2$$

or written as an ordered pair:

$$(2, -2)$$

Example 4: Solve: $2x - 3y = 11$
$$3x + 4y = -9$$

Plan: Solve the
first equation for x

$$2x - 3y = 11$$
$$2x = 11 + 3y$$
$$x = \frac{11 + 3y}{2}$$

Substitute $x = \dfrac{11 + 3y}{2}$ into the second equation at each appearance of x:

$$3x + 4y = -9$$
$$3\left(\frac{11 + 3y}{2}\right) + 4y = -9$$

Multiply the entire equation by 2 to clear the fractions

$$2\left(\frac{33 + 9y}{2} + 4y = -9\right)$$
$$33 + 9y + 8y = -18$$
$$17y = -51$$
$$y = -3$$

Substitute $y = -3$ into one of the original equations:

$$2x - 3y = 11$$
$$2x - 3(-3) = 11$$
$$2x + 9 = 11$$
$$2x = 2$$
$$x = 1$$

Solution:

$$x = 1, y = -3$$

or written as an ordered pair:

$$(1, -3)$$

● **Graphic:** Graph each of the lines on the same set of axes. Any graphical technique can be used, e.g., $y = mx + b$, a chart, or intercepts. The intersection of the two graphs represents the solution.

Example 5:

Solve: $3x + y = 4$
$x - 2y = 6$

Plan: Graph using $y = mx + b$

$3x + y = 4$

Slope-intercept $y = -3x + 4$

$m = \text{slope} = \dfrac{-3}{1} \begin{array}{l} \leftarrow\text{change in } y, 3 \text{ down} \\ \leftarrow\text{change in } x, 1 \text{ to right} \end{array}$

$b = y\text{-intercept} = 4$

$x - 2y = 6$

$-2y = -x + 6$

$y = \dfrac{x}{2} - 3$

$m = \text{slope} = \dfrac{1}{2} \begin{array}{l} \leftarrow\text{change in } y, 1 \text{ up} \\ \leftarrow\text{change in } x, 2 \text{ to right} \end{array}$

$b = y\text{-intercept} = -3$

Example 6:

$2x - 3y = 11$
$3x + 4y = -9$

Plan: Graph by developing a table of values

$2x - 3y = 11$

x	y
1	-3
4	-1
-2	-5

$3x + 4y = -9$

x	y
-3	0
1	-3
-7	3

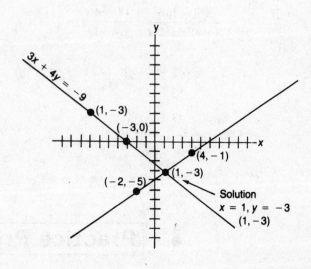

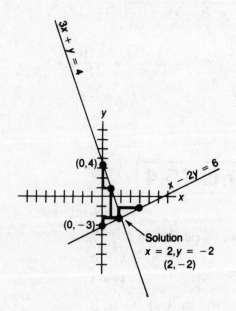

- **Determinants:** Align the equations in the format

$$ax + by = c$$
$$dx + ey = f$$

By Cramer's Rule:

$$x = \frac{\begin{vmatrix} c & b \\ f & e \end{vmatrix}}{\begin{vmatrix} a & b \\ d & e \end{vmatrix}} \qquad y = \frac{\begin{vmatrix} a & c \\ d & f \end{vmatrix}}{\begin{vmatrix} a & b \\ d & e \end{vmatrix}}$$

As a review: $\begin{vmatrix} a & c \\ b & d \end{vmatrix} = ad - bc$

Example 7:

Solve: $3x + y = 4$
$x - 2y = 6$

$$x = \frac{\begin{vmatrix} 4 & 1 \\ 6 & -2 \end{vmatrix}}{\begin{vmatrix} 3 & 1 \\ 1 & -2 \end{vmatrix}} = \frac{4(-2) - (6)(1)}{3(-2) - (1)(1)} = \frac{-8 - 6}{-6 - 1} = 2$$

$$y = \frac{\begin{vmatrix} 3 & 4 \\ 1 & 6 \end{vmatrix}}{\begin{vmatrix} 3 & 1 \\ 1 & -2 \end{vmatrix}} = \frac{3(6) - 1(4)}{3(-2) - (1)(1)} = \frac{18 - 4}{-6 - 1} = -2$$

$$x = 2, y = -2$$

Example 8:

Solve: $2x - 3y = 11$
$3x + 4y = -9$

$$x = \frac{\begin{vmatrix} 11 & -3 \\ -9 & 4 \end{vmatrix}}{\begin{vmatrix} 2 & -3 \\ 3 & 4 \end{vmatrix}} = \frac{11(4) - (-3)(-9)}{2(4) - (-3)(3)}$$

$$= \frac{44 - 27}{8 + 9} = \frac{17}{17} = 1$$

$$y = \frac{\begin{vmatrix} 2 & 11 \\ 3 & -9 \end{vmatrix}}{\begin{vmatrix} 2 & -3 \\ 3 & 4 \end{vmatrix}} = \frac{2(-9) - 3(11)}{2(4) - (-3)(3)}$$

$$= \frac{-18 - 33}{8 + 9} = \frac{-51}{17} = -3$$

$$x = 1, y = -3$$

Practice Problems 15.4

1. Solve by substitution:

$$y - x = -1$$
$$2y + x = 4$$

A) (2, 1) B) (1, 2) C) (−2, −1)
D) (−1, −2), E(−1, −1)

2. Solve by elimination:

$$3x + 4y = 7$$
$$x - 4y = 1$$

A) $(4, \frac{1}{2})$ B) $(\frac{1}{2}, 4)$ C) $(2, \frac{1}{4})$
D) $(\frac{1}{4}, -2)$ E) $(\frac{1}{4}, 2)$

3. Solve by determinants:

$$3x - 2y = -1$$
$$2x + 3y = 8$$

A) (1, 2) B) (−1, 2) C) (1, −2)
D) (−1, −2) E) (1, $\frac{1}{2}$)

4. Solve by substitution:

$$2x - y = 6$$
$$x + y = 9$$

A) (−5, −4) B) (−5, 4) C) (5, −4)
D) (5, 4) E) ($\frac{1}{5}$, $\frac{1}{4}$)

5. Solve by elimination:

$$y - 4 = x$$
$$5x - 20 = y$$

A) (4, 8) B) (1, 5) C) (−4, −8)
D) (6, 10) C) (5, 5)

6. Solve by determinants:

$$4x - y = 10$$
$$x + y = 5$$

A) (3, 2) B) (4, 4) C) (4, 1)
D) (6, −1) E) (−3, 2)

7. Solve graphically:

$$x + y = 3$$
$$2x - y = 6$$

A) (0, 3) B) (0, 0) C) (3, 3)
D) (3, 0) E) (4, −7)

8. Solve by substitution:

$$x + y = 12$$
$$y = 3x$$

A) (1, 3) B) (9, 3) D) (3, 9)
D) ($\frac{4}{3}$, 1) E) (−1, 1)

Solutions on page 263

15.5 Quadratic Equations

There are three algebraic techniques for solving a quadratic equation.

Factoring—the easiest, if it can be applied.
Completing the square—useful to obtain $(x - h)^2$ and $(y - k)^2$ terms for graphing.
Quadratic formula—always works.

The solutions (roots, zeros) of a quadratic equation are the values of the variable which make the equation equal to zero. The roots are graphically represented as the x-intercepts if $y = f(x)$. There are always two roots, not necessarily real.

Factoring

Find the roots of $f(x) = x^2 - 5x - 6$ **by factoring**.

1. To find the roots set $f(x) = 0$.
2. Factor.
3. Set each factor equal to 0.
4. Solve for x.

$$x^2 - 5x - 6 = 0$$
$$(x - 6)(x + 1) = 0$$
$$x - 6 = 0 \mid x + 1 = 0$$
$$x = 6 \mid \quad x = -1$$
The roots are 6, −1.

Completing the Square

Find the roots of $f(x) = x^2 - 5x - 6$ **by completing the square**.

$$x^2 - 5x - 6 = 0$$

$$x^2 - 5x = 6$$

$$x^2 - 5x + \left(\frac{-5}{2}\right)^2 = 6 + \left(\frac{-5}{2}\right)^2$$

$$\left(x - \frac{5}{2}\right)^2 = \frac{49}{4}$$

$$\sqrt{\left(x - \frac{5}{2}\right)^2} = \pm\sqrt{\frac{49}{4}}$$

$$x - \frac{5}{2} = \pm\frac{7}{2}$$

1. To find the roots set $f(x) = 0$.
2. Algebraically move the numerical term to the other side of the equation.
3. Divide the coefficient of the x term by 2 and add this number squared to both sides of the equation.
4. a) Express the left side of the equation as $(x + a)^2$, where a is the value from step 3 found by dividing the coefficient of the x by 2.
   b) Express the right side as a single numerical value.
5. Take the square root of each side of the equation.
6. Separate the equation into two parts and solve for x.

$$x - \frac{5}{2} = \frac{7}{2} \qquad x - \frac{5}{2} = \frac{-7}{2}$$

$$x = 6 \qquad x = -1$$

The roots are -1 and 6.

Quadratic Formula

Find the roots of $f(x) = x^2 - 5x - 6$ **using the quadratic formula**.

Quadratic Formula

$$f(x) = ax^2 + bx + c \qquad\qquad f(y) = ay^2 + by + c$$

$$x = \frac{-b \pm \sqrt{b^2 - 4ac}}{2a} \qquad\qquad y = \frac{-b \pm \sqrt{b^2 - 4ac}}{2a}$$

$$x^2 - 5x - 6 = 0$$

$$a = 1 \quad b = -5 \quad c = -6$$

$$x = \frac{-(-5) \pm \sqrt{(-5)^2 - 4(1)(-6)}}{2(1)}$$

1. To find the roots set $f(x) = 0$.
2. Determine the values of a, b and c.
3. Apply the formula.
4. Solve for the values of the variable.

$$= \frac{5 \pm \sqrt{25 + 24}}{2} = \frac{5 \pm \sqrt{49}}{2}$$

$$= \frac{5 \pm 7}{2}$$

$$x = \frac{5 + 7}{2} = 6 \qquad x = \frac{5 - 7}{2} = -1$$

The roots are 6 and -1.

Graphing

The roots are where the equation crosses the x axis if $y = f(x)$.
Find the roots of $f(x) = x^2 - 5x - 6$ using graphing.

$$f(x) = x^2 - 5x - 6$$

$$x = \frac{-b}{2a} = \frac{5}{2} = \frac{5}{2}$$

1. Arrange the equation into the form $f(x) = ax^2 + bx + c$.

2. Determine the axis of symmetry $x = \frac{-b}{2a}$.

3. Establish a table of values with $x = -\frac{b}{2a}$ the center of 7 values. That is, 3 values above and 4 values below. The values above and below should only be integers.

4. The roots are the x-intercepts.

x	$x^2 - 5x - 6$	y
0	$0^2 - 5(0) - 6$	-6
1	$1^2 - 5(1) - 6$	-10
2	$2^2 - 5(2) - 6$	-12
$\frac{5}{2}$	$(\frac{5}{2})^2 - 5(\frac{5}{2}) - 6$	$-\frac{49}{4}$
3	$3^2 - 5(3) - 6$	-12
4	$4^2 - 5(4) - 6$	-10
5	$5^2 - 5(5) - 6$	-6

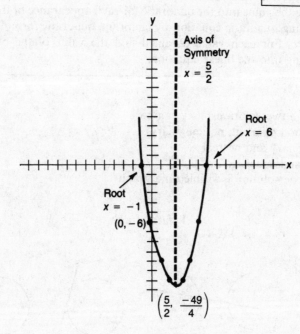

Practice Problems 15.5

1. Find the roots of $f(x) = 2x^2 + 3x - 2$ (use factoring).

A) $\frac{1}{2}$ and -2 B) 2 and -2
C) $\frac{1}{2}$ and $-\frac{1}{2}$ D) $\frac{1}{2}$ and 2
E) $-\frac{1}{2}$ and 2

2. Find the roots of $x^2 - 4x - 7$ (use completing the square).

A) $2 \pm \sqrt{7}$ B) $2 \pm \sqrt{15}$ C) $2 \pm \sqrt{11}$
D) -2 and 2 E) $\pm\sqrt{11}$

3. Find the roots of $x^2 + 4x + 9$ (use the quadratic formula).

   A) $2 \pm \sqrt{5}$ B) $-2 \pm i\sqrt{5}$ C) $2 \pm \sqrt{2}$
   D) $2 \pm 5i\sqrt{2}$ E) $5 \pm 3\sqrt{2}$

4. Find the roots of $x^2 - 4x + 4$ (use graphing).

   A) Double root at 2 B) 2, -2 C) 2, 3
   D) 4, 1 E) 0, 4

Solutions on page 264

15.6 Linear-Quadratic Systems

These systems are solved algebraically by substitution or by graphing. There are **usually two solutions, which can be real or imaginary**.

- **Substitution:** Solve the linear equation for either variable, whichever is easiest. Substitute the value into the quadratic for each appearance of the variable. Solve the resulting quadratic equation by factoring, quadratic formula, or completing the square. For each value obtained find the value of the other variable by substituting into the linear equation.

 If there are two solutions, they can be:
- 2 real and different, no imaginary
- 2 imaginary and no real
- 2 real and equal, no imaginary
  (only one solution is visible on a graph)

Typical graph

| 2 solutions | 2 solutions | 1 solution
(2 the same) | 0 real solutions
(2 imaginary) |

Example 1:

Solve: $x^2 - 4x - y - 5 = 0$

$$y = 3x - 11$$

The linear equation is already solved for *y*. Substitute into the quadratic:

$$x^2 - 4x - (3x - 11) - 5 = 0$$
$$x^2 - 4x - 3x + 11 - 5 = 0$$
$$x^2 - 7x + 6 = 0$$
$$(x - 6)(x - 1) = 0$$
$$x - 6 = 0 \mid x - 1 = 0$$
$$x = 6 \mid x = 1$$

For each *x* value determinine the corresponding *y*:

$x = 6$	$x = 1$
$y = 3x - 11$	$y = 3(1) - 11$
$y = 3(6) - 11$	$= 3 - 11$
$y = 7$	$= -8$
$\therefore x = 6, y = 7$	$\therefore x = 1, y = -8$
$(6, 7)$	$(1, -8)$

Example 2:

$$x^2 + y^2 - 10y - 24 = 0$$
$$y = x - 2$$

The linear equation is already solved for *y*. Substitute into the quadratic:

$$x^2 + (x - 2)^2 - 10(x - 2) - 24 = 0$$
$$x^2 + x^2 - 4x + 4 - 10x + 20 - 24 = 0$$
$$\frac{2x^2}{2} \frac{-14x}{2} = \frac{0}{2}$$
$$x^2 - 7x = 0$$
$$x(x - 7) = 0$$
$$x = 0 \mid x - 7 = 0$$
$$\mid x = 7$$

For each *x*-value determine the corresponding *y* value:

$y = x - 2$	$y = x - 2$
$y = 0 - 2$	$y = 7 - 2$
$y = -2$	$y = 5$
$x = 0, \quad y = -2$	$x = 7, y = 5$
$(0, -2)$	$(7, 5)$

Graphing

> Graph the equations on the same set of axis. The real solutions are at the intersections of the graphs.

Solve: $(x - 2)^2 + (y + 3)^2 = 9$
$$y = 2x - 3$$

$$(x - 2)^2 + (y + 3)^2 = 9$$

This is a circle with center $(2, -3)$ and radius $= 3$.

$$y = 2x - 3$$

This is a line with $m = \text{slope} = \dfrac{2 \leftarrow \text{up}}{1 \leftarrow \text{right}}$ and *y*-intercept of -3.

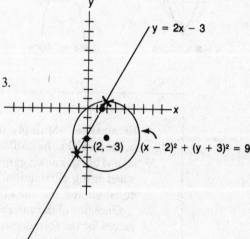

2 real solutions indicated by the *X*'s

Practice Problems 15.6

1. Solve: $y = 5 + 2x$
$x^2 + y^2 = 25$

    A) (1, 7) and (0, 5)
    B) (−4, −3) and (1, 7)
    C) (−4, −3) and (0, 5)
    D) (3, 4) and (−3, −4)
    E) (2, 9) and (2, $\sqrt{21}$)

2. Solve: $x^2 + y^2 - 4x - 2y + 1 = 0$
$y + x = 1$

    A) $(\frac{1}{2}, \frac{1}{2})$ and (0, 1)
    B) (1, 0) and (3, −2)
    C) $(\frac{1}{4}, \frac{3}{4})$ and (−2, 3)
    D) (0, 1) and $(\frac{1}{2}, \frac{1}{2})$
    E) (0, 1) and (2, −1)

3. Solve: $y = 2x^2 - 5x + 5$
$y - x = 5$

    A) (7, 2) and (4, −1)
    B) (0, 5) and (3, 8)
    C) (0, 5) and (−1, 4)
    D) (−3, −8) and (1, −6)
    E) $(\frac{5}{2}, \frac{1}{2})$ and (5, 0)

4. Solve: $y = -x^2 + 4x - 3$
$y + 1 = x$

    A) (1, 0) and (2, 1)
    B) (3, 2) and (−1, 0)
    C) (0, −1) and (2, 3)
    D) (1, 2) and (0, 1)
    E) (5, 4) and (−6, −7)

Solutions on page 265

15.7 Quadratic-Quadratic Systems

In general, either substitution or elimination can be used to solve these systems. Sometimes depending on the particular equations involved only one of these methods may be practical. These systems can have 4, 3, 2, 1 or 0 real solutions.

4 real solutions

3 real solutions

2 real solutions

1 real solution

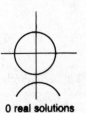
0 real solutions

> - **Elimination:** Multiply one or both of the equations by suitable constants in order to make the coefficients of one of the variables the same—but opposite in sign. This unknown can be eliminated by adding the two equations. Remember, when using elimination be sure the variable terms are all on one side of the equation and the numerical or constant terms are on the other.
>
> Once one of the variables is eliminated solve the remaining equation. Using the values for the solved variable substitute these values back into one of the original equations to find the corresponding values of the eliminated variable.

> • **Substitution:** Solve the easiest of the two equations for the value of either variable or either variable squared. Substitute this value into the other equation and solve. Using the values for the solved variable, substitute these values back into one of the original equations to find the corresponding values of the other variable.

Example 1:

Solve by elimination:

$$2x^2 = 6 + 3y^2$$
$$x^2 = 17 - 2y^2$$

Rearrange the equations.

$$2x^2 - 3y^2 = 6$$
$$x^2 + 2y^2 = 17$$

Multiply the second equation by -2 to eliminate the x^2.

$$-2(x^2 + 2y^2 = 17) \Rightarrow -2x^2 - 4y^2 = -34$$

$$\begin{array}{r} -2x^2 - 4y^2 = -34 \\ 2x^2 - 3y^2 = 6 \\ \hline -7y^2 = -28 \\ y^2 = 4 \\ y = \pm 2 \end{array}$$

For each value of y find the corresponding x values.

$y = 2$	$y = -2$
$x^2 = 17 - 2y^2$	$x^2 = 17 - 2y^2$
$x^2 = 17 - 2(2)^2$	$x^2 = 17 - 2(-2)^2$
$x^2 = 17 - 8 = 9$	$x^2 = 17 - 8 = 9$
$x = \pm 3$	$x = \pm 3$
$x = 3, y = 2$	$x = 3, y = -2$
$x = -3, y = 2$	$x = -3, y = -2$

4 solutions

Example 2:

Solve by substitution:

$$2x^2 = 6 + 3y^2$$
$$x^2 = 17 - 2y^2$$

The second equation is solved for x^2. Substitute into first equation:

$$2x^2 = 6 + 3y^2$$
$$2(17 - 2y^2) = 6 + 3y^2$$
$$34 - 4y^2 = 6 + 3y^2$$
$$7y^2 = 28$$
$$y^2 = 4$$
$$y = \pm 2$$

For each value of y find the corresponding x values:

$y = 2$	$y = -2$
$x^2 = 17 - 2y^2$	$x^2 = 17 - 2y^2$
$x^2 = 17 - 2(2)^2$	$x^2 = 17 - 2(-2)^2$
$x^2 = 17 - 8 = 9$	$x^2 = 17 - 8 = 9$
$x = \pm 3$	$x = \pm 3$
$x = 3, y = 2$	$x = 3, y = -2$
$x = -3, y = 2$	$x = -3, y = -2$

> ● **Graphing:** Plot or sketch each of the curves. Their points of intersection are the real solutions.

Solve graphically: $2x^2 = 6 + 3y^2$
$x^2 = 17 - 2y^2$

$$2x^2 = 6 + 3y^2 \Rightarrow 2x^2 - 3y^2 = 6 \Rightarrow \frac{x^2}{3} - \frac{y^2}{2} = 1 \quad \text{Hyperbola}$$

$$x^2 = 17 - 2y^2 \Rightarrow x^2 + 2y^2 = 17 \Rightarrow \frac{x^2}{17} + \frac{y^2}{\frac{17}{2}} = 1 \quad \text{Ellipse}$$

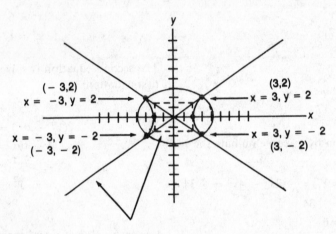

4 Solutions

Practice Problems 15.7

1. How many real solutions are there for the quadratic system: $\quad 4x^2 + y^2 = 20$
$\qquad x^2 + 4y^2 = 20$

A) 0   B) 1   C) 2   D) 3   E) 4

2. How many real solution are there for the quadratic system: $\quad y^2 + 2x = -1$
$\qquad x^2 - y^2 = 16$

A) 0   B) 1   C) 2   D) 3   E) 4

3. How many real solutions are there for the quadratic system: $\quad x^2 + y = 5$
$\qquad x^2 + y^2 = 1$

A) 0   B) 1   C) 2   D) 3   E) 4

4. How many real solutions are there for the quadratic system: $\qquad xy = 12$
$\qquad x^2 + y^2 = 16$

A) 0   B) 1   C) 2   D) 3   E) 4

Solutions on page 266

15.8 Radical Equations

> In solving equations containing radicals it is important to get the radical term alone on one side of the equation. Then square both sides of the equation to eliminate the radical sign. Solve the resulting equation.

☞ $\sqrt{x} \cdot \sqrt{x} = (\sqrt{x})^2 = x \quad x \geq 0$

$\sqrt{4} = 2$ only, $\sqrt{4} \neq -2$

Remember that all solutions to radical equations **must be checked**, as squaring both sides sometimes results in extraneous solutions. It is possible for the solution set of a radical equation to be a null or empty set.

Remember also in squaring each side of an equation, do not make the mistake of simply squaring each term; the entire side of the equation must be multiplied by itself.

$$(x + 2)^2 \neq x^2 + 2^2, \quad (x + 2)^2 = (x + 2)(x + 2) = x^2 + 4x + 4$$

If more than one radical term appears in the equation eliminate one radical term as described above. Then get the second radical term alone and repeat the process. These types of equations involve double squaring operations.

Examples:

1. Solve for x: $\sqrt{x - 3} = 4$

2. Solve for x: $\sqrt{x^2 - 7} - 1 = x$

Solutions:

1. Get the $\sqrt{}$ term alone, then square both sides.

$$\sqrt{x - 3} = 4$$
$$(\sqrt{x - 3})^2 = (4)^2$$
$$x - 3 = 16$$
$$x = 19$$

Check

$$\sqrt{x - 3} \stackrel{?}{=} 4$$
$$\sqrt{19 - 3} \stackrel{?}{=} 4$$
$$\sqrt{16} \stackrel{?}{=} 4$$
$$4 \stackrel{?}{=} 4 \checkmark$$

1 solution
$$x = 4$$

2. Get the $\sqrt{}$ term alone, then square both sides.

$$\sqrt{x^2 - 7} - 1 = x$$
$$\sqrt{x^2 - 7} = x + 1$$
$$(\sqrt{x^2 - 7})^2 = (x + 1)^2 = (x + 1)(x + 1)$$
$$x^2 - 7 = x^2 + 2x + 1$$
$$2x + 8 = 0$$
$$x = -4$$

Check

$$\sqrt{x^2 - 7} - 1 = x$$
$$\sqrt{(-4)^2 - 7} - 1 \stackrel{?}{=} -4$$
$$\sqrt{16 - 7} - 1 \stackrel{?}{=} -4$$
$$\sqrt{9} - 1 \stackrel{?}{=} -4$$
$$3 - 1 \neq -4$$

No solution
$$\{ \ \}$$

3. Solve for x: $\sqrt{x + 2} + \sqrt{x - 3} = 5$

3. First get one of $\sqrt{}$ by itself and square the equation.

$$\sqrt{x + 2} + \sqrt{x - 3} = 5$$
$$\sqrt{x + 2} = 5 - \sqrt{x - 3}$$
$$(\sqrt{x + 2})^2 = (5 - \sqrt{x - 3})^2$$
$$x + 2 = 25 - 10\sqrt{x - 3} + (x - 3)$$

Rearrange and get the $\sqrt{}$ by itself and square the equation.

$$10\sqrt{x - 3} = 20$$
$$\sqrt{x - 3} = 2$$
$$(\sqrt{x - 3})^2 = (2)^2$$
$$x - 3 = 4$$
$$x = 7$$

Check

$$\sqrt{x + 2} + \sqrt{x - 3} = 5$$
$$\sqrt{7 + 2} + \sqrt{7 - 3} \overset{?}{=} 5$$
$$\sqrt{9} + \sqrt{4} \overset{?}{=} 5$$
$$3 + 2 \overset{?}{=} 5$$
$$5 \overset{?}{=} 5 \checkmark$$

1 solution
$$x = 7$$

Practice Problems 15.8

1. Solve for x: $\sqrt{x^2 + 3} = x + 1$

   A) ± 1 B) $+1$ only   C) -1 only
   D) No solution   E) 2

2. Solve for x: $\sqrt{x + 18} + 2 = x$

   A) 7 and -2 B) -2 only   C) 7 only
   D) No solution   E) -7 and -2

3. If $\sqrt{x + 1} + 2 = 0$, what is the solution set for all real values of x?

   A) $\{-1\}$ B) $\{2\}$ C) $\{-1, 2\}$
   D) $\{3\}$ E) $\{\ \}$

4. The equation $\sqrt{2x} - \sqrt{x + 1} = 0$ has

   A) 1 as its only root
   B) 0 as its only root
   C) 0 and 1 as its only roots
   D) 0, 1 and 2 as its roots
   E) no solution

Solutions on page 267

15.9 Absolute Value Equations

Absolute value equations are of the format $|x| = a$. To solve these equations convert the equation into:

$$|x| = a$$
$$\downarrow \quad \downarrow$$
$$x = -a \qquad x = a$$

Solve each equation separately. All possible solutions **must be checked** into the original equation. It is possible for the solution set to an absolute value equation to be the empty or null set.

Examples:

1. Solve for x: $|x - 3| = 9$

2. Solve for x: $|2x - 3| = x + 1$

Solutions:

1.
$$|x - 3| = 9$$
$$\downarrow \quad \downarrow$$

$x - 3 = -9$	$x - 3 = 9$
$x = -6$	$x = 12$

Check

$|x - 3| = 9$	$|x - 3| = 9$
$|-6 - 3| \stackrel{?}{=} 9$	$|12 - 3| \stackrel{?}{=} 9$
$|-9| \stackrel{?}{=} 9$	$|9| = 9$
$9 = 9 \checkmark$	$9 = 9 \checkmark$

2 solutions:
$$x = -6, x = 12$$

2.
$$|2x - 3| = x + 1$$
$$\downarrow \quad \downarrow$$

$2x - 3 = -(x + 1)$	$2x - 3 = x + 1$
$2x - 3 = -x - 1$	$x = 4$
$3x = +2$	
$x = \frac{2}{3}$	$x = 4$

Check

$|2x - 3| = x + 1$	$|2x - 3| = x + 1$
$|2(\frac{2}{3}) - 3| \stackrel{?}{=} \frac{2}{3} + 1$	$|2(4) - 3| \stackrel{?}{=} 4 + 1$
$|-\frac{5}{3}| \stackrel{?}{=} \frac{5}{3}$	$|5| \stackrel{?}{=} 5$
$\frac{5}{3} = \frac{5}{3}$	$\sqrt{5} = 5 \checkmark$

2 solutions:
$$x = \frac{5}{3}, x = 4$$

3. Solve for x: $|x + 4| = -2$

3.

$$|x + 4| = -2$$

$$x + 4 = -(-2) \qquad x + 4 = -2$$
$$x = -2 \qquad\qquad x = -6$$

Check

$$|x + 4| = -2 \qquad\qquad |x + 4| = -2$$
$$|-2 + 4| \overset{?}{=} -2 \qquad |-6 + 4| \overset{?}{=} -2$$
$$|2| \overset{?}{=} -2 \qquad\qquad |-2| \overset{?}{=} -2$$
$$2 \neq -2 \qquad\qquad\quad 2 \neq -2$$

No solutions

Practice Problems 15.9

Solutions on page 268

1. The solution set of $|x - 1| = 3$ is

   A) $\{-2, 4\}$ B) $\{2, -4\}$ C) $\{-2\}$
   D) $\{4\}$ E) $\{\ \}$

2. What is the solution set of the equation

   $|x + 7| = 2x - 1$

   A) $\{-2\}$ B) $\{8\}$ C) $\{-2, 8\}$
   D) $\{2, -8\}$ E) $\{\ \}$

3. The solution set of $|x^2 - 5x| = 6$ is

   A) $\{\ \}$ B) $\{2, 3\}$ C) $\{-1, 6\}$
   D) $\{3, 6\}$ E) $\{-1, 2, 3, 6\}$

4. What is the solution set of the equation

   $$|x - 7| = 2x - 10$$

   A) $\{\frac{17}{3}\}$ B) $\{3\}$ C) $\{-3\}$
   D) $\{\frac{17}{3}, 3\}$ E) $\{\ \}$

15.10 Three Simultaneous Equations

● **Elimination:** This type of system is composed of three variables. Select two different pairs of equations and, using the elimination method, eliminate the same variable from each pair of equations. The result is two linear equations with two variables. Using any of the linear-linear system methods, solve for both variables. Using the two found variables, substitute into any of the three original equations to find the value of the third variable.

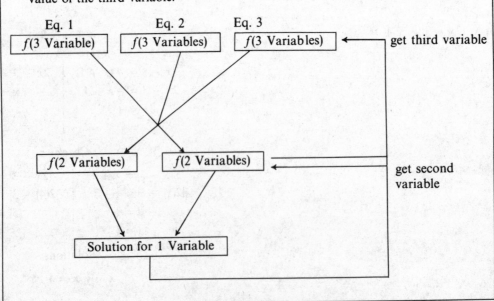

Example:

1. Solve the system of equations:

$$3x + y - 4z = -2$$
$$x + y + 2z = -1$$
$$x + 2y + 6z = -2$$

Solution:

1. Selecting equations 1 with 2 and 1 with 3.

$$3x + y - 4z = -2 \qquad\qquad 3x + y - 4z = -2$$
$$x + y + 2z = -1 \qquad\qquad x + 2y + 6z = -2$$

Multiply first equation by -1 and add it to the second equation

Multiply first equation by -2 and add it to the second equation

$$-3x - y + 4z = 2 \qquad\qquad -6x - 2y + 8z = 4$$
$$\underline{x + y + 2z = -1} \qquad\qquad \underline{x + 2y + 6z = -2}$$
$$\boxed{-2x \quad + 6z = 1} \qquad\qquad \boxed{-5x \quad + 14z = 2}$$

$$-2x + 6z = 1$$
$$-5x + 14z = 2$$

Multiply first equation by 5 and second equation by -2 to eliminate x

$$5(-2x + 6z = 1)$$
$$-2(-5x + 14z = 2)$$
$$\Downarrow$$
$$-10x + 30z = 5$$
$$\underline{10x - 28z = -4}$$
$$2z = 1$$

$$\boxed{z = \tfrac{1}{2}}$$

Substitute $z = \tfrac{1}{2}$ into one of the two 2-variable equations.

$$-2x + 6z = 1$$
$$-2x + 6(\tfrac{1}{2}) = 1$$
$$-2x + 3 = 1$$
$$-2x = -2$$
$$\boxed{x = 1}$$

Substitute $x = 1$, $z = \tfrac{1}{2}$ into one of the three 3-variable equations.

$$x + y + 2z = -1$$
$$1 + y + 2(\tfrac{1}{2}) = -1$$
$$y + 2 = -1$$

$$\boxed{y = -3}$$

$$x = 1, \ y = -3, \ z = \tfrac{1}{2}$$

● **Determinants:** Arrange the equation in the format:

$$ax + by + cz = \boldsymbol{d}$$
$$ex + fy + gz = \boldsymbol{h}$$
$$jx + ky + lz = \boldsymbol{m}$$

If any term is missing insert the variable with a zero coefficient.

$$x = \frac{\begin{vmatrix} d & b & c \\ h & f & g \\ m & k & l \end{vmatrix}}{\begin{vmatrix} a & b & c \\ e & f & g \\ j & k & l \end{vmatrix}} \qquad y = \frac{\begin{vmatrix} a & d & c \\ e & h & g \\ j & m & l \end{vmatrix}}{\begin{vmatrix} a & b & c \\ e & f & g \\ j & k & l \end{vmatrix}} \qquad z = \frac{\begin{vmatrix} a & b & d \\ e & f & h \\ j & k & m \end{vmatrix}}{\begin{vmatrix} a & b & c \\ e & f & g \\ j & k & l \end{vmatrix}}$$

Review:

$$\begin{vmatrix} a & b & c \\ e & f & g \\ j & k & l \end{vmatrix} \begin{matrix} a & b \\ e & f \\ j & k \end{matrix} = [a \cdot f \cdot l + b \cdot g \cdot y + c \cdot e \cdot k] - [j \cdot f \cdot c + k \cdot g \cdot a + l \cdot e \cdot b]$$

$$j \cdot f \cdot c + k \cdot g \cdot a + l \cdot e \cdot b$$
$$a \cdot f \cdot l + b \cdot g \cdot j + c \cdot e \cdot k$$

Example:

2. Using determinants solve the system of equations:

$$3x + y - 4z = -2$$
$$x + y + 2z = -1$$
$$x + 2y + 6z = -2$$

Solutions:

$$x = \frac{\begin{vmatrix} -2 & 1 & -4 \\ -1 & 1 & 2 \\ -2 & 2 & 6 \end{vmatrix}}{\begin{vmatrix} 3 & 1 & -4 \\ 1 & 1 & 2 \\ 1 & 2 & 6 \end{vmatrix}} = \frac{\begin{vmatrix} -2 & 1 & -4 \\ -1 & 1 & 2 \\ -2 & 2 & 6 \end{vmatrix} \begin{matrix} -2 & 1 \\ -1 & 1 \\ -2 & 2 \end{matrix}}{\begin{vmatrix} 3 & 1 & -4 \\ 1 & 1 & 2 \\ 1 & 2 & 6 \end{vmatrix} \begin{matrix} 3 & 1 \\ 1 & 1 \\ 1 & 2 \end{matrix}}$$

$$= \frac{(-12 - 4 + 8) - (8 - 8 - 6)}{(18 + 2 - 8) - (-4 + 12 + 6)}$$

$$= \frac{-2}{-2} = 1$$

$$y = \frac{\begin{vmatrix} 3 & -2 & -4 \\ 1 & -1 & 2 \\ 1 & -2 & 6 \end{vmatrix}}{\begin{vmatrix} 3 & 1 & -4 \\ 1 & 1 & 2 \\ 1 & 2 & 6 \end{vmatrix}} = \frac{\begin{vmatrix} 3 & -2 & -4 \\ 1 & -1 & 2 \\ 1 & -2 & 6 \end{vmatrix} \begin{matrix} 3 & -2 \\ 1 & -1 \\ 1 & -2 \end{matrix}}{\begin{vmatrix} 3 & 1 & -4 \\ 1 & 1 & 2 \\ 1 & 2 & 6 \end{vmatrix} \begin{matrix} 3 & 1 \\ 1 & 1 \\ 1 & 2 \end{matrix}}$$

$$= \frac{(-18 - 4 + 8) - (4 - 12 - 12)}{-2}$$

$$= \frac{6}{-2} = 3$$

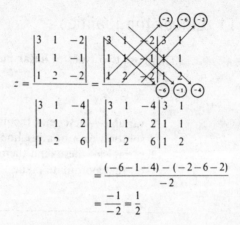

$$z = \frac{\begin{vmatrix} 3 & 1 & -2 \\ 1 & 1 & -1 \\ 1 & 2 & -2 \end{vmatrix}}{\begin{vmatrix} 3 & 1 & -4 \\ 1 & 1 & 2 \\ 1 & 2 & 6 \end{vmatrix}}$$

$$= \frac{(-6-1-4)-(-2-6-2)}{-2}$$

$$= \frac{-1}{-2} = \frac{1}{2}$$

 Note all three denominators are the same. Only calculate the value once.

Practice Problems 15.10

1. Solve by elimination:

$$x + y + 2z = 1$$
$$2x + 3y + 4z = 4$$
$$3x - y - z = 2$$

A) $x=1, y=2, z=-1$ B) $x=2, y=2, z=\frac{3}{2}$
C) $x=4, y=6, z=4$ D) $x=1, y=1, z=\frac{1}{4}$
E) $x=-1, y=2, z=1$

2. Solve by elimination:

$$2x - 3y + 2z = 26$$
$$x - 2y - 3z = -5$$
$$x + y - z = -7$$

A) $x=4, y=3, z=-14$ B) $x=0, y=5, z=-2$
C) $x=-1, y=-2, z=-1$ D) $x=1, y=1, z=1$
E) $x=2, y=-4, z=5$

3. Solve by determinants:

$$2x + 2y - z = 10$$
$$x + 3y + 3z = 4$$
$$5x + z = 3$$

A) $x=1, y=1, z=1$ B) $x=1, y=3, z=-2$
C) $x=1, y=2, z=3$ D) $x=-1, y=2, z=-3$
E) $x=1, y=2, z=-3$

4. Solve by determinants:

$$y + 2z = -6$$
$$2x + z = 11$$
$$x + 2y = 16$$

A) $x=0, y=12, z=3$ B) $x=8, y=4, z=-5$
C) $x=5, y=0, z=1$ D) $x=4, y=-5, z=8$
E) $x=4, y=6, z=0$

Solutions on page 270

5.11 Linear Inequalities

There are two types of **linear inequalities** which we will review in this book.

1 variable—these lend themselves to algebraic solution and a demonstration of the solution on a number line.

2 variables—these lend themselves to graphic representation and represent a 'space' on the coordinate plane.

---

1 variable: The rules to solve a single variable inequality are the same as those for an equality with one exception:

Any multiplication or division by a negative number reverses the direction of the inequality.

The solution can be represented on a number line.

Examples:

1. Find the solution set for $2x - 1 \geqslant 5$ and represent the solution on a number line.

2. If $8 - 2x > 4$, represent the solution set on a number line.

Solutions:

1.
$$2x - 1 \geqslant 5$$
$$2x \geqslant 6$$
$$x \geqslant 3$$

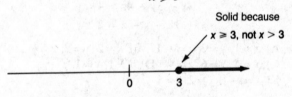

Solid because
$x \geqslant 3$, not $x > 3$

 ● **represents a particular number is included,** e.g., $x \geqslant 3$, $x \leqslant 3$, $x = 3$.
○ **represents a particular value is not included,** e.g., $x > 3$, $x < 3$.

2.
$$8 - 2x > 4$$
$$-2x > -4$$
$$x < 2$$

Division by -2 reverses the direction of the inequality.

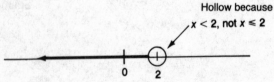

Hollow because
$x < 2$, not $x \leqslant 2$

2 variables: The equation $ax + by = c$ is the boundary between the space $ax + by > c$ and $ax + by < c$. The boundary itself is included if $ax + by \geq c$ or $ax + by \leq c$ and is shown by a solid line (to show inclusion). The boundary is excluded if $ax + by > c$ or $ax + by < c$ and is shown as a dotted line (to show exclusion).

$ax + by \geq c$, $ax + by > c$, $ax + by \leq c$ and $ax + by < c$ represent a space on the coordinate plane.

Examples:

3. Represent $y + 2x \geq 4$.

Solutions:

3. First graph the boundary $y + 2x = 4$. Since the original problem shows $\geq$ the boundary is solid (inclusion).

$$y + 2x = 4$$

x	y
0	4
1	2
2	0

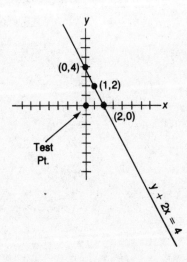

Test a sample point on one side of the boundary line. For example, test the origin $(0, 0)$:

$$y + 2x \geq 4$$

Test point $(0, 0)$:

$$0 + 2(0) \overset{?}{\geq} 4$$

$$0 \overset{?}{\geq} 4 \quad \text{False}$$

∴ The test point is *not* in the space.

Shade the side of the line *not* including the test point.

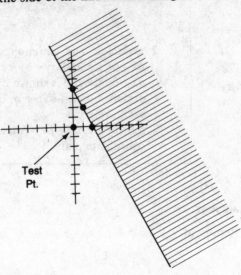

4. Represent $x < 4$.

4. First graph the boundary $x = 4$. Since the original equation involves only $<$, the boundary is dotted (excluded).

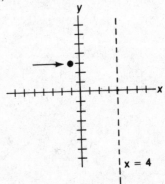

Test a sample point on one side of the boundary line. For example, test point $(-1, 3)$:

$$x < 4$$

Test point $(-1, 3)$:

$$-1 < 4 \quad \text{True}$$

∴ The test point *is* in the sample space.
Shade the side of the line including the test point.

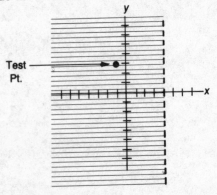

Practice Problems 15.11

1. Graphically represent the solution set to

    $3x + 1 > 7$

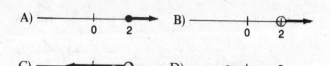

E) ——————+——————●
         0

4. Represent $2x + y \leqslant -6$

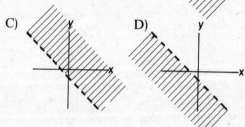

E)

2. Graphically represent the solution set to

    $1 - 3x \leqslant -5$

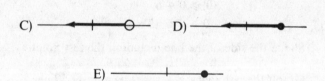

E) ——————+——————●

3. Represent $x - y < 0$

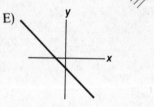

Solutions on page 271

15.12 Linear-Linear Inequalities

> The solution set is the coordinate space in the region overlapped by the two linear graphs. Graph each inequality on the same set of axes and notate the overlap.

Example:

1. On the same set of coordinate axes, graph the following system of inequalities and label the solution set A:

$$2y \geqslant x - 6$$
$$x + y < 2$$

Solution:

1. Graph the boundary $2y = x - 6$ as a solid line:

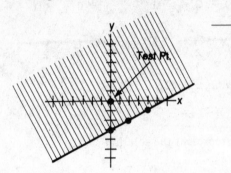

x	y
0	−3
2	−2
4	−1

Test point $(0, 0)$:

$$2y \geqslant x - 6$$
$$2(0) \geqslant 0 - 6$$
$$0 \geqslant -6 \quad \text{True}$$

Shade the side of the line including the test point.

Graph the boundary $x + y = 2$ as a dashed line:

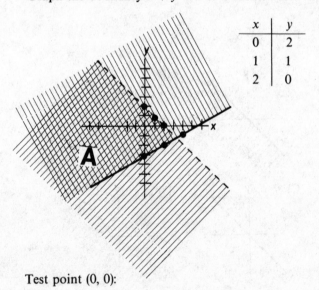

x	y
0	2
1	1
2	0

Test point $(0, 0)$:

$$x + y < 2$$
$$0 + 0 < 2$$
$$0 < 2 \quad \text{True}$$

Shade the side of the boundary including the test point.

The solution set A is the overlapping region.

Practice Problems 15.12

1. On the same set of axes, graph the following system of inequalities and label the solution set S:

$$y > x + 1$$
$$x \leqslant -1$$

2. On the same set of axes, graph the following system of inequalities and label the solution set N:

$$y < x - 2$$
$$y \geqslant -2x + 4$$

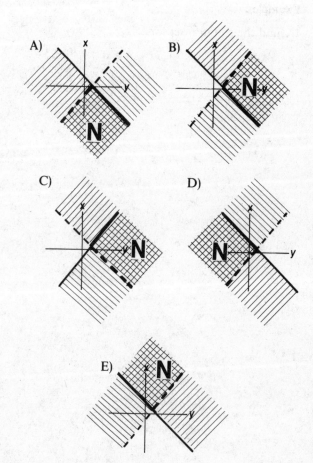

Solutions on page 272

15.13 Quadratic Inequalities

There are two types of quadratic inequalities that we will review in this book.

1 variable—These lend themselves to an algebraic solution and/or analysis on a number line. Using a number analysis, find the values which make the quadratic function equal to 0. Analyze the intervals between these critical values.

2 variables—These lend themselves to graphic representation and represent a 'space' on the coordinate plane.

---

1 variable: Arrange the quadratic function into $ax^2 + bx + c \geqslant 0$, >0, <0, $\leqslant 0$ form. Set the quadratic function equal to 0. The values which satisfy this equality are critical values in a number line analysis. Analyze the interval between these critical values to determine if the inequality is valid. The critical values are included if $\leqslant$ or $\geqslant$; they are excluded if $>$ or $<$. Critical values are represented by ● for inclusion, ○ for exclusion.

Examples:

1. Find the solution set for

$$x^2 + 3x - 4 \geqslant 0$$

Solutions:

1. To find the critical values set $f(x) = 0$.

$$x^2 + 3x - 4 = 0$$
$$(x + 4)(x - 1) = 0$$

$$x + 4 = 0 \quad \bigg| \quad x - 1 = 0$$
$$x = -4 \quad \bigg| \quad \quad x = 1$$

The critical values are -4 and 1. They are included because of the $\geqslant$ in the original problem.

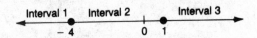

$$x^2 + 3x - 4 \geqslant 0$$

Test the interval $x < -4$, e.g., $x = -10$:

$$(-10)^2 + 3(-10) - 4$$
$$66 \geqslant 0 \quad \text{True} \checkmark$$

Darken $x < -4$ on the number line.

Test the interval $-4 < x < 1$, e.g., $x = 0$:

$$0^2 + 3(0) - 4 \geqslant 0$$
$$-4 \geqslant 0 \quad \text{False}$$

Do *not* darken the number line in this interval.

Test the interval $x > 1$, e.g., $x = 10$:

$$10^2 + 3(10) - 4 \geqslant 0$$
$$126 \geqslant 0 \quad \text{True} \checkmark$$

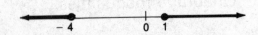

Darken $x > 1$ on the number line.

2. Find the solution set for

$$x^2 - 4 < 0$$

2. To find the critical values set $f(x) = 0$

$$x^2 - 4 = 0$$
$$(x + 2)(x - 2) = 0$$
$$x = -2 \,\big|\, x = 2$$

The critical values are $x = 2$, $x = -2$.
These values are excluded because of the $<$ in the original problem.
Test the interval $x < -2$, e.g., $x = -10$:

$$x^2 - 4 < 0 \quad 10^2 - 4 < 0 \quad \text{False}$$

Do *not* darken the number line in this interval.

Test the interval $-2 < x < 2$, e.g., $x = 0$:

$$x^2 - 4 < 0 \quad 0^2 - 4 < 0 \quad \text{True}$$

Darken the number line in this interval.
Test the interval $x > 2$, e.g., $x = 10$:

$$x^2 - 4 < 0 \quad 10^2 - 4 < 0 \quad \text{False}$$

Do *not* darken the number line in this interval.

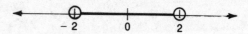

- **2 variables:** The quadratic equation with an $=$ instead of the $\leq$, $\geq$, $>$ or $<$ represents the boundary between the $>$ and $<$ regions. The boundary itself is included if $\geq$ or $\leq$ is represented (the boundary is solid). The boundary is excluded if $<$ or $>$ is represented (the boundary is dashed). The quadratic inequality is basically the "space" on the coordinate axes inside or outside the shape with the boundary either included or excluded as appropriate.

Examples:

3. Represent $x^2 + y^2 < 4$.

Solutions:

3. First graph $x^2 + y^2 = 4$. This is a circle with center $(0, 0)$ and radius $= 2$. The boundary is dashed as the original equation contains a $<$.

Test point $(0, 0)$:

$$x^2 + y^2 < 4$$
$$0^2 + 0^2 < 4$$
$$0 < 4 \quad \text{True}$$

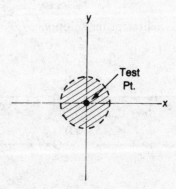

The test point is in the required space. Shade in the inside of the circle.

4. Represent $\dfrac{x^2}{9} - \dfrac{y^2}{4} \leqslant 1$.

4. First sketch $\dfrac{x^2}{9} - \dfrac{y^2}{4} = 1$. This is a hyperbola. The boundary is solid as the original equation has a $\leqslant$.

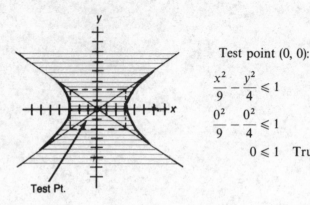

Test point $(0, 0)$:

$$\dfrac{x^2}{9} - \dfrac{y^2}{4} \leqslant 1$$

$$\dfrac{0^2}{9} - \dfrac{0^2}{4} \leqslant 1$$

$$0 \leqslant 1 \quad \text{True}$$

$\therefore$ The test point is in the required space. Shade the space including the test point.

Practice Problems 15.13

1. Graphically represent the solution set to

$x^2 - 5x + 6 < 0$

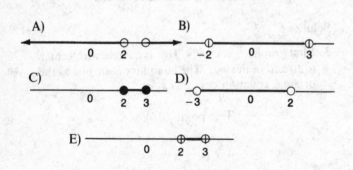

3. Represent $\dfrac{x^2}{9} + \dfrac{y^2}{16} \geqslant 1$.

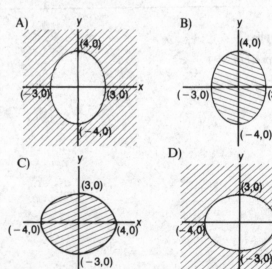

2. Graphically represent the solution set to

$(x + 3)(x - 1) \leqslant 0$

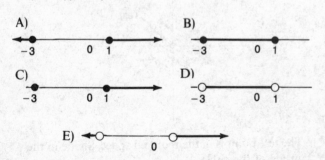

4. Represent $x^2 + y^2 > 16$.

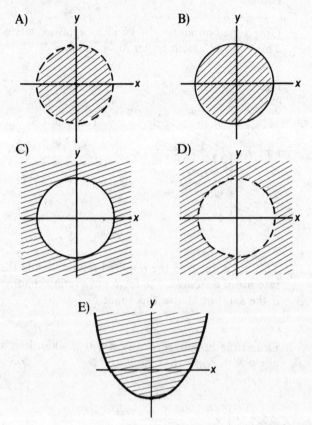

A) B)

C) D)

E)

Solutions on page 273

15.14 Quadratic-Linear and Quadratic-Quadratic Inequality Systems

> The solution set is the coordinate space in the region overlapped by the two graphs. Graph each inequality on the same set of axes and notate the overlapped region.

Examples:

1. On the same set of coordinate axes graph the following system of inequalities and label the solution set N.

$$x - y > 0$$
$$x^2 + y^2 < 9$$

Solutions:

1. Graph the boundary $x - y = 0$ as a dashed line

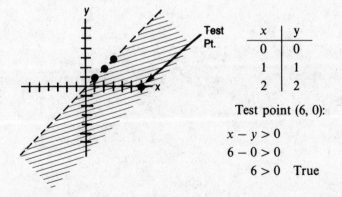

x	y
0	0
1	1
2	2

Test point (6, 0):

$$x - y > 0$$
$$6 - 0 > 0$$
$$6 > 0 \quad \text{True}$$

Shade the side of the boundary including the test point.

Graph the boundary $x^2 + y^2 = 9$ as a dashed circle. This is a circle with center $(0, 0)$, radius $= 3$.

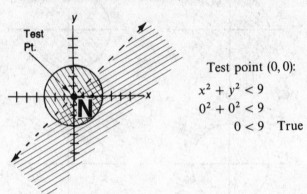

Test point $(0, 0)$:

$x^2 + y^2 < 9$

$0^2 + 0^2 < 9$

$0 < 9$ True

The test point *is* in the required region. Shade the interior of the circle. The region of overlap shading is the solution. Label this space N.

2. On the same set of coordinate axes graph the following system of inequalities.

$$x^2 + y^2 \geqslant 4$$
$$x^2 + y^2 \leqslant 9$$

2. Graph the boundary $x^2 + y^2 = 4$. It is solid. It is a circle with *center*: $(0, 0)$, *radius* $= 2$.

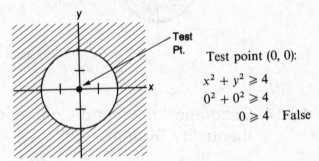

Test point $(0, 0)$:

$x^2 + y^2 \geqslant 4$

$0^2 + 0^2 \geqslant 4$

$0 \geqslant 4$ False

The test point is not in the required space. Shade outside the circle.

Graph the boundary $x^2 + y^2 = 9$. It is solid. It is a circle with *center*: $(0, 0)$, *radius* $= 3$.

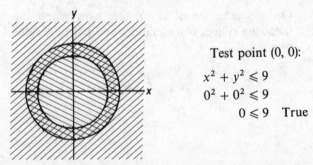

Test point $(0, 0)$:

$x^2 + y^2 \leqslant 9$

$0^2 + 0^2 \leqslant 9$

$0 \leqslant 9$ True

The test point *is* in the required space. Shade inside the circle.

The overlapping region is the solution set. It is the space between the circles including the boundaries.

Practice Problems 15.14

1. On the same set of axes, graph the solution set to the system of inequalities given below.

$$x \leqslant 1$$
$$y < x^2 + 3x$$

2. Graphically display the solution set to the following system of inequalities:

$$\frac{x^2}{9} + \frac{y^2}{4} < 1$$

$$\frac{x^2}{9} - \frac{y^2}{4} \geqslant 1$$

A)
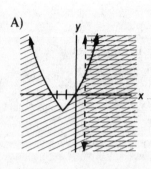

B)
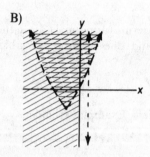

A)
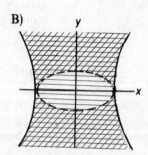

B)
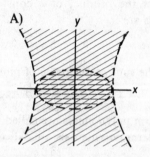

C)
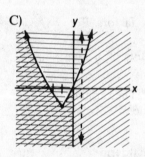

D)
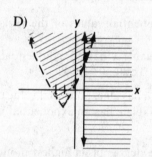

A)
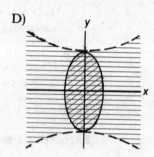

B)
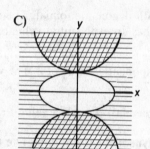

E)
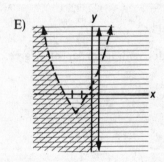

C)

D)

E)
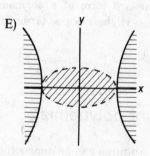

No solution set

Solutions on page 274

16. Polynomials

16.1 Fundamental Facts

Monomial—An algebraic expression with one term. Monomials are either a constant, a variable or the product of a constant and one or more variables. Examples of monomials are:

$$5 \quad -5 \quad x \quad 5x \quad 5x^2 \quad 5x^2y \quad xyz \quad xy^2z^3$$

Polynomial—The sum or difference of monomials. Examples include:

$$5 + x \quad x^2 + 3x + 2 \quad 3x^5 + 2x^4 - 3x^3 + xy - 6 \quad x^{75} - 1$$

—Two-term polynomials are called binomials, e.g., $x^2 - 1$.
—Three-term polynomials are called trinomials, e.g., $x^2 - 3x + 1$.

Degree of a monomial—The sum of the exponential values of the variable factors. For example,
$3x^3y$ is a 4th degree monomial. The variable x has an exponential value of 3. The variable y has an exponential value of 1. Total: $3 + 1 = 4$

$6a^2b^2c^3$ is a 7th degree monomial. The variable a has an exponential value of 2. The variable b has an exponential value of 2. The variable c has an exponential value of 3. Total: $2 + 2 + 3 = 7$

Degree of a polynomial—Equal in value to the degree of its highest monomial term

Simplifying a polynomial—Combining like terms. For example,

$$3xy - 2xy = xy \quad 7x^4 + 4x^4 = 11x^4$$

Standard form of a polynomial—Arranging the terms in descending order of degree. Highest degree terms first. For example,

$$P(x) = x^5 - 3x^4 + 3x^3 + x^2 + x - 1$$

Addition of polynomials

Addition can be done either horizontally or vertically.

> **Addition:**
>
> - Arrange the polynomials in standard form.
> - Associate like terms.
> - Add like terms by adding their coefficients.

Example:

1. Add $6x^3 + 4x^2 + 2x - 7$ and $8x^2 - 3x + 4$.

Solution:

1. The polynomials are already in standard form.

Horizontal Addition

$$(6x^3 + 4x^2 + 2x - 7) + (8x^2 - 3x + 4)$$
$$= 6x^3 + \underbrace{4x^2 + 8x^2}_{+12x^2} + \underbrace{2x - 3x}_{-x} \underbrace{-7 + 4}_{-3}$$
$$= 6x^3 + \quad +12x^2 \quad -x \quad -3$$
$$= 6x^3 + 12x^2 - x - 3$$

Vertical Addition

$$6x^3 + 4x^2 + 2x - 7$$
$$\underline{ 8x^2 - 3x + 4}$$
$$6x^3 + 12x^3 - x - 3$$

Subtraction of polynomials

Subtraction is the addition of the opposite (used when subtracting vertically) or the distribution of the minus sign (used when subtracting horizontally).

Subtraction

- Arrange the polynomials in standard form.
- For vertical calculations change all the signs of the polynomial being subtracted and then add.
- For horizontal calculations distribute the minus sign.
- Associate like terms.
- Add like terms by adding their coefficients.

Example:

2. From $6x^3 + 4x^2 - 2x - 7$ subtract $8x^2 - 3x - 4$.

Solution:

2. Polynomials are already in standard form.

Horizontal Subtraction

$$(6x^3 + 4x^2 - 2x - 7) - (8x^2 - 3x - 4)$$
$$6x^3 + 4x^2 - 2x - 7 - 8x^2 + 3x + 4$$
$$6x^3 + \underbrace{4x^2 - 8x^2}_{-4x^2} \underbrace{-2x + 3x}_{+x} \underbrace{-7 + 4}_{-3}$$
$$6x^3 \qquad -4x^2 \qquad +x \qquad -3$$
$$6x^3 - 4x^2 + x - 3$$

Vertical Subtraction

$$6x^3 + 4x^2 - 2x - 7$$
$$- \qquad 8x^2 - 3x - 4$$
$$\Downarrow$$
$$6x^3 + 4x^2 - 2x - 7$$
$$\underline{+ \qquad -8x^2 + 3x + 4}$$
$$6x^3 - 4x^2 + x - 3$$

Multiplying polynomials

Multiplication is best done horizontally.

Multiplication:

A polynomial by a monomial

- **Distribute** the monomial to each term of the polynomial.

$$a(b + c + d) = ab + ac + ad$$

- **Associate** and add like terms.

A binomial by a binomial

- **Distribute** each term of the first binomial to each term of the second binommial.

$$(a + b)(c + d) = ac + ad + bc + bd$$

This procedure is sometimes called **FOIL**. Referring to the terms of the binomials, multiply the First terms, Outside terms, Inside and Last terms.

$$(a + b)(c + d) = \frac{ac}{\text{First}} + \frac{ad}{\text{Outside}} + \frac{bc}{\text{Inside}} + \frac{bd}{\text{Last}}$$

- **Associate** and **add** like terms.

A polynomial by a polynomial

- **Distribute** each term of the first polynomial to each term of the second polynomial.

$$(a + b + c)(x + y + z) = a(x + y + z) + b(x + y + z) + c(x + y + z) \text{ or}$$
$$= ax + ay + az + bx + by + bz + cx + cy + cz$$

- **Associate** and **add** like terms.

☞ Recall $x^a \cdot x^b = x^{a+b}$ and $x \cdot y = xy$.

Examples:

3. $6x(3x^2 - 2x - 4xy + 5) =$

Solutions:

3. $6x(3x^2 - 2x - 4xy + 5)$

$= 18x^3 - 12x^2 - 24x^2y + 30x$

4. Simplify $3x^2 + 2(x+1)(x-3) - (x+1)(x-1)$.

4. $3x^2 + 2\underbrace{(x+1)(x-3)} \quad -\underbrace{(x+1)(x-1)}$

$= 3x^2 + 2(x^2 - 3x + x - 3) - (x^2 - x + x - 1)$

$= 3x^2 + 2\underbrace{(x^2 - 2x - 3)} - \underbrace{(x^2 - 1)}$

$= 3x^2 + 2x^2 - 4x - 6 - x^2 + 1$

$= \underbrace{3x^2 + 2x^2 - x^2}_{4x^2} \underbrace{-4x}_{-4x} \underbrace{-6+1}_{-5}$

$= 4x^2 - 4x - 5$

5. Multiply $(x^2 - 3x + 1)(x^2 - 2x - 3)$.

5. $(x^2 - 3x + 1)(x^2 - 2x - 3)$

$= x^2(x^2 - 2x - 3) - 3x(x^2 - 2x - 3) + 1(x^2 - 2x - 3)$ or

$= x^4 - 2x^3 - 3x^2 - 3x^3 + 6x^2 + 9x + x^2 - 2x - 3$

$= x^4 \underbrace{-2x^3 - 3x^3}_{-5x^3} \underbrace{-3x^2 + 6x^2 + x^2}_{+4x^2} \underbrace{+9x - 2x}_{+7x} -3$

$= x^4 - 5x^3 + 4x^2 + 7x - 3$

Division of polynomials

Division is best expressed in fraction form.

Division:

A polynomial by a polynomial

- Divide each term of the polynomial by the monomial

$$\frac{ab + ac}{a} = \frac{ab}{a} + \frac{ac}{a} = b + c$$

- This is done similiar to long division for numbers

Steps

Example

divide $x^2 - 5x - 7$ by $x - 1$

$x - 1 \overline{)x^2 - 5x - 7}$

1. **Divide** the left term of the divisor into the left term of the dividend to obtain the first term of the quotient. Place the result in the quotient over the corresponding term of the dividend.

1. divide x into $x^2 \Rightarrow x$

$$x - 1 \overline{)x^2 - 5x - 7}^{\quad x}$$

---

2. **Multiply** the divisor by the first term of the quotient and

Subtract this result from the dividend

☞ subtraction is done by changing the signs of the polynomial being subtracted and then add

2. $x(x-1)=x^2-x$

$$\begin{array}{r} x \\ x-1\overline{\smash{)}\ {-x^2-5x-7}} \\ \boldsymbol{x^2-x} \end{array}$$

☞
$$\begin{array}{r} x \\ x-1\overline{\smash{)}\ x^2-5x-7} \\ + \\ \underline{-x^2+x} \\ -4x-7 \end{array}$$

3. **Divide** the left term of the divisor into the left term of the **new** dividend to obtain the next term of the quotient.

3. divide x into $-4x \Rightarrow -4$

$$\begin{array}{r} x-4 \\ x-1\overline{\smash{)}\ x^2-5x-7} \\ \underline{-x^2+\ \ x} \\ -4x-7 \end{array}$$

4. **Multiply** the divisor by the new term of the quotient and

Subtract this result from the new dividend (Add the opposite)

4. $-4(x-1)=-4x+4$

$$\begin{array}{r} x-4 \\ x-1\overline{\smash{)}\ x^2-5x-7} \\ \underline{-x^2+\ \ x} \\ \\ -4x-7 \\ -4x+4 \end{array}$$

becomes

$$\begin{array}{r} x-4 \\ x-1\overline{\smash{)}\ x^2-5x-7} \\ \underline{-x^2+\ \ x} \\ -4x-7 \\ + \\ \underline{4x-4} \\ -11 \end{array}$$

5. **Continue** this process until the subtraction has a remainder of 0 or is of a lower power than the divisor.

If the remainder $=0$, the quotient is complete. If the remainder $\neq 0$, express as a fraction $\dfrac{\text{remainder}}{\text{divisor}}$

5. The remainder -11 is a lower power than the divisor $x-1$

Solution: $x-4+\dfrac{-11}{x-1}$

$$\text{Divisor}\overline{\smash{)}\ \underset{\text{Dividend}}{\text{Quotient}+\dfrac{\text{Remainder}}{\text{Divisor}}}} \quad \text{or} \quad \dfrac{\text{Dividend}}{\text{Divisor}}=\text{Quotient}+\dfrac{\text{Remainder}}{\text{Divisor}}$$

Replace any missing degree terms in the dividend before dividing. For example, if the dividend is x^2-1 use x^2+0x-1; if it is x^3-3x+2 use x^3+0x^2-3x+2.

Examples:

6. Divide $\dfrac{3x^3+6x^2+3x}{3x}$.

7. Divide $2x^3+x-1$ by $x+3$.

Solutions:

6. $\dfrac{3x^3+6x^2+3x}{3x}$

$= \dfrac{3x^3}{3x} + \dfrac{6x^2}{3x} + \dfrac{3x}{3x}$

$= x^2 + 2x + 1$

⚡ Common error:

$$\frac{3x^3+6x^2+3x}{3x} = x^2 + 2x$$

Note: $\dfrac{3x}{3x} = 1$ not 0

7.

$$x+3\,\overline{\smash{\big)}\,2x^3+\mathbf{0x^2}+x-1}$$

Insert the missing term:

$$\begin{array}{r} \mathbf{2x^2} \\ x+3\,\overline{\smash{\big)}\,2x^3+0x^2+x-1} \end{array} \qquad \begin{array}{r} 2x^2 \\ x\,\overline{\smash{\big)}\,2x^3} \end{array}$$

Multiply $2x^2(x+3)$ and subtract from the dividend:

$$\begin{array}{r} 2x^2 \\ x+3\,\overline{\smash{\big)}\,-2x^3+0x^2+x-1} \\ \underline{\mathbf{2x^3+6x^2}} \end{array} \Rightarrow$$

$$\begin{array}{r} 2x^2 \\ x+3\,\overline{\smash{\big)}\,2x^3+0x^2+x-1} \\ +\ \mathbf{-2x^3-6x^2} \\ \hline -6x^2+x-1 \end{array}$$

$$\begin{array}{r} -6x \\ x\,\overline{\smash{\big)}\,-6x^2} \end{array}$$

$$\begin{array}{r} 2x^2-6x \\ x+3\,\overline{\smash{\big)}\,2x^3+0x^2+x-1} \\ +\ -2x^3-6x^2 \\ \hline -6x^2+x-1 \end{array}$$

Multiply $-6x(x+3)$ and subtract from the new dividend:

$$\begin{array}{r} 2x^2-6x \\ x+3\,\overline{\smash{\big)}\,2x^3+0x^2+x-1} \\ +\ -2x^3-6x^2 \\ \hline -6x^2+x-1 \\ -\ \mathbf{-6x^2-18x} \end{array}$$

$$\begin{array}{r} 2x^2-6x \\ x+3{\overline{\smash{\big)}\,2x^3+0x^2+x-1}} \\ +\ \underline{-2x^3-6x^2} \\ -6x^2+x-1 \\ +\ \underline{6x^2+18x} \\ 19x-1 \end{array}$$

$$\begin{array}{r} 19 \\ x{\overline{\smash{\big)}\,19x}} \end{array}$$

$$\begin{array}{r} 2x^2-6x+19 \\ x+3{\overline{\smash{\big)}\,2x^3+0x^2+x-1}} \\ +\ \underline{-2x^3-6x^2} \\ -6x^2+x-1 \\ +\ \underline{6x^2+18} \\ 19x-1 \end{array}$$

Multiply $19(x+3)$ and subtract from the new dividend:

$$\begin{array}{r} 2x^2-6x+19 \\ x+3{\overline{\smash{\big)}\,2x^3+0x^2+x-1}} \\ +\ \underline{-2x^3-6x^2} \\ -6x^2+x-1 \\ +\ \underline{6x^2+18x} \\ -19x-1 \\ 19x+57 \end{array}$$

$$\begin{array}{r} 2x^2-6x+19 \\ x+3{\overline{\smash{\big)}\,2x^3+0x^2+x-1}} \\ +\ \underline{-2x^3-6x^2} \\ -6x^2+x-1 \\ \underline{+6x^2+18x} \\ -19x-1 \\ +\ \underline{19x-57} \\ -58 \end{array}$$

$$\text{Quotient}=2x^2-6x+19+\frac{-58}{x+3}$$

Synthetic division

If a polynomial is divided by a binomial of the form $x - a$, synthetic division simplifies the division process.

Synthetic Division:

Division of a polynomial by $x - a$

Example

Divide $2x^3 + x + 1$ by $x + 3$

Steps

- Insert any missing terms.

- Represent the coefficients of the dividend. Place the value of a in the bracket on the left side.

$$x + 3 \overline{\smash{\big)}\, 2x^3 + 0x^2 + x - 1}$$

$$
\begin{array}{ccccc}
a & x^3 & x^2 & x & \# \\
\downarrow & \downarrow & \downarrow & \downarrow & \downarrow \\
-3| & 2 & 0 & 1 & -1
\end{array}
$$

- Bring down the first number.

- Multiply the brought down number by the bracket number and place under the second coefficient number.

$$-3(2) = -6$$

$$
\begin{array}{ccc}
 & & -6 \\
\hline
2 & -6 &
\end{array}
$$

- Add the new number to the coefficient number and bring down the new sum.

- Multiply the new brought down number by the bracket number and place under the 3rd coefficient number.

$$
\begin{array}{ccccc}
-3| & 2 & 0 & 1 & -1 \\
 & & -6 & 18 & \\
\hline
 & 2 & -6 & 19 &
\end{array}
$$

$$-3(-6) = 18$$

- Add the new number to the coefficient number and bring down the new sum.

Continue until all places are filled.

$$
\begin{array}{ccccc}
-3| & 2 & 0 & 1 & -1 \\
 & & -6 & 18 & -57 \\
\hline
 & 2 & -6 & 19 & -58
\end{array}
$$

$$-3(19) = -57$$

- Box the last number. This represents the remainder. The numbers to the left of the remainder represent the quotient (with each term one power less than the original dividend.

$$
\begin{array}{c c c c c}
-3\ | & 2 & 0 & 1 & -1 \\
 & & -6 & 18 & -57 \\
\hline
 & 2 & -6 & 19 & \boxed{-58} \\
 & \uparrow & \uparrow & \uparrow & \\
 & x^2 & x & \# &
\end{array}
$$

$$\text{Quotient} = 2x^2 - 6x + 19 + \frac{-58}{x + 3}$$

Examples:

8. Divide $x^2 - 5x - 7$ by $x - 1$.

9. Divide $x^2 - 5x + 6$ by $x - 5$.

Solutions:

8.
$$
\begin{array}{ccccc}
a & x^2 & x & \# \\
\downarrow & \downarrow & \downarrow & \downarrow \\
\underline{1} & 1 & -5 & -7 \\
& & 1 & -4 \\
\hline
& 1 & -4 & \boxed{-11} \\
& \uparrow & \uparrow \\
& x & \#
\end{array}
$$

Quotient $= x - 4 + \dfrac{-11}{x-1}$

9.
$$
\begin{array}{ccccc}
a & x^2 & x & \# \\
\downarrow & \downarrow & \downarrow & \downarrow \\
\underline{5} & 1 & -5 & +6 \\
& & 5 & 0 \\
\hline
& 1 & 0 & \boxed{6} \\
& \uparrow & \uparrow \\
& x & \#
\end{array}
$$

Quotient $= x + \dfrac{6}{x-5}$

Practice Problems 16.1

1. Multiply $(2y-5)(y-4)$.

    A) $3y+20$ B) $-7y-9$
    C) $2y^2+20$ D) $2y^2-9y+20$
    E) $2y^2-13y+20$

2. From $6x^2-8x+6$ subtract x^2-3x-6.

    A) $3x^2-5x$ B) $5x^2-11x+12$
    C) $5x^2-5x+12$ D) $-5x^2+5x-12$
    E) $5x-5$

3. Divide $9x^6+3x^2$ by $3x^2$.

    A) $3x^3$ B) $3x^4+1$ C) $3x^3+1$
    D) $3x^4$ E) $6x^4+3x^2$

4. $(x-2)(x^2+4x-4)=$

    A) $2x^3+5x+8$ B) x^2+5x-6
    C) x^3+8 D) x^3-8
    E) $x^3+2x^2-12x+8$

5. $(1+a^3b^3)(1-a^3b^3)=$

    A) $1-a^6b^6$ B) 2   C) $1-a^9b^9$
    D) $1-2a^3b^3$ E) $1-2a^6b^6-a^9b^9$

6. $x^3+3x^2-6x-3(2x+3)+2x(-3x+1)=$

    A) $-8x^3+5x^2-6x-9$
    B) $x^3-3x^2-10x-9$
    C) x^3-3x^2+2x-9
    D) x^3-2x^2-2x+1
    E) x^8+4

7. Divide x^3+2x^2-4x-8 by $x+2$.

    A) x^2+2 B) x^2-3x+2 C) x^2-4
    D) $x-2$ E) x^2+2x-2

8. Divide $x^2+9x-10$ by $x+10$.

    A) $x-1$ B) $x+1$ C) $x-9$
    D) $x+9$ E) $x-10$

Solutions on page 275

16.2 Asymptotes

Asymptotes—These are lines which a graph approaches (but does not get to) at some finite value of x or when $x = \pm \infty$.

There are three types of asymptotes: **Vertical**
Horizontal
Slant

Given: $f(x) = \dfrac{g(x)}{h(x)}$			
Type of Asymptote	**Criterion**	**Equation of Asymptote**	**Graphic Illustration**
Vertical	$h(a) = 0$	$x = a$	vertical
Horizontal	$\lim\limits_{x \to \infty} f(x) = a$ $a \neq \infty$	$y = a$	vertical, Horiz, Horiz
Slant	$\dfrac{g(x)}{h(x)} = Q(x) + R$ $g(x)$ is of a higher degree than $h(x)$	$y = Q(x)$	slant

Examples:

1. Find any vertical, horizontal or slant asymptotes for $f(x) = \dfrac{4x}{x-3}$.

Solutions:

1. $f(x) = \dfrac{g(x)}{h(x)} = \dfrac{4x}{x-3}$

- $h(x) = 0$, $x - 3 = 0$, $x = 3$ There is a vertical asymptote at $x = 3$.

- $\lim\limits_{x \to \infty} \dfrac{4x}{x-3} = \lim\limits_{x \to \infty} \dfrac{\dfrac{4x}{x}}{\dfrac{x}{x} - \dfrac{3}{x}}$

$= \lim\limits_{x \to \infty} \dfrac{4}{1 - \dfrac{3}{x}}$

$= 4$

There is a horizontal asymptote at $y = 4$.

- There are no slant asymptotes; the degree of $g(x)$ is not greater than the degree of $h(x)$.

2. Find any horizontal, vertical or slant asymptotes of

$$f(x) = \frac{x^2 + 1}{x - 1}$$

2. $f(x) = \dfrac{g(x)}{h(x)} = \dfrac{x^2 + 1}{x - 1}$

$h(x) = 0$, $x - 1 = 0$, $x = 1$ There is a vertical asymptote at $x = 1$.

$$\lim_{x \to \infty} \frac{x^2 + 1}{x - 1} = \lim_{x \to \infty} \frac{\dfrac{x^2}{x^2} + \dfrac{1}{x^2}}{\dfrac{x}{x^2} - \dfrac{1}{x^2}} = \lim_{x \to \infty} \frac{1 + \dfrac{1}{x^2}}{\dfrac{1}{x} - \dfrac{1}{x^2}}$$

$$= \frac{1 + 0}{0 - 0} = \frac{1}{0} = \infty$$

No horizontal asymptote

Since $\lim_{x \to \infty} f(x) = \infty$, there is no horizontal asymptote.

Divide $\dfrac{x^2 + 1}{x - 1}$ using synthetic division.

$$
\begin{array}{c}
\quad x^2 \quad x \quad \# \\
\quad \downarrow \quad \downarrow \quad \downarrow \\
\underline{1}| \quad 1 \quad 0 \quad 1 \\
\quad\quad\quad 1 \quad 1 \\
\overline{\quad\quad 1 \quad 1 \quad \boxed{2}} \\
\quad \uparrow \quad \uparrow \\
\quad x \quad \#
\end{array}
$$

$$\frac{x^2 + 1}{x - 1} = x + 1 + \frac{2}{x - 1}$$

There is a slant asymptote of $y = x + 1$.

Practice Problem 16.2

1. Determine any vertical asymptote(s) for

$$f(x) = \frac{x^3 + 3x^2 + 4x - 3}{x}.$$

A) $x = 1$ B) $y = 0$ C) $x = 0$
D) $y = 1$ E) No vertical asymptote

2. Determine any horizontal asymptote(s) for

$$f(x) = \frac{x^2}{x^2 + 4}$$

A) $x = 1$ B) $x = -2$ C) $x = 2$
D) $y = 1$ E) $y = -2$

3. Determine the vertical asymptote(s) for

$$f(x) = \frac{x^2}{x^2 - 4}$$

A) $y = 1$ B) $x = 2$ only   C) $x = -2$ only
D) $x = \pm 2$ E) No vertical asymptotes

4. Determine the slant asymptote(s) for

$$x^2 - yx + y + 2 = 0$$

A) $y = 2$ B) $x = 2$ C) $y = x + 1$
D) $x = y + 1$ E) No slant asymptote

Solutions on page 276

16.3 Properties of Polynomials

$P(x) = a_n x^n + a_{n-1} x^{n-1} \cdots + a_1 x + a_0$	
Roots or Zeros of a Polynomial	The values of x for which $P(x) = 0$.
The Number of Roots	A polynomial of degree n has n roots, some of which may be real or complex, and some of which may have a multiplicity greater than 1.
Fundamental Theorem of Algebra	Every polynomial of degree $\geqslant 1$ has at least one real or complex root.
Remainder Theorem	If $\dfrac{P(x)}{x-a} = Q(x) + R$ The remainder $R = P(a)$.
Factor Theorem	If $\dfrac{P(x)}{x-a} = Q(x) + 0$ or $P(a) = 0$, then $(x-a)$ is a factor of $P(x)$ and a is a root.
Imaginary Roots	If $P(x)$ has real coefficients and if $a + bi$ is a root, then $a - bi$ is a root.
Irrational Roots	If $P(x)$ has rational coefficients and if $a + \sqrt{b}$ is root, then $a - \sqrt{b}$ is a root.
Rational Root Theorem	If a_n to a_0 are integral coefficients, and if $\dfrac{p}{q}$ reduced to lowest terms is a rational number and a root of $P(x)$, then p is a factor of a_0 and q is a factor of a_n.
Descartes' Rule of Sign	If $P(x)$ has real coefficients and $a_0 \neq 0$, the number of positive real roots is equal to the number of sign variations of $P(x)$ or less than that number by a multiple of two. The number of negative real roots is equal to the number of sign variations of $P(-x)$ or less than that number by a multiple of two.
Upper and Lower Bounds	If $P(x)$ has real coefficients and a_n is positive, there is a lower limit to the roots, that is, no root is smaller than L if for $L \leqslant 0$ the third row in the synthetic division of $\dfrac{P(x)}{x-L}$ has values that alternate in sign. (Zero is considered $+$ or $-$ as required.) There is an upper limit to the roots, that is, no root is larger than U if for $U \geqslant 0$ the third row of the synthetic division of $\dfrac{P(x)}{x-U}$ has only non-negative numbers.

The Sums and Products of the Roots

$\dfrac{a_{n-1}}{a_n}$ = Sum of the roots

$-\dfrac{a_{n-2}}{a_n}$ = Sum of the roots taken two at a time

$\dfrac{a_{n-3}}{a_n}$ = Sum of the roots taken three at a time

$-\dfrac{a_{n-4}}{a_n}$ = Sum of the roots taken four at a time

$\vdots \qquad \vdots$

$\left. \begin{array}{l} \dfrac{a_0}{n} \text{ (if } n \text{ is even)} \\[2em] -\dfrac{a_0}{n} \text{ (if } n \text{ is odd)} \end{array} \right\}$

Sum of the roots taken n at a time

or

Product of the roots

Procedure for **locating the roots (zeros) of a polynomial** with real/rational coefficients

Step	Description	Rule Utilized
1	Display the possible rational roots in the form $\dfrac{p}{q}$.	Rational Root Theorem
2	Eliminate impossible roots. For example, there may be no positive roots.	Descartes' Rule of Signs
3	Test possible values, looking for $P(a) = 0$. Then a is a root and $(x - a)$ is a factor.	Factor Theorem
4	Upon finding a root, depress the equation by removing the factor and lowering the degree of the equation.	Synthetic Division
5	Continue to search for roots using the depressed equation. Remember: Look for upper and lower bounds. : A root may have multiplicity > 1. : Try the easier numbers first.	
6	If possible, continue locating roots until a quadratic equation is achieved. Solve the quadratic equation to obtain the last two roots/factors.	Quadratic Formula

Examples:

1. Find the roots (zeros) of

$$P(x) = x^5 + 4x^4 - 4x^3 - 34x^2 - 45x - 18$$

Solutions:

1. List the possible roots: $\dfrac{\text{factors of } 18}{\text{factors of } 1}$

$$\frac{p}{q} = \frac{\pm 18, \ \pm 9, \ \pm 6, \ \pm 3, \ \pm 2, \ \pm 1}{\pm 1}$$

Descarte's rule of signs: 1 sign change $\therefore$ 1 positive root

$$P(-x) = -x^5 + x^4 + 4x^3 - 34x^2 + 45x - 18$$

4 sign changes

$\therefore$

+	−	imag.
1	4	0
1	2	2
1	0	4

Look for the positive root:
$$P(x) = x^5 + 4x^4 - 4x^3 - 34x^2 - 45x - 18$$
$$P(1) = 1 + 4 - 4 - 34 - 45 - 18 \neq 0$$
$$P(2) = 32 + 64 - 32 - 136 - 90 - 18 \neq 0$$
$$P(3) = 243 + 324 - 108 - 306$$
$$\qquad\quad - 135 - 18 = 0$$

$$x^5 + 4x^4 - 4x^3 - 34x^2 - 45x - 18$$

$$
\begin{array}{r|rrrrrr}
3] & 1 & 4 & -4 & -34 & -45 & -18 \\
   &   & 3 & 21 & 51 & 51 & 18 \\
\hline
   & 1 & 7 & 17 & 17 & 6   & |0| \\
\end{array}
$$

$$(x - 3)(x^4 + 7x^3 + 17x^2 + 17x + 6)$$

There are no more positive roots. Try for any negative roots using the depressed equation:

$$D_1(x) = x^4 + 7x^3 + 17x^2 + 17x + 6$$
$$D_1(-1) = 1 - 7 + 17 - 17 + 6 = 0 \ \sqrt{}$$

$$
\begin{array}{r|rrrrr}
-1] & 1 & +7 & +17 & +17 & +6 \\
    &   & -1 & -6 & -11 & -6 \\
\hline
    & 1 & 6 & 11 & 6   & |0| \\
\end{array}
$$

$$(x - 3)(x + 1)(x^3 + 6x^2 + 11x + 6)$$

Using the depressed equation $x^3 + 6x^2 + 11x + 6$:

$$D_2(x) = x^3 + 6x^2 + 11x + 6$$
$$D_2(-1) = -1 + 6 - 11 + 6 = 0 \ \sqrt{}$$

(-1 was used again looking for a possible multiple root.)

$$
\begin{array}{r|rrrr}
-1] & 1 & +6 & +11 & +6 \\
    &   & -1 & -5 & -6 \\
\hline
    & 1 & 5 & 6   & |0| \\
\end{array}
$$

$$(x - 3)(x + 1)(x + 1)\underbrace{(x^2 + 5x + 6)}_{(x + 3)(x + 2)}$$

$\therefore P(x) = (x - 3)(x + 1)(x + 1)(x + 3)(x + 2)$
Roots are $3, -1, -2, -3$.
-1 is a double root.

Practice Problems 16.3

1. What is the multiplicity of the root $+1$ for the equation $x^3 - 3x^2 + 3x - 1 = 0$?

   A) 4   B) 3   C) 2   D) 1   E) 0

2. The equation $2x^3 + 3x^2 - 11x - 6 = 0$ has how many fractional roots?

   A) 4   B) 3   C) 2   D) 1   E) 0

3. What is the remainder when $x^{98} + 3$ is divided by $x - 1$?

   A) $(98)^{98}$ B) 98   C) 3   D) 2   E) 4

4. What is the sum of the roots of the equation $4x^3 - 2x^2 - x + 7 = 0$?

   A) 2   B) $\frac{1}{2}$ C) $\frac{7}{4}$ D) $\frac{4}{7}$ E) $-\frac{1}{2}$

5. What is the highest multiplicity of a root of the equation $x^3 - 3x^2 + 4 = 0$?

   A) 4   B) 3   C) 2   D) 1   E) 0

6. How many roots of $x^4 - 16 = 0$ are not real?

   A) 4   B) 3   C) 2   D) 1   E) 0

7. Which of the following is not a possible rational root of $f(x) = 4x^3 + 6x^2 + 8x + 9$?

   A) -3 B) $-\frac{3}{2}$ C) $-\frac{3}{4}$ D) 2   E) 9

8. If a polynomial with real/rational coefficients has one root $3 + 2i$ and another $1 - 3i$ what is the lowest degree equation possible?

   A) 5th   B) 4th   C) 3rd   D) 2nd
   E) 1st

Solutions on page 277

16.4 Sketching a Polynomial

$$f(x) = \frac{g(x)}{h(x)}$$

Step	Details	Comment
1	Locate y-intercept	Let $x = 0$, solve for y
2	Locate x-intercepts (roots)	Let $y = 0$, solve for x
3	Note any single, double, triple roots, etc.	 Factors of the form $(x - a)$ cross the x-axis at a as shown by either the solid or dotted line.

Factors of the form $(x-a)^2$, double roots at $x=a$, do not cross the axis at a they are tangent to it as shown by the solid or dotted curve.

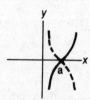

Factors of the form $(x-a)^3$, triple roots cross the x-axis as shown by either the solid or dotted line.

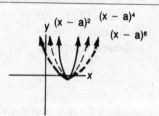

Factors of the form $(x-a)^n$, **where n is even**, vary as they **touch** the x-axis by their degree of flatness.

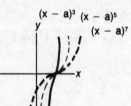

Factors of the form $(x-a)^n$, **where n is odd**, vary as they **cross** the x-axis by their degree of flatness

| 4 | Determine when the graph is above and below the x-axis. | Test the region between the roots and between the lowest root and $-\infty$, and the largest root and $+\infty$. A number line can prove useful. In each interval test one value of x to determine if $f(x)$ is + or −. |

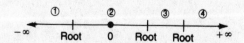

In this case there are four intervals to test. If there are vertical asymptotes use these as

additional points on the number line to achieve additional intervals.

5	Evaluate asymptotes: Vertical Horizontal Slant	$h(a)=0$ $x=a$ is a vertical asymptote $\lim\limits_{x\to\infty} f(x)=a$ $y=a$ is a horizontal asymptote $f(x)=\dfrac{g(x)}{h(x)}=Q(x)+R$ $y=Q(x)$ is a slant asymptote
6*	Determine the relative min/max points. Test the critical values.	Use the first derivative $=0$ to determine possible min/max points. Determine if the points are/are not min/max points.
7*	Determine the points of inflection. Determine concavity.	Use the second derivative $=0$ to determine possible points of inflection. Determine if the curve is concave or convex, etc.

*These are further topics discussed in a calculus context which aid in sketching the graph of $f(x)$.

Examples:

1. Sketch the graph of

$$f(x)=x^3-7x^2+11x-5$$

Solutions:

1. $y=x^3-7x^2+11x-5$

Let $x=0$

$y=0^3-7(0)^2+11(0)-5=-5$

The y-intercept is $(0, -5)$

Let $y=0$

$0=x^3-7x^2+11x-5$

$\dfrac{p}{q}=\dfrac{\pm5, \pm1}{\pm1}$

$P(x)=x^3-7x^2+11x-5$

$P(1)=1^3-7(1)^2+11(1)-5=0 \checkmark$

$$\underline{1|}\quad 1\quad -7\quad 11\quad -5$$
$$\qquad\quad\; 1\quad -6\quad\;\; 5$$
$$\overline{\quad\;\; 1\quad -6\quad\;\; 5\quad\; \underline{|0}}$$

$P(x)=(x-1)(x^2-6x+5)$

$\quad\;\;\;=(x-1)(x-5)(x-1)$

$\quad\;\;\;=(x-1)^2(x-5)$

There is a root at $x=5$ and a double root at $x=1$.

There are no asymptotes since there is no division.

Test the intervals between the roots.

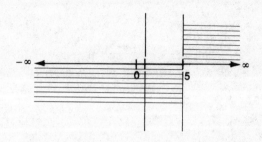

Test the intervals:

$-\infty$ to 1 $f(x) = $ "$-$"
  1 to 5 $f(x) = $ "$-$"
  5 to ∞ $f(x) = $ "$+$"

a "$-$" indicates the graph is below the x-axis in this interval.

a "$+$" indicates the graph is above the x-axis in this interval.

The shaded area on the number line above shows the regions the graph passes through.

Recall that a single root at $x = 5$ passes through the axis as a straight line.

Recall: a double root at $x = 1$ indicates the curve is tangent to the axis.

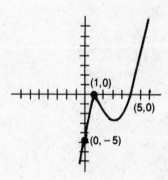

2. Sketch the graph of

$$f(x) = x^3 + x^2 - 5x + 3$$

(Use the first and second derivatives to aid in sketching.)

2. $y = x^3 + x^2 - 5x + 3$

Let $x = 0$

$y = 0^3 + 0^2 - 5(0) + 3 = 3$

The y-intercept is $(0, 3)$

Let $y = 0$

$0 = x^3 + x^2 - 5x + 3$

$\dfrac{p}{q} = \dfrac{\pm 3, \pm 1}{\pm 1}$

$P(x) = x^3 + x^2 - 5x + 3$

$$P(1) = 1^3 + 1^2 - 5(1) + 3 = 0 \checkmark$$

$$
\begin{array}{r|rrrr}
1 & 1 & 1 & -5 & 3 \\
  &   & 1 & 2 & -3 \\
\hline
  & 1 & 2 & -3 & \underline{0}
\end{array}
$$

$$= (x-1)(x^2 + 2x - 3)$$
$$= (x-1)(x+3)(x-1)$$
$$= (x-1)^2(x+3)$$

There is a root at $x = -3$ and a double root at $x = 1$.

___

There are no asymptotes since there is no division.

___

Test the intervals between the roots.

Evaluate the intervals:

$-\infty$ to -3 $f(x) = $ "$-$"
-3 to 1 $f(x) = $ "$+$"
1 to ∞ $f(x) = $ "$+$"

"$-$" means the graph is below the x-axis.
"$+$" means the graph is above the x-axis.
Also recall that the single root at $x = -3$ crosses the x axis as a line at $x = -3$.
The double root at $x = 1$, is tangent to the axis.

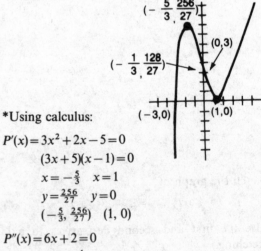

*Using calculus:

$$P'(x) = 3x^2 + 2x - 5 = 0$$
$$(3x+5)(x-1) = 0$$
$$x = -\tfrac{5}{3} \quad x = 1$$
$$y = \tfrac{256}{27} \quad y = 0$$
$$(-\tfrac{5}{3}, \tfrac{256}{27}) \quad (1, 0)$$

$$P''(x) = 6x + 2 = 0$$
$$x = -\tfrac{1}{3}$$
$$(-\tfrac{1}{3}, \tfrac{128}{27})$$

is a point of inflection

$P''(1) = 6(1) + 2 = 8 \quad \therefore (1, 0)$ is a minimum point
$P''(-\tfrac{5}{3}) = 6(-\tfrac{5}{3}) + 2 = -8 \quad \therefore (-\tfrac{5}{3}, \tfrac{256}{27})$ is a maximum point.

Practice Problems 16.4

1. Sketch a graph of $f(x) = \dfrac{x^2}{x^2 - 16}$.

2. Sketch a graph of $f(x) = x^3 - 4x^2 - 3x + 18$.

Solutions on page 278

Cumulative Review Test

1. If $5^{3x-1} = 25$, find x.

   A) 1   B) $\frac{2}{3}$ C) $\frac{3}{2}$ D) -1 E) $\frac{1}{3}$

2. The graph of the equation $x^2 + 3x + 4y^2 = 36$ is a(n)

   A) line   B) parabola   C) ellipse
   D) circle   E) hyperbola

3. Factor completely $6x^2 - 7x - 3$.

   A) $(6x-1)(x+3)$ B) $(3x+1)(2x-3)$
   C) $(3x-3)(2x+1)$ D) $3(2x-1)(2x+1)$
   E) $3(x-1)(2x+1)$

4. If the measures of the three angles of a triangle are represented by x, $2x-20$ and $3x-10$, then the triangle is

   A) right   B) obtuse   C) acute
   D) equilateral   E) isosceles

5. Solve the following system of equations
   for x: $4x - 3y = 8$
    $2x + y = -1$

   A) -2 B) -1 C) 0   D) $\frac{1}{2}$ E) $\frac{3}{4}$

6. If $x^4 - 7x^3 + 9x^2 - 4 = 0$, find the sum of the roots.

   A) 9   B) -9 C) 7   D) -7 E) 4

7. Factor $y^{2n} - 16$.

   A) $(y^n + 4)(y^n - 4)$ B) $(y^n + 1)(y^n - 16)$
   C) $(y + 4)(y^{2n-1} - 4)$ D) $2(y^n - 8)$
   E) $2(y + 4)(y - 4)$

8. The graph of the equation $y = x^2 + 3x + k$ passes through the point $(2, 0)$. Find k.

   A) $-\frac{3}{2}$ B) 2   C) -2 D) -10
   E) 10

9. Find the equation of the line which passes through the points $(1, 6)$ and $(-1, -2)$.

   A) $y - 6 = 4(x - 1)$ B) $y + 6 = -4(x + 1)$
   C) $y - 2 = -4(x - 1)$ D) $y - 1 = 4(x - 6)$
   E) $y - 6 = -\frac{1}{4}(x - 1)$

10. Express 6.5×10^4 as an integer.

    A) .00065   B) .0065   C) 6.54
    D) 65000   E) 65400

11. Express $\sqrt{-81} + \sqrt{-25} + 3\sqrt{-49}$ in terms of i.

    A) 35   B) $35i$ C) -35
    D) $-35i$ E) $21i$

12. Solve for x: $2 + 3\sqrt{x} = 8$.

    A) 0   B) 2   C) 4   D) 6   E) 8

13. Factor $12x^2 - 5x - 3$.

    A) $(3x+1)(4x-3)$ B) $3(x-1)(4x+1)$
    C) $4(3x-1)(x-1)$ D) $(12x+1)(x-3)$
    E) $(6x-1)(2x+3)$

14. If $f(x) = \dfrac{x-3}{x+2}$, find $f^{-1}(x)$.

    A) $\dfrac{x+2}{x-3}$ B) $\dfrac{x+3}{x-2}$ C) $\dfrac{2x+3}{x+1}$

    D) $\dfrac{2x+3}{1-x}$ E) $\dfrac{3x+2}{x+1}$

15. What is the locus of points 5 units from a given line and 3 units from a point on the line?

    A) 0 points   B) 1 point   C) 2 points
    D) 3 points   E) 4 points

16. Solve for r: $S = \dfrac{rl - a}{r - 1}$.

    A) $\dfrac{S-a}{l}$ B) $\dfrac{S-a}{S-l}$ C) $\dfrac{l-a}{S}$

    D) $\dfrac{S-1}{l+1}$ E) $\dfrac{S-l}{S-a}$

17. Find the roots of $f(x) = 3x^2 + 2x - 6$.

    A) $-1 \pm \sqrt{19}$ B) $-2 \pm \sqrt{38}$ C) $\dfrac{-2 \pm \sqrt{19}}{3}$

    D) $\dfrac{-1 \pm \sqrt{19}}{3}$ E) $\dfrac{-2 \pm \sqrt{19}}{2}$

18. If $1 + i$ and $1 - i$ are the roots of the quadratic equation $x^2 + bx + c$, what is the value of c?

    A) 0 B) 1 C) 2 D) 3 E) 4

19. The solution set for the equation $2x^3 - 3x^2 - 23x + 12 = 0$ is

    A) $\{-4, \frac{1}{2}, -3\}$ B) $\{-4, -2, -1\}$
    C) $\{4, 2, -1\}$ D) $\{3, -1, \frac{1}{2}\}$
    E) $\{4, \frac{1}{2}, -3\}$

20. The solution set for the equation $3 + |x - 1| = 2x$ is

    A) $\{2\}$ B) $\{2, \frac{4}{3}\}$ C) $\{\frac{4}{3}\}$
    D) $\{\ \}$ E) $\{2, -\frac{4}{3}\}$

Solutions on page 279

Diagnostic Chart for Cumulative Review Test

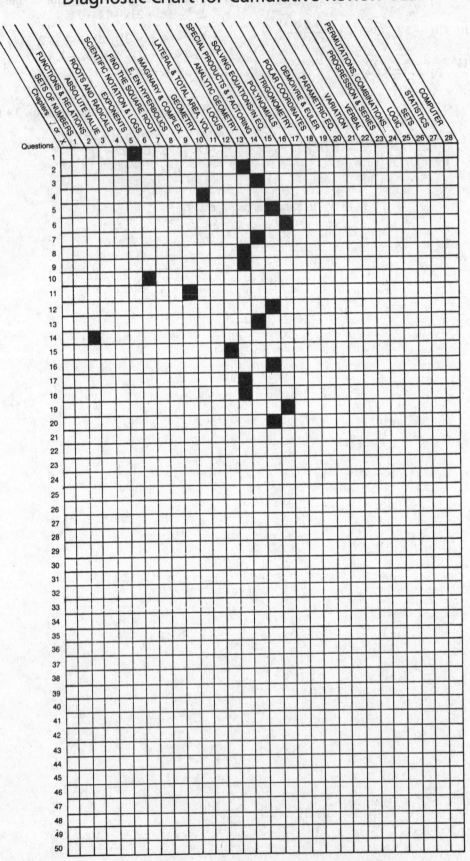

Solutions to Problems in Unit Four

Practice Problems 13.1

1. **(A)** $N(-3, -10), L(4, -4)$

$$m_{NL} = \frac{\Delta y}{\Delta x} = \frac{-4-(-10)}{4-(-3)}$$

$$= \frac{-4+10}{4+3} = \frac{6}{7}$$

2. **(B)** $N(-7, 0), L(-4, -2)$

$$d_{NL} = \sqrt{(x_2-x_1)^2 + (y_2-y_1)^2}$$

$$= \sqrt{[-4-(-7)]^2 + [-2-0]^2}$$

$$= \sqrt{(-4+7)^2 + (-2)^2}$$

$$= \sqrt{3^2 + (-2)^2} = \sqrt{9+4} = \sqrt{13}$$

3. **(E)**

$$m_{RS} = \frac{\Delta y}{\Delta x} = \frac{y_2-y_1}{x_2-x_1} = \frac{-2-2}{-5-5} = \frac{-4}{-10}$$

$$= \frac{2}{5}$$

4. **(C)**

$$M_{JK} = \left(\frac{x_1+x_2}{2}, \frac{y_1+y_2}{2}\right) = \left(\frac{-6+(-8)}{2}, \frac{-2+4}{2}\right)$$

$$= \left(\frac{-14}{2}, \frac{2}{2}\right) = (-7, 1)$$

5. **(B)**

$$x \text{ midpoint} = 3 = \frac{8+x}{2}$$

$$6 = 8+x, \quad x = -2$$

$$y \text{ midpoint} = 2 = \frac{3+y}{2}$$

$$4 = 3+y, \quad y = 1$$

$$B(-2, 1)$$

6. **(A)** The three points N, J, and L are colinear if $m_{NJ} = m_{NL} = m_{JL}$.

$$N(-2, -11) \; J(0, -7) \quad L(4, 1)$$

$$m_{NJ} = \frac{y_2-y_1}{x_2-x_1} = \frac{-7-(-11)}{0-(-2)} = \frac{-7+11}{2}$$

$$= \frac{4}{2} = 2$$

$$m_{JL} = \frac{y_2-y_1}{x_2-x_1} = \frac{1-(-7)}{4-(0)} = \frac{1+7}{4}$$

$$= \frac{8}{4} = 2$$

$$m_{NL} = \frac{y_2-y_1}{x_2-x_1} = \frac{1-(-11)}{4-(-2)} = \frac{1+11}{4+2}$$

$$= \frac{12}{6} = 2$$

The points are colinear.

7. **(D)**

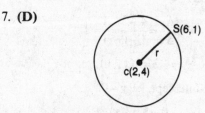

r is the distance from the center to point S.

$$d = \sqrt{(x_1-x_2)^2 + (y_2-y_1)^2}$$

$$= \sqrt{(2-6)^2 + (4-1)^2} = \sqrt{(-4)^2 + (3)^2}$$

$$= \sqrt{16+9} = \sqrt{25} = 5$$

8. **(D)** Perpendicular lines have slopes that are negative reciprocals.

$$m_1 = -\frac{5}{2}, \text{ then } m_2 = +\frac{2}{5}$$

☞ Negative reciprocals are two numbers with
a) opposite signs
b) one upside down of the other
c) multiply to -1

Practice Problems 13.2

1. (D) First find the slope of the given line:

$$m = \tfrac{1}{2}$$

A line perpendicular to the given line has a slope which is the negative reciprocal.

$$\therefore m = -\frac{2}{1}$$

Using $m = -2$ and the point $(-1, 5)$

$$y - 5 = -2(x - (-1))$$
$$y - 5 = -2(x + 1)$$

2. (C) $N(5, 0)$ $L(0, 5)$

These are the x- and y-intercepts.

$$\frac{x}{5} + \frac{y}{5} = 1 \quad \text{or} \quad x + y = 5$$

3. (C) Rearrange the equation into $y = mx + b$ form:

$$7 + 3y = -4x$$
$$3y = -4x - 7$$
$$y = -\tfrac{4}{3}x - \tfrac{7}{3}$$
$$m = -\tfrac{4}{3}$$

4. (A) $N(0, 3)$ $L(1, 4)$

$$m = \frac{\Delta y}{\Delta x} = \frac{4 - 3}{1 - 0} = \frac{1}{1} = 1$$

$$y - 3 = 1(x - 0)$$

To find the x-intercept, let $y = 0$:

$$0 - 3 = 1(x)$$
$$-3 = x$$
$$(-3, 0)$$

5. (B) First find the slope of the given line:

$$2x - 3y = 6$$

$$\frac{-3y}{-3} = \frac{-2x}{-3} + \frac{6}{-3}$$

$$y = \frac{2}{3}x - 2$$

$$m = \frac{2}{3}$$

A line parallel to the given line has the same slope.

$$\therefore m = \frac{2}{3} \text{ and using the point } (1, 3)$$

$$(y - 3) = \tfrac{2}{3}(x - 1)$$

☞ If a point (x_1, y_1) is on a line, then the coordinates x_1 and y_1, when substituted into the equation of the line, will satisfy it.

6. (C) Test the five answer choices:

$$2y = 4 - x$$

A) $(2, 1)$ $2(1) = 4 - (-2)$
$\qquad\qquad 2 = 2$ ✓

B) $(-4, 4)$ $2(4) = 4 - (-4)$
$\qquad\qquad 8 = 8$ ✓

C) $(-2, -3)$ $2(-3) = 4 - (-2)$
$\qquad\qquad\quad -6 \neq 6$ **X**

D) $(0, 2)$ $2(2) = 4 - 0$
$\qquad\qquad 4 = 4$ ✓

E) $(4, 0)$ $2(0) = 4 - 4$
$\qquad\qquad 0 = 0$ ✓

7. (E) To be colinear $m_{NJ} = m_{JL} = m_{NL}$:

$$m_{NJ} = \frac{6 - 0}{3 - 1} = \frac{6}{2} = \frac{3}{1}$$

$$m_{NL} = \frac{k - 0}{-4 - 1} = \frac{k}{-5}$$

$$\frac{3}{1} = \frac{k}{-5} \qquad k = -15$$

8. (D) $y = 3x + b$

If a point $(-3, 1)$ is on a line, it can be substituted into the equation of the line to make a true sentence:

$$1 = 3(-3) + b$$
$$1 = -9 + b$$
$$b = 10$$

Practice Problems 13.4

1. (D) $r_1 = 1$ $r_2 = 2$
Factors were $(x-1)(x-2) = 0$
$$x^2 - 3x + 2 = 0$$

2. (D) Real and equal roots of Discriminant $= 0$.
$$x^2 + 4x + c = 0$$
$$\therefore a = 1 \quad b = 4 \quad c = 4$$
$$b^2 - 4ac = 0$$
$$4^2 - 4(1)(c) = 0$$
$$16 = 4c$$
$$c = 4$$

3. (A) $y = x^2 - 4x$
Complete the square:
$$y + (-2)^2 = x^2 - 4x + (-2)^2$$
$$y + 4 = (x-2)^2$$
$$(x-2)^2 = y + 4$$

$\begin{cases} V:(2, -4) \\ p = 1 \\ F: (2, -3) \end{cases}$

4. (B) $x^2 = 12(y-4)$

$\begin{cases} V: (0, 4) \\ 4p = 12 \quad p = 3 \end{cases}$

5. (C)
$$x^2 - 4x - 5 = 0$$
$$(x-5)(x+1) = 0$$
$$x - 5 = 0 \mid x + 1 = 0$$
$$x = 5 \mid x = -1$$
$$\{-1, 5\}$$

6. (A) $f(\text{root}) = 0$ $r_1 = 2$
$$y = x^2 + kx - 2$$
$$y = (2)^2 + k(2) - 2 = 0$$
$$4 + 2k - 2 = 0$$
$$2k = -2$$
$$k = -1$$

7. (B) A parabola tangent to the x-axis has equal roots. The Discriminant $= 0$.
$$kx^2 - 4x + k = 0$$
$$a = k \quad b = -4 \quad c = k$$
$$b^2 - 4ac = 0$$
$$(-4)^2 - 4(k)(k) = 0$$
$$16 - 4k^2 = 0$$
$$4k^2 = 16$$
$$k^2 = 4$$
$$k = \pm 2$$

8. (B) $2x^2 - 5x - 4 = 0$
$$a = 2 \quad b = -5 \quad c = -4$$
$$x = \frac{-b \pm \sqrt{b^2 - 4ac}}{2a}$$
$$= \frac{-(-5) \pm \sqrt{(-5)^2 - 4(2)(-4)}}{2(2)}$$
$$= \frac{5 \pm \sqrt{25 + 32}}{4}$$
$$= \frac{5 \pm \sqrt{57}}{4}$$

Practice Problems 13.5

1. (E) $(x-h)^2 + (y-k)^2 = r^2$ *center*: $(-1, 1)$
$$(x+1)^2 + (y-1)^2 = 3^2 \quad r = 3$$
$$(x+1)^2 + (y-1)^2 = 9$$

2. (B)
$$x^2 + y^2 = r^2 \quad \textit{center}: (0, 0)$$
$$x^2 + y^2 = 8 \quad r = \sqrt{8} = 2\sqrt{2}$$

3. **(A)** Complete the square.

$$x^2 - 6x + (-3)^2 + y^2 + 4y + (+2)^2$$
$$= -10 + (-3)^2 + (2)^2$$
$$(x-3)^2 + (y+2)^2 = 3$$
$$center: (3, -2) \quad r = \sqrt{3}$$

4. **(E)** *center*: $(-1, -1)$ P: $(4, -5)$

$$(x-h)^2 + (y-k)^2 = r^2$$
$$(x+1)^2 + (y+1)^2 = r^2$$

Substitute the point on the circle:

$$(4+1)^2 + (-5+1)^2 = r^2$$
$$5^2 + (-4)^2 = r^2$$
$$25 + 16 = r^2$$
$$r^2 = 41$$
$$(x+1)^2 + (y+1)^2 = 41$$

Practice Problems 13.6

1. **(B)**

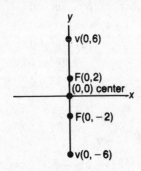

form: $\dfrac{y^2}{a^2} + \dfrac{x^2}{b^2} = 1$

$c =$ distance focus to center $= 2$
$a =$ distance vertex to center $= 6$

$$b^2 = a^2 - c^2 = 36 - 4 = 32$$
$$\frac{y^2}{36} + \frac{x^2}{32} = 1$$

2. **(E)** To set the equation equal to 1, divide each term by 3

$$\frac{(x-1)^2}{\dfrac{4}{3}} + \frac{(y-2)^2}{\dfrac{3}{3}} = \frac{3}{3}$$

$$\frac{(x-1)^2}{12} + \frac{(y-2)^2}{9} = 1$$

$a =$ distance from center to major vertex
$= \sqrt{12}$ or $2\sqrt{3}$

3. **(C)**

$$\frac{4x^2}{36} + \frac{9y^2}{36} = \frac{36}{36} \quad \text{Divide each term by 36.}$$

$$\frac{x^2}{9} + \frac{y^2}{4} = 1 \qquad \begin{array}{l} a = 3 \\ b = 2 \end{array}$$

Length of latus rectum

$$= \left| \frac{2b^2}{a} \right| = \frac{2(2)^2}{3} = \frac{8}{3}$$

4. **(A)** Rewrite the given equation in the form:

$$\frac{x^2}{a^2} + \frac{y^2}{b^2} = 1$$

☞ $\dfrac{a}{b} = \dfrac{1}{\dfrac{b}{a}}$,   e.g., $\dfrac{25}{9} = \dfrac{1}{\dfrac{9}{25}}$

$$\frac{25x^2}{9} + 4y^2 = 1$$

$$\frac{x^2}{\dfrac{9}{25}} + \frac{y^2}{\dfrac{1}{4}} = 1$$

$$a = \sqrt{\frac{9}{25}} = \frac{3}{5}, \quad b = \sqrt{\frac{1}{4}} = \frac{1}{2}$$

$$b^2 + c^2 = a^2$$
$$\frac{1}{4} + c^2 = \frac{9}{25}$$
$$c^2 = \frac{9}{25} - \frac{1}{4} = \frac{26-25}{100} = \frac{11}{100}$$
$$c = \frac{\sqrt{11}}{10}$$

Practice Problems 13.7

1. (B) $a = 5$ $b = 3$

latus rectum length $= \dfrac{2b^2}{a} = \dfrac{2(3)^2}{5} = \dfrac{18}{5}$

2. (B) center: $(1, -11)$

$a = \sqrt{2}, \quad b = \sqrt{2}$

$b^2 + a^2 = c^2$

$2 + 2 = c^2$

$\qquad c = 2$

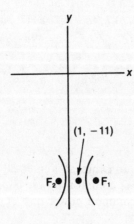

(1, −11)

The foci are each a distance 2 from the center.

$F_1: (3, -11) \quad F_2: (-1, -11)$

3. (A) Complete the square

$16(x^2 + 6x + (3)^2) - 9(y^2 - 4y + (-2)^2)$
$$= 36 + 16(3)^2 - 9(-2)^2$$

$16(x + 3)^2 - 9(y - 2)^2 = 144$

Divide each term by 144 to make the equation equal to 1

$$\frac{16(x + 3)^2}{144} - \frac{9(y - 2)^2}{144} = \frac{144}{144}$$

$$\frac{(x + 3)^2}{9} - \frac{(y - 2)^2}{16} = 1$$

center: $(-3, 2)$ $a = 3, \quad b = 4$

Asymptotes: $y = \pm \frac{4}{3}(x + 3) + 2$

$\therefore y = \frac{4}{3}(x + 3) + 2 \Rightarrow 3y - 4x - 18 = 0$

$\quad y = -\frac{4}{3}(x + 3) + 2 \Rightarrow 3y + 4x + 6 = 0$

4. (C) $xy = -6$

Make a table of values.

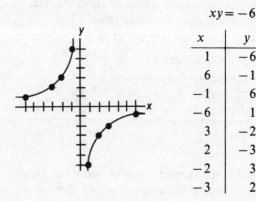

$xy = -6$

x	y
1	−6
6	−1
−1	6
−6	1
3	−2
2	−3
−2	3
−3	2

II and IV only

Practice Problems 13.8

1. (E) The x^2 and y^2 terms have the same sign, different coefficients.

Ellipse

2. (A) The x^2 and y^2 terms have the same sign and the same coefficients.

Circle

3. (D) The x^2 and y^2 terms have different signs.

Hyperbola

4. (B) One variable is squared, the other is linear.

Parabola

Practice Problems 14.1

1. (A) Using the model
$$(x-y)(x-y)=x^2-2xy+y^2$$
where
$$x=3f \text{ and } y=6g^2$$
$$(3f-6g^2)(3f+6g^2)=(3f)^2-2(3f)(6g^2)$$
$$+(6g^2)^2$$
$$=9f^2-36fg^2+36g^4$$

2. (A) Using the model
$$(x-y)(x+y)=x^2-y^2$$
where $x=5m^3$ and $y=3$
$$(5m^3-3)(5m^3+3)=(5m^3)^2-(3)^2$$
$$=25m^6-9$$

3. (D) Using the model
$$(x-y)(x^2+xy+y^2)=x^3-y^3$$
where $x=5$ and $y=3x$
$$(5-3x)(25+15x+9x^2)=(5)^3-(3x)^3$$
$$=125-27x^3$$

4. (B) Using the model
$$(x+y)(x-y)=x^2-y^2$$
where $x=7\sqrt{a}$ and $y=3\sqrt{b}$
$$(7\sqrt{a}+3\sqrt{b})(7\sqrt{a}-3\sqrt{b})=(7\sqrt{a})^2-(3\sqrt{b})^2$$
$$=49a-9b$$

Practice Problems 14.2

1. (B) 9 is the largest common numerical factor.
x^2 is the highest common power of x.
There is no common power of y.

1. The GCF $=9x^2$

2. $\dfrac{9x^2-9x^2y}{9x^2}=\dfrac{9x^2}{9x^2}-\dfrac{9x^2y}{9x^2}=1-y$

3. $9x^2-9x^2y=9x^2(1-y)$

2. (C) 3 is the largest common numerical factor.
m^2 is the highest common power of m.
n^2 is the highest common power of n.

1. The GCF $=3m^2n^2$

2. $\dfrac{3m^2n^2+3m^3n^3}{3m^2n^2}=\dfrac{3m^2n^2}{3m^2n^2}+\dfrac{3m^3n^3}{3m^2n^2}$
$$=1+mn$$

3. $3m^2n^2+3m^3n^3=3m^2n^2(1+mn)$

3. (A) There is no common numerical factor.
a is the highest common power of a.
There is no common power of b.

1. The GCF $=a$

2. $\dfrac{4a^2+20a+5a^3b}{a}=\dfrac{4a^2}{a}+\dfrac{20a}{a}+\dfrac{5a^3b}{a}$
$$=4a+20+5a^2b$$

3. $4a^2+20a+5a^3b=a(4a+20+5a^2b)$

4. (D) 3 is the greatest common numerical factor.
n is the highest common power of n.
l is the highest common power of l.

1. The GCF $=3nl$

2. $\dfrac{3n^2l^2+6nl+9n^3l^3}{3nl}=\dfrac{3n^2l^2}{3nl}+\dfrac{6nl}{3nl}+\dfrac{9n^3l^3}{3nl}$
$$=nl+2+3n^2l^2$$

3. $3n^2l^2+6nl+9n^3l^3=3nl(nl+2+3n^2l^2)$

Practice Problems 14.3

1. (E) Using the format x^2-y^2, where $x=3a$ and $y=4$,
$$x^2-y^2=(x+y)(x-y)$$
$$9a^2-16=(3a+4)(3a-4)$$

2. (B) Using the format x^2-y^2, where $x=4$ and $y=3a$,
$$x^2-y^2=(x+y)(x-y)$$
$$16-9a^2=(4+3a)(4-3a)$$

3. (D) Using the format $x^2 + 2xy + y^2$, where $x = a$ and $y = 5$,

$$x^2 + 2xy + y^2 = (x+y)(x+y)$$
$$a^2 + 10a + 25 = (a+5)(a+5)$$

4. (D) Using the format $x^3 - y^3$, where $x = 3a$ and $y = 2$,

$$x^3 - y^3 = (x-y)(x^2 + xy + y^2)$$
$$27a^3 - 8 = (3a-2)(9a^2 + 6a + 4)$$

5. (B) Using the format $x^3 + y^3$, where $x = 3a$ and $y = 2$,

$$x^3 + y^3 = (x+y)(x^2 - xy + y^2)$$
$$27a^3 + 8 = (3a+2)(9a^2 - 6a + 4)$$

6. (C) Using the format $x^2 - 2xy + y^2$, where $x = 3a$ and $y = 2$,

$$x^2 - 2xy + y^2 = (x-y)(x-y)$$
$$9a^2 - 12a + 4 = (3a-2)(3a-2)$$

Practice Problems 14.4

1. (C) $(x \quad)(x \quad)$ $a = 1$

$c = 10$, middle term $= -7x$

$+10, +1$
$-10, -1$
$+5, +2$
$-5, -2$

$(x + 10)(x + 1)$ $10x + 1x = 11x \neq -7x$

$(x - 10)(x - 1)$ $-10x - 1x = -11x \neq -7x$

$(x + 5)(x + 2)$ $5x + 2x = 7x \neq -7x$

$(x - 5)(x - 2)$ $-5x - 2x = -7x = -7x \checkmark$

Factors are: $(x-5)(x-2)$

2. (D) $3x^2 - 27 = 3(x^2 - 9)$ GCF: 3
$\qquad\qquad = 3(x+3)(x-3)$

$x^2 - 9$ is the difference of two squares.

3. (A) $(3x \quad)(x \quad)$

$a = 3$ a prime number

$c = 4$, middle term $= -4x$

$+4, +1$
$-4, -1$
$+2, -2$

$(3x + 4)(x + 1)$ $4x + 3x = 7x \neq -4x$

$(3x + 1)(x + 4)$ $x + 12x = 13x \neq -4x$

$(3x - 4)(x - 1)$ $-4x + (-3x) = -7x \neq -4x$

$(3x - 1)(x - 4)$ $-x + (-12x) = -13x \neq -4x$

$(3x + 2)(x - 2)$ $2x + (-6x) = -4x = -4x \checkmark$

Factors are: $(3x+2)(x-2)$

4. (C) $15y^2 - 5y = 5y(3y - 1)$ GCF: $5y$

5. (E) $2x^3 - 16 = 2(x^3 - 8)$ GCF: 2

$x^3 - 8$ is the difference of two cubes $x^3 - y^3$, where $x = x$, $y = 2$.

$$2x^3 - 16 = 2(x-2)(x^2 + 2x + 4)$$

6. (B)

$4x^2 + 12x - 16$ GCF: 4
$4(x^2 + 3x - 4)$ a is 1
$4(x \quad)(x \quad)$
$c = -4$, middle term $= +3x$

$+4, -1$
$-1, +4$
$-2, +2$

$(x + 4)(x - 1)$ $4x + (-x) = 3x = 3x \checkmark$

Factors are: $4(x+4)(x-1)$

7. **(D)**

$8x^8 - 8$ GCF: 8

$= 8(x^8 - 1)$

$x^8 - 1$ is the difference of two squares.

$= 8(x^4 + 1)(x^4 - 1)$

$x^4 - 1$ is the difference of two squares.

$= 8(x^4 + 1)(x^2 + 1)(x^2 - 1)$

$x^2 - 1$ is the difference of two squares.

$= 8(x^4 + 1)(x^2 + 1)(x + 1)(x - 1)$

8. **(A)** $a = 4$ a non-prime number

$\left.\begin{matrix} 4, 1 \\ 2, 2 \end{matrix}\right\}$ factors of a

$(4x \quad)(x \quad)$

$c = 3$, middle term $= -8x$

$\begin{matrix} 3, 1 \\ -3, -1 \end{matrix}$ factors of c

$\underset{4x}{(4x + \overset{3x}{3})(x + 1)}$

$3x + 4x = 7x \neq -8x$

$\underset{12x}{(4x + \overset{x}{1})(x + 3)}$ $x + 12x = 13x \neq -8x$

$\underset{-4x}{(4x - \overset{-3x}{3})(x - 1)}$ $-3x + (-4x) = -7x \neq -8x$

$\underset{-12x}{(4x - \overset{-x}{1})(x - 3)}$ $-x + (-12x) = -13x \neq -8x$

$\underset{2x}{(2x + \overset{6x}{3})(2x + 1)}$ $6x + 2x = 8x \neq -8x$

$\underset{-2x}{(2x - \overset{-6x}{3})(2x - 1)}$ $-6x + (-2x) = -8x = -8x \checkmark$

Factors are: $(2x - 3)(2x - 1)$

9. **(B)** $(7x \quad)(x \quad)$

$a = 7$ a prime number

$c = -2$, middle term $= -13x$

$\begin{matrix} -2, +1 \\ +2, -1 \end{matrix}$

$\underset{7x}{(7x - \overset{-2x}{2})(x + 1)}$ $-2x + 7x = 5x \neq -13x$

$\underset{-14x}{(7x + \overset{x}{1})(x - 2)}$ $x + (-14x) = -13x = -13x \checkmark$

Factors are: $(7x + 1)(x - 2)$

10. **(D)** $(x \quad)(x \quad)$ $a = 1$

$c = 6$, middle term $= -5x$

$\begin{matrix} 6 & 1 \\ -6 & -1 \\ 3 & 2 \\ -3 & -2 \end{matrix}$

$\underset{x}{(x + \overset{6x}{6})(x + 1)}$ $6x + 1x = 7x \neq -5x$

$\underset{-x}{(x - \overset{-6x}{6})(x - 1)}$ $-6x - x = -7x \neq -5x$

$\underset{2x}{(x + \overset{3x}{3})(x + 2)}$ $3x + 2x = 5x \neq -5x$

$\underset{-2x}{(x - \overset{-3x}{3})(x - 2)}$ $-3x - 2x = -5x \neq -5x \checkmark$

Factors are: $(x - 3)(x - 2)$

11. **(C)** $(2x \quad)(x \quad)$

$a = 2$ a prime number

$c = -3$, middle term $= +1x$

$\begin{matrix} -3, 1 \\ 3, -1 \end{matrix}$

$\underset{2x}{(2x - \overset{-3x}{3})(x + 1)}$ $-3x + 2x = -x \neq x$

$\underset{-6x}{(2x + \overset{x}{1})(x - 3)}$ $x + (-6x) = -5x \neq x$

$\underset{-2x}{(2x + \overset{3x}{3})(x - 1)}$ $3x + (-2x) = x = x \checkmark$

Factors are: $(2x + 3)(x - 1)$

12. **(E)** $4x^3 - 49x$ GCF: x

$= x(4x^2 - 49)$

$4x^2 - 49$ is the difference of two squares.

$= x(2x + 7)(2x - 7)$

Factors are: $x(2x + 7)(2x - 7)$

Practice Problems 14.5

1. (B)

$$\frac{x+3}{2x+4}\cdot\frac{x^2-x-6}{2x+6}$$

$$=\frac{\cancel{x+3}^{(1)}}{2(\cancel{x+2})}\cdot\frac{(x-3)(\cancel{x+2})^{(1)}}{2(\cancel{x+3})}=\frac{x-3}{4}$$

2. (E)

$$\frac{a^2+8a+15}{a^2+5a}\div\frac{a^2-9}{a^2+2a-3}$$

$$=\frac{a^2+8a+15}{a^2+5a}\cdot\frac{a^2+2a-3}{a^2-9}$$

$$=\frac{(\cancel{a+3})^{(1)}(\cancel{a+5})^{(1)}}{a(\cancel{a+5})}\cdot\frac{(a+3)(a-1)}{(\cancel{a+3})(a-3)}$$

$$=\frac{(a+3)(a-1)}{a(a-3)}$$

3. (A)

$$\frac{x^2+3x+2}{x^2-x-2}\cdot\frac{x^2+x-6}{x^2-x-12}$$

$$=\frac{(x+2)(\cancel{x+1})^{(1)}}{(\cancel{x+1})(\cancel{x-2})}\cdot\frac{(\cancel{x-2})^{(1)}(\cancel{x+3})^{(1)}}{(\cancel{x+3})(x-4)}$$

$$=\frac{x+2}{x-4}$$

4. (C)

$$\frac{a+2}{2a}\cdot\frac{2a-4}{2+a}\cdot\frac{a^2-4}{12-6a}\div\frac{2+a}{12a^2}$$

$$=\frac{a+2}{2a}\cdot\frac{2a-4}{2+a}\cdot\frac{a^2-4}{12-6a}\cdot\frac{12a^2}{2+a}$$

$$=\frac{\cancel{a+2}^{(1)}}{\cancel{2a}}\cdot\frac{\cancel{2}(a-2)}{\cancel{a+2}}\cdot\frac{(\cancel{a+2})^{(1)}(\cancel{a-2})^{(-1)}}{\cancel{6}(2-a)}\cdot\frac{\cancel{12a^2}^{(2)\,(a)}}{\cancel{2+a}}$$
$$\qquad\qquad\qquad\quad_{(1)}$$

$$=-2a(a-2)$$

Practice Problems 14.6

1. (C)

$$x^2-4=0$$
$$(x+2)(x-2)=0$$
$$x+2=0\ \Big|\ x-2=0$$
$$x=-2\ \Big|\ \ x=2$$

$$x=-2\text{ or }2$$

2. (A)

$$x^2+9=10x$$
$$x^2-10x+9=0$$
$$(x-1)(x-9)=0$$
$$x-1=0\ \Big|\ x-9=0$$
$$x=1\ \Big|\ \ x=9$$

$$x=1\text{ or }x=9$$

3. (E) $2n^3-2n=0$

$$2n(n^2-1)=0\qquad\text{GCF: }2n$$
$$2n(n+1)(n-1)=0$$
$$2n=0\ \Big|\ n+1=0\ \Big|\ n-1=0$$
$$n=0\ \Big|\ \ n=-1\ \Big|\ \ n=1$$
$$\{0,\,1,\,-1\}$$

4. (B)

$$8x^2+10x-3=0$$
$$(4x-1)(2x+3)=0$$
$$4x-1=0\ \Big|\ 2x+3=0$$
$$x=\tfrac{1}{4}\ \Big|\ \ x=-\tfrac{3}{2}$$
$$\{\tfrac{1}{4},\,-\tfrac{3}{2}\}$$

Practice Problems 15.1

1. (A) $3x+5=5x-3$

$$5+3=5x-3x$$
$$2x=8$$
$$x=4$$

2. (B) $2\left(\dfrac{d}{2}+6=4\right)$

$$d+12=8$$
$$d=-4$$

3. **(C)** $10(x - .2 = 1.8)$
$$10x - 2 = 18$$
$$10x = 20$$
$$x = 2$$

4. **(D)** $4(b - 1) - 3 = 17$
$$4b - 4 - 3 = 17$$
$$4b - 7 = 17$$
$$4b = 24$$
$$b = 6$$

5. **(E)** $12\left(\dfrac{2n}{3} - \dfrac{5n}{12} = \dfrac{5}{4}\right)$
$$8n - 5n = 15$$
$$3n = 15$$
$$n = 5$$

6. **(A)** $5 + 3(a + 2) = 14$
$$5 + 3a + 6 = 14$$
$$3a + 11 = 14$$
$$3a = 3$$
$$a = 1$$

Practice Problems 15.2

1. **(B)** $\dfrac{3a + 1}{4} = \dfrac{5}{2}$
$$2(3a + 1) = 20$$
$$6a + 2 = 20$$
$$6a = 18$$
$$a = 3$$

2. **(C)** $\dfrac{m}{4} = \dfrac{5}{2}$
$$2m = 20$$
$$m = 10$$

3. **(D)** $\dfrac{x + 1}{8} = \dfrac{2x + 1}{4}$
$$4(x + 1) = 8(2x + 1)$$
$$4x + 4 = 16x + 8$$
$$12x = -4$$
$$x = -\tfrac{4}{12} = -\tfrac{1}{3}$$

4. **(A)** $\dfrac{3l - 1}{5l + 2} = \dfrac{2}{7}$
$$7(3l - 1) = 2(5l + 2)$$
$$21l - 7 = 10l + 4$$
$$11l = 11$$
$$l = 1$$

Practice Problems 15.3

1. **(D)** $\dfrac{cx + 1}{a} = \dfrac{b}{3}$
$$3(cx + 1) = ab$$
$$3cx + 3 = ab$$
$$\dfrac{ab}{b} = \dfrac{3cx + 3}{b}$$
$$a = \dfrac{3cx + 3}{b}$$

2. **(A)** $3(x + 2) = a(b + x)$
$$3x + 6 = ab + ax$$
$$3x - ax = ab - 6$$
$$\dfrac{x(3 - a)}{3 - a} = \dfrac{ab - 6}{3 - a}$$
$$x = \dfrac{ab - 6}{3 - a}$$

3. **(C)**
$$\dfrac{a}{1} = \dfrac{t + b}{t - c}$$
$$a(t - c) = 1(t + b)$$
$$at - ac = t + b$$
$$at - t = b + ac$$
$$\dfrac{t(a - 1)}{a - 1} = \dfrac{b + ac}{a - 1}$$
$$t = \dfrac{b + ac}{a - 1}$$

4. **(C)** $abn\left(\dfrac{1}{a} = \dfrac{1}{n} + \dfrac{1}{b}\right)$
$$bn = ab + an$$
$$bn - an = ab$$
$$\dfrac{n(b - a)}{b - a} = \dfrac{ab}{b - a}$$
$$n = \dfrac{ab}{b - a}$$

Practice Problems 15.4

1. (A)

$$y - x = -1$$
$$2y + x = 4$$

Solve the first equation for x and substitute into the second equation:

$$y - x = -1$$
$$-x = -1 - y$$
$$x = 1 + y$$

Substitute $x = 1 + y$ into the second equation.

$$2y + x = 4$$
$$2y + (1 + y) = 4$$
$$2y + 1 + y = 4$$
$$3y = 3$$
$$y = 1$$

Replace y with 1 in one of the original equations:

$$2y + x = 4$$
$$2(1) + x = 4$$
$$2 + x = 4$$
$$x = 2$$
$$x = 2, \quad y = 1$$
$$(2, 1)$$

2. (C) Add the equations to eliminate y:

$$
\begin{array}{l}
3x + 4y = 7 \\
\underline{x - 4y = 1} \\
4x \quad\quad = 8 \\
x \quad\quad = 2
\end{array}
$$

Replace $x = 2$ into one of the original equations:

$$x - 4y = 1$$
$$2 - 4y = 1$$
$$-4y = -1$$
$$y = \tfrac{1}{4}$$
$$x = 2 \quad y = \tfrac{1}{4}$$
$$(2, \tfrac{1}{4})$$

3. (A) $3x - 2y = -1$
$\quad\quad\quad 2x + 3y = 8$

$$x = \frac{\begin{vmatrix} -1 & -2 \\ 8 & 3 \end{vmatrix}}{\begin{vmatrix} 3 & -2 \\ 2 & 3 \end{vmatrix}} = \frac{-1(3) - (-2)(8)}{9 - (-2)(2)} = \frac{13}{13} = 1$$

$$y = \frac{\begin{vmatrix} 3 & -1 \\ 2 & 8 \end{vmatrix}}{\begin{vmatrix} 3 & -2 \\ 2 & 3 \end{vmatrix}} = \frac{3(8) - (-1)(2)}{3(3) - (-2)(2)} = \frac{26}{13} = 2$$

$$x = 1 \quad y = 2$$
$$(1, 2)$$

4. (D) $2x - y = 6$
$\quad\quad\quad x + y = 9$

Solve the second equation for y and substitute into the first equation:

$$x + y = 9$$
$$y = 9 - x$$

Substitute $9 - x$ for y in the first equation.

$$2x - y = 6$$
$$2x - (9 - x) = 6$$
$$2x - 9 + x = 6$$
$$3x = 15$$
$$x = 5$$

Replace $x = 5$ into one of the original equations:

$$x + y = 9$$
$$5 + y = 9$$
$$y = 4$$
$$x = 5 \quad y = 4$$
$$(5, 4)$$

5. (D) $\quad y - 4 = x$
$\quad\quad\quad 5x - 20 = y$

Rearrange the equations:

$$x - y = -4$$
$$5x - y = 20$$

Multiply the first equation by -1 to eliminate y:

$$-1(x - y = -4) \Rightarrow -x + y = 4$$

$$
\begin{array}{l}
-x + y = \ 4 \\
\underline{5x - y = 20} \\
4x \quad\quad = 24 \\
x \quad\quad = 6
\end{array}
$$

Substitute $x = 6$ and solve for y:

$$y - 4 = x$$
$$y - 4 = 6$$
$$y = 10$$
$$x = 6 \quad y = 10$$
$$(6, 10)$$

6. **(A)** $4x - y = 10$
$\qquad x + y = 5$

$$x = \frac{\begin{vmatrix} 10 & -1 \\ 5 & 1 \end{vmatrix}}{\begin{vmatrix} 4 & -1 \\ 1 & 1 \end{vmatrix}} = \frac{10(1) - (5)(-1)}{4(1) - (-1)(1)} = \frac{15}{5} = 3$$

$$y = \frac{\begin{vmatrix} 4 & 10 \\ 1 & 5 \end{vmatrix}}{\begin{vmatrix} 4 & -1 \\ 1 & 1 \end{vmatrix}} = \frac{4(5) - (10)(1)}{4(1) - (-1)(1)} = \frac{10}{5} = 2$$

$$x = 3 \quad y = 2$$
$$(3, 2)$$

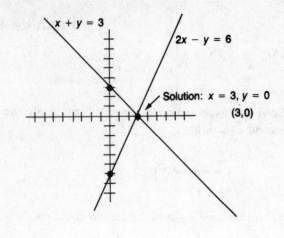

7. **(D)** Put into intercept form:

$$x + y = 3$$
$$2x - y = 6$$

$$\frac{x}{3} + \frac{y}{3} = 1 \qquad \begin{array}{l} x \text{ intercept} = 3 \\ y \text{ intercept} = 3 \end{array}$$

$$\frac{2x}{6} - \frac{y}{6} = \frac{6}{6}$$
$$\downarrow$$
$$\frac{x}{3} - \frac{y}{6} = 1 \qquad \begin{array}{l} x \text{ intercept} = 3 \\ y \text{ intercept} = -6 \end{array}$$

8. **(C)** $x + y = 12$
$\qquad y = 3x$

Substitute the second equation into the first:

$$x + y = 12$$
$$x + (3x) = 12$$
$$4x = 12$$
$$x = 3$$

Replace x with 3, solve for y:

$$y = 3x$$
$$y = 3(3) = 9$$
$$x = 3 \quad y = 9$$
$$(3, 9)$$

Practice Problems 15.5

1. **(A)**

$$2x^2 + 3x - 2 = 0$$
$$(2x - 1)(x + 2) = 0$$
$$2x - 1 = 0 \mid x + 2 = 0$$
$$x = \tfrac{1}{2} \mid \quad x = -2$$

2. **(C)**

$$x^2 - 4x - 7 = 0$$
$$x^2 - 4x \quad = 7$$
$$x^2 - 4x + (-2)^2 = 7 + (-2)^2$$
$$(x - 2)^2 = 11$$
$$\sqrt{(x - 2)^2} = \sqrt{11}$$
$$x - 2 = \pm\sqrt{11}$$
$$x = 2 \pm \sqrt{11}$$

3. **(B)**

$$f(x) = x^2 + 4x + 9 \quad a = 1, \, b = 4, \, c = 9$$

$$x = \frac{-b \pm \sqrt{b^2 - 4ac}}{2a}$$

$$= \frac{-4 \pm \sqrt{4^2 - 4(1)(9)}}{2(1)}$$

$$= \frac{-4 \pm \sqrt{16 - 36}}{2}$$

$$= \frac{-4 \pm \sqrt{-20}}{2} = \frac{-4 \pm 2i\sqrt{5}}{2}$$

$$= -2 \pm i\sqrt{5}$$

4. (A) axis of symmetry $x = \dfrac{-(-4)}{2} = 2$

x	$x^2 - 4x + 4$	y
-1	$(-1)^2 - 4(-1) + 4$	9
0	$0^2 - 4(0) + 4$	4
1	$1^2 - 4(1) + 4$	1
2	$2^2 - 4(2) + 4$	0
3	$3^2 - 4(3) + 4$	1
4	$4^2 - 4(4) + 4$	4
5	$5^2 - 4(5) + 4$	9

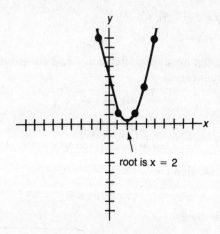

root is x = 2

root is $x = 2$. When curve is tangent to the axis there are really two roots of $x = 2$.

$\therefore$ There is a double root at $x = 2$.

Practice Problems 15.6

1. (C) $y = 5 + 2x$
$x^2 + y^2 = 25$

Substitute $y = 2x + 5$ into the quadratic:

$$x^2 + (2x + 5)^2 = 25$$
$$x^2 + 4x^2 + 20x + 25 - 25 = 0$$

Divide by 5:

$$5x^2 + 20x = 0$$
$$x^2 + 4x = 0$$
$$x(x + 4) = 0$$

$x = 0$	$x + 4 = 0$
	$x = -4$
$y = 5 + 2x$	$y = 5 + 2x$
$y = 5 + 2(0)$	$y = 5 + 2(-4)$
$y = 5$	$y = -3$
$(0, 5)$	$(-4, -3)$

2. (E) $x^2 + y^2 - 4x - 2y + 1 = 0$
$y + x = 1$

Solve the linear equation for x and substitute:

$$y + x = 1$$
$$x = 1 - y$$
$$x^2 + y^2 - 4x - 2y + 1 = 0$$
$$(1 - y)^2 + y^2 - 4(1 - y) - 2y + 1 = 0$$
$$1 - 2y + y^2 + y^2 - 4 + 4y - 2y + 1 = 0$$

combine terms

$$2y^2 - 2 = 0$$

Divide by 2

$y^2 - 1 = 0$	
$(y + 1) \quad (y - 1) = 0$	
$y + 1 = 0$	$y - 1 = 0$
$y = -1$	$y = 1$
$y + x = 1$	$y + x = 1$
$-1 + x = 1$	$1 + x = 1$
$x = 2$	$x = 0$
$(2, -1)$	$(0, 1)$

3. (B) $y = 2x^2 - 5x + 5$
$y - x = 5$

Solve the linear equation for y and substitute into the quadratic:

$$y - x = 5 \Rightarrow y = x + 5$$

$$y = 2x^2 - 5x + 5$$

$$x + 5 = 2x^2 - 5x + 5$$

combine like terms

$$2x^2 - 6x = 0$$

Divide by 2:

$$x^2 - 3x = 0$$

$$x \quad (x - 3) = 0$$

$x = 0$	$x - 3 = 0$
	$x = 3$
$y - x = 5$	$y - x = 5$
$y - 0 = 5$	
$y = 5$	$y = 8$
$(0, 5)$	$(3, 8)$

4. (A) $y = -x^2 + 4x - 3$
$y + 1 = x$

Solve the linear equation for y and substitute:

$$y + 1 = x \Rightarrow y = x - 1$$

$$y = -x^2 + 4x - 3$$

$$x - 1 = -x^2 + 4x - 3$$

combine like terms

$$-x^2 + 3x - 2 = 0$$

Multiply by -1: $\quad x^2 - 3x + 2 = 0$

$$(x - 2) \quad (x - 1) = 0$$

$x - 2 = 0$	$x - 1 = 0$
$x = 2$	$x = 1$
$y + 1 = x$	$y + 1 = x$
$y + 1 = 2$	$y + 1 = 1$
$y = 1$	$y = 0$
$(2, 1)$	$(1, 0)$

Practice Problems 15.7

1. (E) $4x^2 + y^2 = 20$
$x^2 + 4y^2 = 20$

Multiply the second equation by -4 and add it to the first equation to eliminate the x^2:

$$-4(x^2 + 4y^2 = 20) \Rightarrow -4x - 16y = -80$$
$$-4x^2 - 16y^2 = -80$$
$$4x^2 + y^2 = 20$$

$$-15y^2 = -60$$
$$y^2 = 4$$
$$y = \pm 2$$

For each value of y determine the corresponding values of x:

$y = 2$	$y = -2$
$x^2 + 4y^2 = 20$	$x^2 + 4y^2 = 20$
$x^2 + 4(2)^2 = 20$	$x^2 + 4(-2)^2 = 20$
$x^2 = 4$	$x^2 = 4$
$x = \pm 2$	$x = \pm 2$
$(2, 2)$	$(2, -2)$
$(-2, 2)$	$(-2, -2)$

4 solutions

2. (C) Add the equations to eliminate the y^2.

$$y^2 + 2x = -1$$
$$-y^2 + x^2 = 16$$

$$x^2 + 2x - 15 = 0$$

Solve the quadratic for x.

$(x + 5)$	$(x - 3) = 0$
$x + 5 = 0$	$x - 3 = 0$
$x = -5$	$x = 3$

For the corresponding y values

$y^2 - 2x = -1$	$y^2 - 2x = -1$
$y^2 - 2(-5) = -1$	$y^2 - 2(3) = -1$
$y^2 = -11$	$y^2 = 5$
$y = \pm\sqrt{11}i$	$y = \pm\sqrt{5}$
$(-5, \sqrt{11}i)$	$(3, \sqrt{5})$
$(-5, -\sqrt{11}i)$	$(3, -\sqrt{5})$

2 real solutions

3. (A)

$$x^2 + y = 5$$
$$x^2 + y^2 = 1$$

Multiply the second equation by -1 and add to the first equation to eliminate x.

$$-1(x^2 + y^2 = 1) \Rightarrow -x^2 - y^2 = -1$$
$$\underline{} -x^2 - y^2 = -1$$
$$x^2 + y = 5$$
$$\overline{-y^2 + y = 4}$$

Multiply by -1: $-y^2 + y - 4 = 0$
$$y^2 - y + 4 = 0$$

Use the quadratic equation.

$$y = \frac{-b \pm \sqrt{b^2 - 4ac}}{2a} \quad \begin{array}{l} a = 1 \\ b = -1 \\ c = 4 \end{array}$$

$$= \frac{1 \pm \sqrt{(-1)^2 - 4(1)(4)}}{2(1)}$$

$$= \frac{1 \pm \sqrt{1 - 16}}{2} = \frac{1 \pm \sqrt{-15}}{2}$$

All values of y are imaginary. There are no real solutions.

4. (A) Solved graphically.

$xy = 12$ is a hyperbola

x	y
1	12
12	1
-1	-12
-12	-1
2	6
6	2
-2	-6
-6	-2
4	3
-4	-3
-3	-4
-3	4

$x^2 + y^2 = 16$ is a circle with center at $(0, 0)$ and radius $= 4$.

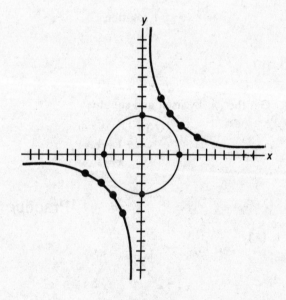

No real solutions.

Practice Problems 15.8

1. (B) $\sqrt{x^2 + 3} = x + 1$

Square both sides:

$$(\sqrt{x^2 + 3})^2 = (x + 1)^2$$
$$x^2 + 3 = x^2 + 2x + 1$$
$$2x - 2 = 0$$
$$x = 1$$

Check

$$\sqrt{x^2 + 3} = x + 1$$
$$\sqrt{1^2 + 3} \overset{?}{=} 1 + 1$$
$$\sqrt{4} \overset{?}{=} 2$$
$$2 = 2 \checkmark$$

1 solution $x = 1$

2. (C) Get the $\sqrt{}$ by itself and square the equation

$$\sqrt{x+18}+2=x$$
$$\sqrt{x+18}=x-2$$
$$(\sqrt{x+18})^2=(x-2)^2$$
$$x+18=x^2-4x+4$$
$$x^2-5x-14=0$$
$$(x-7)\quad(x+2)=0$$
$$x-7=0 \mid x+2=0$$
$$x=7 \mid \quad x=-2$$

Check

$x=7$	$x=-2$
$\sqrt{x+18}+2=x$	$\sqrt{x+18}+2=x$
$\sqrt{7+18}+2\overset{?}{=}7$	$\sqrt{-2+18}+2\overset{?}{=}-2$
$\sqrt{25}+2\overset{?}{=}7$	$\sqrt{16}+2\overset{?}{=}-2$
$5+2\overset{?}{=}7$	$4+2\overset{?}{=}-2$
$7=7$	$6\neq-2$
$\checkmark$	X

1 solution

$$x=7$$

3. (E)

$$\sqrt{x+1}+2=0$$

Get the $\sqrt{}$ by itself and square:

$$(\sqrt{x+1})^2=(-2)^2$$
$$x+1=4$$
$$x=3$$

Check

$$\sqrt{x+1}+2=0$$
$$\sqrt{3+1}+2\overset{?}{=}0$$
$$\sqrt{4}+2\overset{?}{=}0$$
$$2+2\overset{?}{=}0$$
$$4\neq0$$

No solution

Solution set $\varnothing$ or { }

4. (A)

$$\sqrt{2x}-\sqrt{x+1}=0$$

Get one $\sqrt{}$ by itself and square:

$$(\sqrt{2x})^2=(\sqrt{x+1})^2$$
$$2x=x+1$$
$$x=1$$

Check

$$\sqrt{2x}-\sqrt{x+1}=0$$
$$\sqrt{2(1)}-\sqrt{1+1}\overset{?}{=}0$$
$$\sqrt{2}-\sqrt{2}\overset{?}{=}0$$
$$0=0\checkmark$$

1 solution

$$x=1$$

Practice Problems 15.9

1. (A)

$$|x-1|=3$$

$$x-1=-3 \qquad x-1=3$$
$$x=-2 \qquad x=4$$

Check

$|x-1|=3$	$|x-1|=3$
$|-2-1|\overset{?}{=}3$	$|4-1|\overset{?}{=}3$
$|-3|\overset{?}{=}3$	$|3|\overset{?}{=}3$
$3=3\checkmark$	$3=3\checkmark$

2 solutions

$$\{-2,4\}$$

2. (B)

$$|x+7|=2x-1$$

$$x+7=-(2x-1) \qquad x+7=2x-1$$
$$x+7=-2x+1 \qquad\qquad 8=x$$
$$3x=-6$$
$$x=-2$$

Check

$|x+7|=2x-1$	$|x+7|=2x-1$
$|-2+7|\overset{?}{=}2(-2)-1$	$|8+7|\overset{?}{=}2(8)-1$
$|5|\overset{?}{=}-5$	$|15|\overset{?}{=}16-1$
$5\neq-5$	$15=15\checkmark$

1 solution

$$x=8$$
$$\{8\}$$

3. (E)

$$|x^2 - 5x| = 6$$
$$\downarrow$$
$$x^2 - 5x = -6$$
$$x^2 - 5x + 6 = 0$$
$$(x-3)(x-2) = 0$$

$$x - 3 = 0 \mid x - 2 = 0$$
$$x = 3 \mid \quad x = 2$$

Check

$\|x^2 - 5x\| = 6$	$\|x^2 - 5x\| = 6$
$\|3^2 - 5(3)\| \overset{?}{=} 6$	$\|2^2 - 5(2)\| \overset{?}{=} 6$
$\|9 - 15\| \overset{?}{=} 6$	$\|4 - 10\| \overset{?}{=} 6$
$\|-6\| \overset{?}{=} 6$	$\|-6\| \overset{?}{=} 6$
$6 = 6 \checkmark$	$6 = 6 \checkmark$

$$|x^2 - 5x| = 6$$
$$\downarrow$$
$$x^2 - 5x = 6$$
$$x^2 - 5x - 6 = 0$$
$$(x-6)(x+1) = 0$$

$$x - 6 = 0 \mid x + 1 = 0$$
$$x = 6 \mid \quad x = -1$$

Check

$\|x^2 - 5x\| = 6$	$\|x^2 - 5x\| = 6$
$\|6^2 - 5(6)\| \overset{?}{=} 6$	$\|(-1)^2 - 5(-1)\| \overset{?}{=} 6$
$\|36 - 30\| \overset{?}{=} 6$	$\|1 + 5\| \overset{?}{=} 6$
$\|6\| \overset{?}{=} 6$	$\|6\| \overset{?}{=} 6$
$6 = 6 \checkmark$	$6 = 6$

There are 4 solutions
$$\{-1, 2, 3, 6\}$$

4. (A)

$$|x - 7| = 2x - 10$$

$$x - 7 = -(2x - 10) \mid x - 7 = 2x - 10$$
$$x - 7 = -2x + 10 \mid 3 = x$$
$$3x = 17$$
$$x = \tfrac{17}{3}$$

Check

$\|x - 7\| = 2x - 10$	$\|x - 7\| = 2x - 10$
$\|\tfrac{17}{3} - 7\| \overset{?}{=} 2(\tfrac{17}{3}) - 10$	$\|3 - 7\| \overset{?}{=} 2(3) - 10$
$\|-\tfrac{4}{3}\| \overset{?}{=} \tfrac{34}{3} - \tfrac{30}{3}$	$\|-4\| \overset{?}{=} 6 - 10$
$\tfrac{4}{3} = \tfrac{4}{3} \checkmark$	$\|-4\| = -4$
	$4 \neq -4$

1 solution
$$\{\tfrac{17}{3}\}$$

Practice Problems 15.10

1. **(A)** Using Equations 1 with 2 and 1 with 3 eliminate y:

$$x+y+2z=1 \qquad\qquad x+y+2z=1$$
$$2x+3y+4z=4 \qquad\qquad 3x-y-z=2$$

Multiply the first equation by -3 and add it to second equation to eliminate y

Add the equations to eliminate y

$$-3x-3y-6z=-3 \qquad\qquad x+y+2z=1$$
$$\underline{2x+3y+4z=4} \qquad\qquad \underline{3x-y-z=2}$$
$$\boxed{-x-2z=1} \qquad\qquad \boxed{4x+z=3}$$

$$-x-2z=1$$
$$4x+z=3$$

Multiply second equation by 2 and add it to first equation to eliminate z

$$-x-2z=1$$
$$\underline{8x+2z=6}$$
$$7x=7$$
$$\boxed{x=1}$$

Substitute $x=1$ into one of the two variable equations

$$4x+z=3$$
$$4(1)+z=3$$
$$4+z=3$$
$$\boxed{z=-1}$$

Substitute $x=1$ and $z=-1$ into one of the 3-variable equations

$$3x-y-z=2$$
$$3(1)-y-(-1)=2$$
$$3-y+1=2$$
$$\boxed{y=2}$$

2. **(E)** Using equations 1 with 2 and 2 with 3 eliminate x:

$$2x-3y+2z=26 \qquad\qquad x-2y-3z=-5$$
$$x-2y-3z=-5 \qquad\qquad x+y-z=-7$$

Multiply second equation by -2 and add to first equation to eliminate x

Multiply first equation by -1 and add to second equation to eliminate x

$$2x-3y+2z=26 \qquad\qquad -x+2y+3z=5$$
$$\underline{-2x+4y+6z=10} \qquad\qquad \underline{x+y-z=-7}$$
$$\boxed{y+8z=36} \qquad\qquad \boxed{3y+2z=-2}$$

$$y+8z=36$$
$$3y+2z=-2$$

Multiply first equation by -3 and add to second equation to eliminate y

$$-3y-24z=-108$$
$$\underline{3y+2z=-2}$$
$$-22z=-110$$
$$\boxed{z=5}$$

Substitute $z=5$ into one of the 2-variable equations

$$y+8z=36$$
$$y+8(5)=36$$
$$y+40=36$$
$$\boxed{y=-4}$$

Substitute $z=5$ and $y=-4$ into one of the 3-variable equations

$$x+y-z=-7$$
$$x+(-4)-(5)=-7$$
$$x-4-5=-7$$
$$x-9=-7$$
$$\boxed{x=2}$$

3. **(B)**

$$x = \frac{\begin{vmatrix} 10 & 2 & -1 \\ 4 & 3 & 3 \\ 3 & 0 & 1 \end{vmatrix}}{\begin{vmatrix} 2 & 2 & -1 \\ 1 & 3 & 3 \\ 5 & 0 & 1 \end{vmatrix}} =$$

$$= \frac{[30+18+0]-[-9+8+0]}{[6+30+0]-[-15+2+0]} = \frac{49}{49} = 1$$

$$y = \frac{\begin{vmatrix} 2 & 10 & -1 \\ 1 & 4 & 3 \\ 5 & 3 & 1 \end{vmatrix}}{\begin{vmatrix} 2 & 2 & -1 \\ 1 & 3 & 3 \\ 5 & 0 & 1 \end{vmatrix}}$$

$$= \frac{[8+150-3]-[-20+10+18]}{49} = \frac{147}{49} = 3$$

$$z = \frac{\begin{vmatrix} 2 & 2 & 10 \\ 1 & 3 & 4 \\ 5 & 0 & 3 \end{vmatrix}}{\begin{vmatrix} 2 & 2 & -1 \\ 1 & 3 & 3 \\ 5 & 0 & 1 \end{vmatrix}}$$

$$= \frac{[18+40+0]-[150+6+0]}{49} = \frac{-98}{49} = -2$$

4. **(B)**

$$x = \frac{\begin{vmatrix} -6 & 1 & 2 \\ 11 & 0 & 1 \\ 16 & 2 & 0 \end{vmatrix}}{\begin{vmatrix} 0 & 1 & 2 \\ 2 & 0 & 1 \\ 1 & 2 & 0 \end{vmatrix}}$$

$$= \frac{[0+16+44]-[0+0-12]}{[0+1+8]-[0+0+0]} = \frac{72}{9} = 8$$

$$y = \frac{\begin{vmatrix} 0 & -6 & 2 \\ 2 & 11 & 1 \\ 1 & 16 & 0 \end{vmatrix}}{\begin{vmatrix} 0 & 1 & 2 \\ 2 & 0 & 1 \\ 1 & 2 & 0 \end{vmatrix}}$$

$$= \frac{[0-6+64]-[22+0+0]}{9} = \frac{36}{9} = 4$$

$$z = \frac{\begin{vmatrix} 0 & 1 & -6 \\ 2 & 0 & 11 \\ 1 & 2 & 16 \end{vmatrix}}{\begin{vmatrix} 0 & 1 & 2 \\ 2 & 0 & 1 \\ 1 & 2 & 0 \end{vmatrix}}$$

$$= \frac{[0+11-24]-[0+32+0]}{9} = \frac{-45}{9} = -5$$

Practice Problems 15.11

1. **(B)** $3x+1>7$
$$3x>6$$
$$x>2$$

hollow, excludes x = 2

2. **(A)** $1-3x \leqslant -5$
$$-3x \leqslant -6$$
$$x \geqslant 2$$

The inequality reverses when division is by -3.

solid, x = 2 is included

3. **(E)** Graph the boundary: $\quad x-y=0$
$$x=y$$

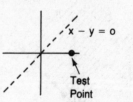

The line is dashed as $x-y<0$ does not include the $=$.

Test point $(6, 0)$:

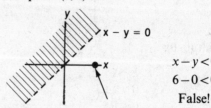

$$x-y=0$$
$$x-y<0$$
$$6-0<0$$
$$\text{False!}$$

Shade the side of the line *not* including the test point.

4. (B) Graph the boundary $2x + y = -6$

x	y
-3	0
-4	2
-2	-2

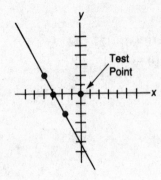

The line is solid as $2x + y \leq -6$ (includes the =)

Test point $(0, 0)$:

$$2x + y \leq -6$$
$$2(0) + 0 \leq -6$$
$$0 \leq -6 \quad \text{False!}$$

Shade the side of the line *not* including the test point.

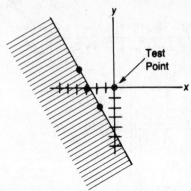

Practice Problems 15.12

1. (D)

$$y > x + 1$$
$$x \leq -1$$

Graph the boundary $y = x + 1$ as a dashed line:

x	y
0	1
1	2
2	3

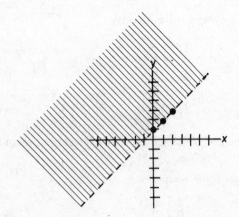

Test point $(0, 0)$:

$$y > x + 1$$
$$0 > 0 + 1$$
$$0 > 1$$

$$\text{False}$$

Shade the side of the boundary *not* including the test point.

Graph the boundary $x = -1$ as a solid line:

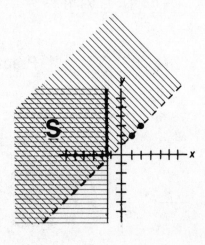

Test point $(0, 0)$:

$$x \leq -1$$
$$0 \leq -1$$
$$\text{False}$$

Shade the side of the boundary *not* including the test point.

Label the overlap shaded areas **S**.

2. **(B)** $y < x - 2$

$y \geqslant -2x + 4$

Graph the boundary $y = x - 2$ as a dashed line:

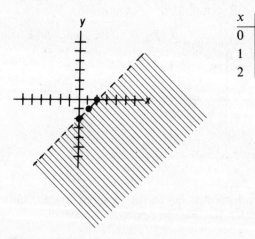

x	y
0	-2
1	-1
2	0

Graph the boundary $y = -2x + 4$ as a solid line.

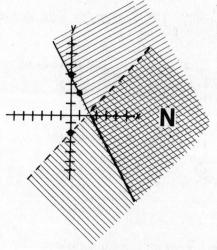

x	y
0	4
1	2
2	0

Test point (0, 0):

$$y < x - 2$$
$$0 < 0 - 2$$
$$0 < -2$$

False

Shade the side of the boundary *not* including the test point.

Test point (0, 0)

$$y \geqslant -2x + 4$$
$$0 \geqslant -2(0) + 4$$
$$0 \geqslant 4$$

False

Shade the side of the boundary *not* including the test point.

Label the overlap shaded area **N**.

Practice Problems 15.13

1. **(E)** To find the critical values set $f(x) = 0$:

$$x^2 - 5x + 6 = 0$$
$$(x - 2)(x - 3) = 0$$
$$x = 2 \mid x = 3$$

The critical values are excluded.
Test the interval $x < 2$, e.g., $x = 0$:
$x^2 - 5x + 6 < 0$ $0^2 - 5(0) + 6 < 0$ False

Do *not* darken the number line in this interval.

Test the interval $2 < x < 3$, e.g., $x = \frac{5}{2}$:
$(\frac{5}{2})^2 - 5(\frac{5}{2}) + 6 < 0$ True

Darken the number line in this interval.

Test the interval $x > 3$, e.g., $x = 10$:
$10^2 - 5(10) + 6 < 0$ False

Do *not* darken the number line in this interval.

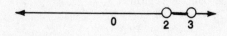

2. **(B)** To find the critical values set $f(x) = 0$:

$$(x + 3) \quad (x - 1) = 0$$
$$x = -3 \mid x = 1$$

The critical values are included.
Test the interval $x < -3$, e.g., $x = -10$:
$(x + 3)(x - 1) \leqslant 0$
$(-10 + 3)(-10 - 1) \leqslant 0$ False

Do *not* darken the number line in this interval.

Test the interval $-3 < x < 1$, e.g., $x = 0$:
$(0 + 3)(0 - 1) \leqslant 0$ True

Darken the number line in this interval.

Test the interval $x > 1$, e.g., $x = 10$:
$(10 + 3)(10 - 1) \leqslant 0$ False

Do *not* darken the number line in this interval.

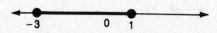

3. **(A)** First sketch $\dfrac{x^2}{9}+\dfrac{y^2}{16}=1$.

This is an ellipse. The boundary is solid as the original equation has $a \geqslant 1$.

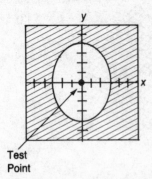

Test Point

Test point (0, 0):

$$\frac{x^2}{9}+\frac{y^2}{16} \geqslant 1$$

$$\frac{0^2}{9}+\frac{0^2}{16} \geqslant 1$$

$$0 \geqslant 1$$

False

The test point is *not* in the required space. Shade in the region outside the ellipse.

4. **(D)** First sketch $x^2+y^2=16$.

This is a circle. The boundary is dashed as the original equation does not have a $\geqslant$.

center (0, 0)   radius $=4$

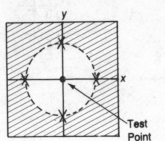

Test Point

Test point (0, 0):

$$x^2+y^2 > 16$$

$$0^2+0^2 > 16$$

$$0 > 16$$

False

The test point is *not* in the required space. Shade the region outside the circle.

Practice Problems 15.14

1. **(E)** Graph the boundary $x=1$. It is solid. It is a vertical line through $x=1$.

Test point (0, 0):

$$x \leqslant 1$$

$$0 \leqslant 1 \qquad \text{True}$$

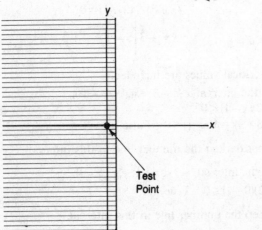

Test Point

The test point is in the included region.

Shade the side of the boundary including the test point.

Graph the boundary $y=x^2+3x$. It is dashed. It is a parabola with roots $x=0$, $x=-3$.

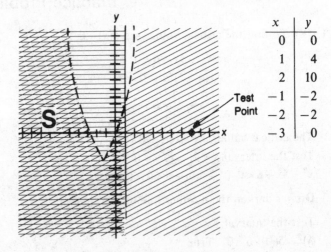

Test Point

x	y
0	0
1	4
2	10
-1	-2
-2	-2
-3	0

Test point (8, 0):

$$y < x^2+3x$$

$$0 < 8^2+3(8) \quad \text{True}$$

The test point is in the required space.

Shade the space outside the parabola.

Label the overlap region **S**.

2. **(E)** Graph the elliptical boundary $\dfrac{x^2}{9}+\dfrac{y^2}{4}=1$. It is dashed.

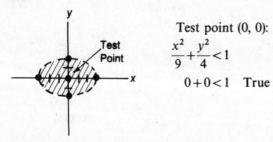

Test point (0, 0):

$$\frac{x^2}{9}+\frac{y^2}{4}<1$$

$$0+0<1 \quad \text{True}$$

The test point is in the required region.

Shade the inside of the ellipse

Graph the hyperbolic boundary $\dfrac{x^2}{9}-\dfrac{y^2}{4}=1$. It is solid.

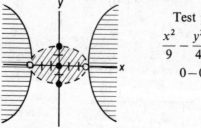

Test point (0, 0):

$$\frac{x^2}{9}-\frac{y^2}{4}\geq 1$$

$$0-0\geq 1 \quad \text{False}$$

The test point is not in the required region.

Shade as shown.

There is no solution. There is no area or point of overlap.

Practice Problems 16.1

1. **(E)**

$$(2y-5)(y-4)=2y^2-8y-5y+20$$
$$=2y^2-13y+20$$

2. **(C)**

$$6x^2-8x+6-(x^2-3x-6)$$
$$=6x^2-8x+6-x^2+3x+6$$
$$=\underbrace{6x^2-x^2}_{5x^2}-\underbrace{8x+3x}_{-5x}+\underbrace{6+6}_{+12}$$
$$5x^2-5x+12$$

3. **(B)**

$$\frac{9x^6+3x^2}{3x^2}=\frac{9x^6}{3x^2}+\frac{3x^2}{3x^2}=3x^4+1$$

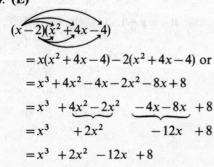 $\left(\dfrac{x^6}{x^2}\right)=x^{6-2}=x^4$

4. **(E)**

$$(x-2)(x^2+4x-4)$$
$$=x(x^2+4x-4)-2(x^2+4x-4) \text{ or}$$
$$=x^3+4x^2-4x-2x^2-8x+8$$
$$=x^3+\underbrace{4x^2-2x^2}_{+2x^2}\underbrace{-4x-8x}_{-12x}+8$$
$$=x^3+2x^2-12x+8$$

5. **(A)**

$$(1+a^3b^3)(1-a^3b^3)$$
$$=1(1-a^3b^3)+a^3b^3(1-a^3b^3) \text{ or}$$
$$=1-a^3b^3+a^3b^3-a^6b^6$$
$$=1-a^6b^6$$

6. **(B)**

$$x^3+3x^2-6x\underbrace{-3(2x+3)}\underbrace{+2x(-3x+1)}$$
$$=x^3+3x^2-6x\underbrace{-6x-9}\underbrace{-6x^2+2x}$$
$$=x^3\underbrace{+3x^2-6x^2}\underbrace{-6x-6x+2x}-9$$
$$=x^3\underbrace{-3x^2}\underbrace{-10x}-9$$
$$x^3-3x^2-10x-9$$

7. **(C)**

$$\frac{x^3+2x^2-4x-8}{x+2}$$

a	x^3	x^2	x	#
↓	↓	↓	↓	↓
$-2\rfloor$	1	$+2$	-4	-8
		-2	0	$+8$
	1	0	-4	$\lfloor 0$
	↑	↑	↑	
	x^2	x	#	

$$x^2+0x-4=x^2-4$$

8. (A)

$$\frac{x^2+9x-10}{x+10}$$

$$
\begin{array}{c|ccc}
 & x^2 & x & \# \\
 & \downarrow & \downarrow & \downarrow \\
-10 & 1 & +9 & -10 \\
 & & -10 & +10 \\
\hline
 & 1 & -1 & \boxed{0} \\
 & \uparrow & \uparrow & \\
 & x & \# &
\end{array}
$$

$$x-1$$

Practice Problems 16.2

1. (C)

$$f(x)=\frac{g(x)}{h(x)}=\frac{x^3+2x^2+4x-3}{x}$$

$h(x)=0$ for vertical asymptotes

$\therefore x=0$ is a vertical asymptote

2. (D)

$$f(x)=\frac{x^2}{x^2+4}$$

$$\lim_{x\to\infty}\frac{x^2}{x^2+4}=\lim_{x\to\infty}\frac{\dfrac{x^2}{x^2}}{\dfrac{x^2}{x^2}+\dfrac{4}{x^2}}=\lim_{x\to\infty}\frac{1}{1+\dfrac{4}{x^2}}=1$$

There is a horizontal asymptote at $y=1$.

3. (D)

$$f(x)=\frac{g(x)}{h(x)}=\frac{x^2}{x^2-4}$$

$h(x)=x^2-4=0\Rightarrow x=\pm2$

There are two vertical asymptotes.

$$x=2 \qquad x=-2$$

4. (C) Rearrange the equation to obtain $f(x)$:

$$x^2-yx+y+2=0$$

$$x^2+2-y(x-1)=0$$

$$y(x-1)=x^2+2$$

$$y=\frac{x^2+2}{x-1}$$

$$y=f(x)=\frac{x^2+2}{x-1}$$

Since degree $g(x)>$ degree $h(x)$ divide using synthetic division:

$$\frac{x^2+0x+2}{x-1}$$

$$
\begin{array}{c|ccc}
1 & 1 & 0 & 2 \\
 & & 1 & 1 \\
\hline
 & 1 & 1 & \boxed{3}
\end{array}
$$

$$\frac{x^2+2}{x-1}=x+1+\frac{3}{x-1}$$

Slant asymptote at $y=x+1$.

Practice Problems 16.3

1. (B) $P(x) = x^3 - 3x^2 + 3x - 1$

Factor out the root $+1$ using synthetic division:

Root
↓

original equation coefficient

$$1 \mid \quad 1 \quad -3 \quad +3 \quad -1$$
$$\quad\quad\quad\quad 1 \quad -2 \quad 1$$
$$\overline{\quad\quad 1 \quad -2 \quad 1 \quad \boxed{0}}$$

depressed equation coefficients   remainder

$$P(x) = x^3 - 3x^2 + 3x - 1$$
$$= (x-1)(x^2 - 2x - 1)$$
$$(x-1)(x-1)$$
$$= (x-1)(x-1)(x-1)$$

There is a multiplicity of 3 for the root $x = +1$.

2. (D) $P(x) = 2x^3 + 3x^2 - 11x - 6$

$$\frac{p}{q} = \frac{\pm 6, \pm 3, \pm 2, \pm 1}{\pm 2, \pm 1}$$

By Descartes' Rule: 1 positive root
$P(-x) = -2x^3 + 3x^2 + 11x - 6$:
Either 2 negative and 0 complex or 0 negative and 2 complex

Try to find the positive root:

$P(x) = 2x^3 + 3x^2 - 11x - 6$
$P(1) = 2 + 3 - 11 - 6 \neq 0$
$P(2) = 16 + 12 - 22 - 6 = 0 \checkmark$

$x = 2$ is a root, $x - 2$ is a factor

$$2 \mid \quad 2 \quad 3 \quad -11 \quad -6$$
$$\quad\quad\quad\quad 4 \quad 14 \quad 6$$
$$\overline{\quad 2 \quad 7 \quad 3 \quad \boxed{0}}$$

$$P(x) = 2x^2 + 3x^2 - 11x - 6$$
$$= (x-2)(2x^2 + 7x + 3)$$
$$(2x+1)(x+3)$$

$$= (x-2)(2x+1)(x+3) = 0$$
$$x = 2 \mid x = -\tfrac{1}{2} \mid x = -3$$

There is one fractional root.

3. (E)

$$\frac{P(x)}{x-a} = Q(x) + R$$

$R = P(a)$ by the remainder theorem
$R = P(1) = x^{98} + 3 = 1^{98} + 3 = 4$

4. (B)

$$\text{Sum} = \frac{-a_{n-1}}{a_n} = \frac{-(-2)}{4} = \frac{1}{2}$$

5. (C) $P(x) = x^3 - 3x^2 + 4$

$$\frac{p}{q} = \frac{\pm 4, \pm 2, \pm 1}{\pm 1}$$

$P(x)$ has 2 sign changes
$P(-x) = -x^3 + 3x^2 + 4$ has one sign change

Possibilities

+	2	0
−	1	1
Complex	0	2

Try for the negative root:

$$1 \mid \quad 1 \quad -3 \quad 0 \quad 4$$
$$\quad\quad\quad\quad -1 \quad 4 \quad -4$$
$$\overline{\quad 1 \quad -4 \quad 4 \quad \boxed{0}}$$

$$P(x) = x^3 - 3x^2 + 4$$
$$= (x+1)(x^2 - 4x + 4)$$
$$(x-2)(x-2)$$
$$= (x+1)\ (x-2)\ (x-2)$$
$$x = -1 \mid x = 2 \mid x = 2$$

Note: The root $x = 2$ has multiplity 2
$x = -1$ is a root, $x + 1$ is a factor

6. (C) $P(x) = x^4 - 16$

By Descartes' Rule $P(x)$ has 1 sign change.
∴ 1 positive root.
$P(-x) = x^4 - 16$ has 1 sign change.
∴ 1 negative root.

Since there is only 1 positive real root and only 1 negative real root the other two roots must be complex.

7. **(D)** $P(x) = 4x^3 + 6x^2 + 8x + 9$

$$\frac{p}{q} = \frac{\pm 9, \pm 3, \pm 1}{\pm 4, \pm 2, \pm 1}$$

None of the $\frac{p}{q}$ values $= 2$.

8. **(B)** Since the coefficients are real, if $3 + 2i$ is a root, $3 - 2i$ is a root; if $1 - 3i$ is a root, $1 + 3i$ is a root. There is a minimum of four roots.

Hence a 4th degree equation is the minimum possible.

Practice Problems 16.4

1. Let $x = 0$, $y = \dfrac{0}{-16} = 0$

$(0, 0)$ is the y-intercept.

Let $y = 0$, $0 = \dfrac{x^2}{x^2 - 16}$

$x^2 = 0 \quad x = 0$; the only root is at the origin.

Asymptotes

$$P(x) = \frac{g(x)}{h(x)} = \frac{x^2}{x^2 - 16}$$

$h(x) = x^2 - 16 = 0 \quad \begin{aligned} x &= +4 \\ x &= -4 \end{aligned}$

Two vertical asymptotes

$$\lim_{x \to \infty} \frac{x^2}{x^2 - 16} = \lim_{x \to \infty} \frac{\dfrac{x^2}{x^2}}{\dfrac{x^2}{x^2} - \dfrac{16}{x^2}}$$

$$= \lim_{x \to \infty} = \frac{1}{1 - 0} = 1$$

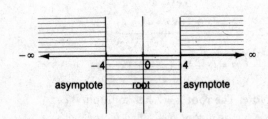

A horizontal asymptote at $y = 1$.

Test the intervals:

$-\infty$ to $-4 \quad P(x) = +$
-4 to $\ \ 0 \quad P(x) = -$
$\ \ 0$ to $\ \ 4 \quad P(x) = -$
$\ \ 4$ to $\infty \quad P(x) = +$

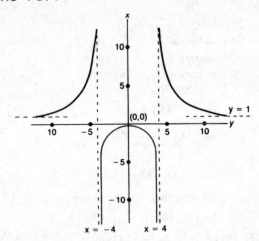

These results are reflected by the shaded areas on the number line above.

2. Let $x = 0$.

$$y = 0^3 - 4(0)^2 - 3(0) + 18 = 18$$

The y intercept is $(0, 18)$.

Let $y = 0$.

$$0 = x^3 - 4x^2 - 3x + 18$$

$$\frac{p}{q} = \frac{\pm 18, \pm 9, \pm 6, \pm 3, \pm 2, \pm 1}{\pm 1}$$

By Descarte's Law of Signs: Either 2 positive or 0 positive roots

$P(-x) = -x^3 - 4x^2 + 3x + 18$: 1 sign change

$\therefore$ 1 negative root

Possibilities

+	2	0
−	1	1
Complex	0	2

Look for the negative root:

$$P(x) = x^3 - 4x^2 - 3x + 18$$
$$P(-1) = -1 - 4 + 3 + 18 \neq 0$$
$$P(-2) = -8 - 16 + 6 + 18 = 0 \checkmark$$

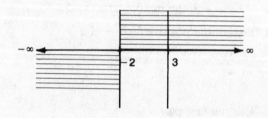

$$P(x) = (x+2)(x^2 - 6x + 9)$$
$$= (x+2)(x-3)(x-3)$$

A single root at $x = -2$.
The graph passes through the x axis at $x = -2$ as a straight line.

A double root at $x = 3$.
The graph is tangent to the axis.

Test the intervals:

$$-\infty \text{ to } -2 \quad f(x) = -$$
$$-2 \text{ to } 3 \quad f(x) = +$$
$$3 \text{ to } \infty \quad f(x) = +$$

The results are reflected by the shaded regions on the number line above.

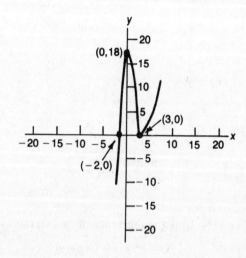

Cumulative Review 1-16

1. **(A)** $5^{3x-1} = 25$
   $$5^{3x-1} = 5^2$$
   $$3x - 1 = 2$$
   $$x = 1$$

2. **(C)** Both the x and y variables are squared. They both have coefficients of the same sign, but different in value. The relation is an ellipse.

3. **(B)** $6x^2 - 7x - 3$
   $a = 6$, $c = -3$, middle term $= 7x$
   1. List all the positive factors of a
      3, 2
      6, 1
   2. Use the first pair
      $(3x \quad)(2x \quad)$
   3. Pairs of factors of c
      $-3, \quad 1$
      $3, \quad -1$

4. Insert and test (be sure to use each pair reversed).

   $$(3x - 3)(2x + 1) \quad -6x + 3x = -3x \neq -7x$$

   $$(3x + 1)(2x - 3) \quad 2x + (-9x) = -7x = -7x \checkmark$$

   The factors are: $(3x + 1)(2x - 3)$

4. **(B)**

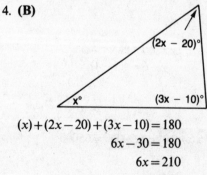

$$(x) + (2x - 20) + (3x - 10) = 180$$
$$6x - 30 = 180$$
$$6x = 210$$
$$x = 35$$
$$2x - 20 = 50$$
$$3x - 10 = 95$$

The triangle is obtuse.

5. **(D)**

$$4x - 3y = 8$$
$$(2x + y = -1) - 2 \Rightarrow -4x - 2y = 2$$

$$\begin{array}{r} 4x - 3y = 8 \\ -4x - 2y = 2 \\ \hline -5y = 10 \\ y = -2 \end{array}$$

$$\therefore 4x - 3y = 8$$
$$4x - 3(-2) = 8$$
$$4x = 2$$
$$x = \tfrac{1}{2}$$

6. **(C)**

$$\text{Sum} = -\frac{b}{a} = 7$$

7. **(A)** $y^{2n} - 16$ is the difference of two squares.

$$x^2 - y^2 = (x + y)(x - y)$$
$$y^{2n} - 16 = (y^n + 4)(y^n - 4)$$

8. **(D)**

$$y = x^2 + 3x + k$$
$$0 = (2)^2 + 3(2) + k$$
$$0 = 4 + 6 + k$$
$$k = -10$$

9. **(A)**

$$m = \frac{\Delta y}{\Delta x} = \frac{6 - (-2)}{1 - (-1)} = \frac{8}{2} = 4$$

Using point (1, 6)

$$(y - y_1) = m(x - x_1)$$
$$y - 6 = 4(x - 1)$$

10. **(D)**

$$6.5 \times 10^4 = 6.50000 = 65000$$

11. **(B)**

$$\sqrt{-81} = 9i \quad \sqrt{-25} = 5i \quad \sqrt{-49} = 7i$$
$$\sqrt{-81} + \sqrt{-25} + 3\sqrt{-49} = 9i + 5i + 3(7i) = 35i$$

12. **(C)**

$$2 + 3\sqrt{x} = 8$$
$$3\sqrt{x} = 6$$
$$\sqrt{x} = 2$$
$$(\sqrt{x})^2 = (2)^2$$
$$x = 4$$

Check

$$2 + 3\sqrt{x} = 8$$
$$2 + 3\sqrt{4} \overset{?}{=} 8$$
$$2 + 3(2) \overset{?}{=} 8$$
$$8 = 8 \checkmark$$

13. **(A)** $12x^2 - 5x - 3$

$a = 12$, $c = -3$, the middle term $= -5x$

1. List all the positive factors of a

   12, 1
   3, 4
   6, 2

2. Use the first pair

   $(12x \quad)(x \quad)$

3. List the factors of c

   $c = -3$, middle term $= -5x$

   $-3, 1$
   $3, -1$

4. Insert and try

4. $\underset{12x}{(12x \overset{-3x}{-} 3)(x + 1)}$ $-3x + 12x = 9x \neq -5x$

   $\underset{-36x}{(12x \overset{x}{+} 1)(x - 3)}$ $x + (-36x) = -35x \neq -5x$

   $\underset{-12x}{(12x \overset{3x}{+} 3)(x - 1)}$ $3x + (-12x) = -9x \neq -5x$

   $\underset{36x}{(12x \overset{-x}{-} 1)(x + 3)}$ $36x + (-x) = 35x \neq -5x$

   $\underset{+3x}{(3x \overset{-12x}{-} 3)(4x + 1)}$ $-12x + 3x = -9x \neq -5x$

   $\underset{-9x}{(3x \overset{4x}{+} 1)(4x - 3)}$ $4x + (-9x) = -9x = -5x \checkmark$

The factors are:

   $(3x + 1)(4x - 3)$

14. **(D)**

$$y = \frac{x-3}{x+2}$$

$$x = \frac{y-3}{y+2} \Rightarrow xy + 2x = y - 3$$

$$xy - y = -2x - 3$$

$$y(x-1) = -2x - 3$$

$$y = \frac{-2x-3}{x-1}$$

$$y = \frac{2x+3}{1-x}$$

15. **(A)**

There are no points that share the individual loci. There are 0 points which satisfy the given conditions.

16. **(B)**

$$S = \frac{rl-a}{r-1}$$

$$Sr - S = rl - a$$

$$Sr - rl = S - a$$

$$r(S-l) = S - a$$

$$r = \frac{S-a}{S-l}$$

17. **(D)**

$$x = \frac{-b \pm \sqrt{b^2-4ac}}{2a}$$

$$x = \frac{-2 \pm \sqrt{(2)^2-4(3)(-6)}}{2(3)}$$

$$x = \frac{-2 \pm \sqrt{4+72}}{6}$$

$$x = \frac{-2 \pm \sqrt{76}}{6} = \frac{-2 \pm 2\sqrt{19}}{6}$$

$$= \frac{-1 \pm \sqrt{19}}{3}$$

18. **(C)** $f(x) = x^2 + bx + c$

Product of the roots $= \dfrac{c}{a} = \dfrac{c}{1} = c$

$$= (1+i)(1-i) = 1 - i^2 = 2$$

$$c = 2$$

19. **(E)**

$$P(x) = 2x^3 - 3x^2 - 23x + 12$$

$$\frac{p}{q} = \frac{\pm 12,\ \pm 6,\ \pm 4,\ \pm 3,\ \pm 2,\ \pm 1}{\pm 2,\ \pm 1}$$

$$P(1) = 2 - 3 - 23 + 12 \neq 0$$

$$P(2) = 16 - 12 - 46 + 12 \neq 0$$

$$P(3) = 54 - 27 - 69 + 12 \neq 0$$

$$P(4) = 128 - 48 - 92 + 12 = 0$$

$$
\begin{array}{r|rrrr}
4 & 2 & -3 & -23 & 12 \\
  &   & 8 & 20 & -12 \\
\hline
  & 2 & 5 & -3 & \boxed{0}
\end{array}
$$

$$P(x) = 2x^3 - 3x^2 - 23x + 12$$

$$(x-4)\underbrace{(2x^2 + 5x - 3)}_{(2x-1)(x+3)}$$

$$= (x-4)(2x-1)(x+3) = 0$$

$$x = 4 \mid x = \tfrac{1}{2} \mid x = -3$$

$$\{4,\ \tfrac{1}{2},\ -3\}$$

20. **(A)**

$$3 + |x-1| = 2x$$

$$|x-1| = 2x - 3$$

$-(x-1) = 2x - 3$	$x - 1 = 2x - 3$
$-x + 1 = 2x - 3$	$x = 2$
$3x = 4$	
$x = \tfrac{4}{3}$	
check	*check*
$3 + |x-1| = 2x$	$3 + |x-1| = 2x$
$3 + |\tfrac{4}{3} - 1| = 2(\tfrac{4}{3})$	$3 + |2-1| = 2(2)$
$3 + |\tfrac{1}{3}| = \tfrac{8}{3}$	$3 + |1| = 2(2)$
$\tfrac{10}{3} \neq \tfrac{8}{3}$	$3 + 1 = 2(2)$
	$4 = 4 \ \checkmark$

Unit Five

17. Trigonometry

17.1 Basics and Beyond

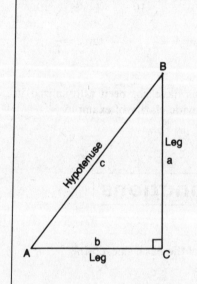

$\sin \theta = \dfrac{\text{leg Opposite angle } \theta}{\text{Hypotenuse}} = \dfrac{\text{Opp}}{\text{Hyp}}$	Sine is abbreviated sin.
$\cos \theta = \dfrac{\text{leg Adjacent to angle } \theta}{\text{Hypotenuse}} = \dfrac{\text{Adj}}{\text{Hyp}}$	Cosine is abbreviated cos.
$\tan \theta = \dfrac{\text{leg Opposite angle } \theta}{\text{leg Adjacent to angle } \theta} = \dfrac{\text{Opp}}{\text{Adj}}$	Tangent is abbreviated tan.
A mnemonic to remember these rules is: **SOH CAH TOA**	

The **opposite** leg is always directly across (opposite) from the angle.
The **adjacent** leg is next to (adjacent) to the angle. It is *never* the hypotenuse.

Examples:

1. Find $\sin \phi$, $\cos \phi$ and $\tan \phi$.

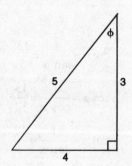

Solutions:

1. $\sin \phi = \dfrac{\text{leg opposite } \angle \phi}{\text{hypotenuse}} = \dfrac{4}{5}$

   $\cos \phi = \dfrac{\text{leg adjacent to } \angle \phi}{\text{hypotenuse}} = \dfrac{3}{5}$

   $\tan \phi = \dfrac{\text{leg opposite } \angle \phi}{\text{leg adjacent to } \angle \phi} = \dfrac{4}{3}$

2. Find: sin *N*, cos *N*, tan *N*, sin *J*, cos *J* and tan *J*.

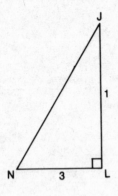

2. First find the hypotenuse using the pythagorean theorem:

$$(\text{leg})^2 + (\text{leg})^2 = (\text{hypotenuse})^2$$
$$1^2 + 3^2 = x^2$$
$$x^2 = 10$$
$$x = \sqrt{10}$$

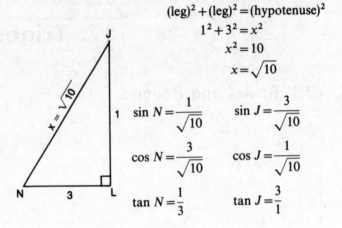

$$\sin N = \frac{1}{\sqrt{10}} \qquad \sin J = \frac{3}{\sqrt{10}}$$

$$\cos N = \frac{3}{\sqrt{10}} \qquad \cos J = \frac{1}{\sqrt{10}}$$

$$\tan N = \frac{1}{3} \qquad \tan J = \frac{3}{1}$$

☞ The fractions have *not* been rationalized in order to provide clarity of example.

Reciprocal Trigonometric Functions

The **inverse or reciprocal trigonometric functions** are the reciprocals of the basic sine, cosine and tangent functions.

$\cot \theta = \dfrac{1}{\tan \theta} = \dfrac{\text{leg adjacent}}{\text{leg opposite}}$	Cotangent is abbreviated cot.
$\sec \theta = \dfrac{1}{\cos \theta} = \dfrac{\text{hypotenuse}}{\text{leg adjacent}}$	Secant is abbreviated sec.
$\csc \theta = \dfrac{1}{\sin \theta} = \dfrac{\text{hypotenuse}}{\text{leg opposite}}$	Cosecant is abbreviated csc.

Examples:

3. Find sin α, cos α, tan α, cot α, sec α, and csc α.

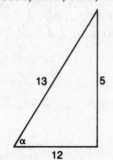

Solutions:

3. $\sin \alpha = \dfrac{\text{opp}}{\text{hyp}} = \dfrac{5}{13} \qquad \csc \alpha = \dfrac{1}{\sin \alpha} = \dfrac{13}{5}$

$\cos \alpha = \dfrac{\text{adj}}{\text{hyp}} = \dfrac{12}{13} \qquad \sec \alpha = \dfrac{1}{\cos \alpha} = \dfrac{13}{12}$

$\tan \alpha = \dfrac{\text{opp}}{\text{adj}} = \dfrac{5}{12} \qquad \cot \alpha = \dfrac{1}{\tan \alpha} = \dfrac{12}{5}$

4. a is an acute angle such that $\sin a = \frac{8}{17}$. Find the value of the other five trigonometric functions.

4. First draw a right triangle such that
$$\sin a = \frac{\text{opp}}{\text{hyp}} = \frac{8}{17}.$$

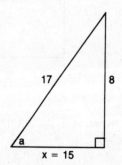

Then find the third side of the triangle.
$$8^2 + x^2 = 17^2$$
$$x^2 = 225$$
$$x = 15$$

$$\sin a = \frac{\text{opp}}{\text{hyp}} = \frac{8}{17}$$

$$\csc a = \frac{1}{\sin a} = \frac{\text{hyp}}{\text{opp}} = \frac{17}{8}$$

$$\cos a = \frac{\text{adj}}{\text{hyp}} = \frac{15}{17}$$

$$\sec a = \frac{1}{\cos a} = \frac{\text{hyp}}{\text{adj}} = \frac{17}{15}$$

$$\tan a = \frac{\text{opp}}{\text{adj}} = \frac{8}{15}$$

$$\cot a = \frac{1}{\tan a} = \frac{\text{adj}}{\text{opp}} = \frac{15}{8}$$

Complementary Functions (Cofunctions)

Complementary trigonometric functions are those trigonometric functions which are equal when their angles are complementary (add to 90°)

Cosine is the complementary function of sine.   sin and cos are cofunctions.
Cotangent is the complementary function of tangent.   tan and cot are cofunctions.
Cosecant is the complementary function of secant.   sec and csc are cofunctions.

$$\sin \theta = \cos(90 - \theta) \quad \Rightarrow \quad \cos \theta = \sin(90 - \theta)$$
$$\tan \theta = \cot(90 - \theta) \quad \Rightarrow \quad \cot \theta = \tan(90 - \theta)$$
$$\sec \theta = \csc(90 - \theta) \quad \Rightarrow \quad \csc \theta = \sec(90 - \theta)$$

Example: cofunctions
$$\sin 60° = \cos 30°$$
angles add to 90°

Examples:

5. Express csc 3° in terms of its cofunction.

6. If $\sin(3x - 26)° = \cos(5x - 60)°$, find x.

7. If $\sin(3x - 26)° = \dfrac{1}{\csc(5x - 60)°}$, find x.

Solutions:

5.   cofunction
$$\csc 3° = \sec 87°$$
adds to 90

6. If $\sin x = \cos y$, then $x + y = 90°$.

If $\sin a = \cos b$, this is a **cofunction** relationship.

$$(3x - 26) + (5x - 60) = 90$$
$$8x - 86 = 90$$
$$8x = 176$$
$$x = 22$$

7. If $\sin x = \dfrac{1}{\csc y}$, this is an **inverse** relationship, not a cofunction relationship.

If $\sin x = \dfrac{1}{\csc y}$, then $x = y$.

$$(3x - 26) = (5x - 60)$$
$$34 = 2x$$
$$x = 17$$

$$\tan \theta = \frac{\sin \theta}{\cos \theta} \qquad \cot \theta = \frac{\cos \theta}{\sin \theta}$$

Pythagorean Identities

$$\sin^2 \theta + \cos^2 \theta = 1 \qquad \tan^2 \theta + 1 = \sec^2 \theta \qquad \cot^2 \theta + 1 = \csc^2 \theta$$

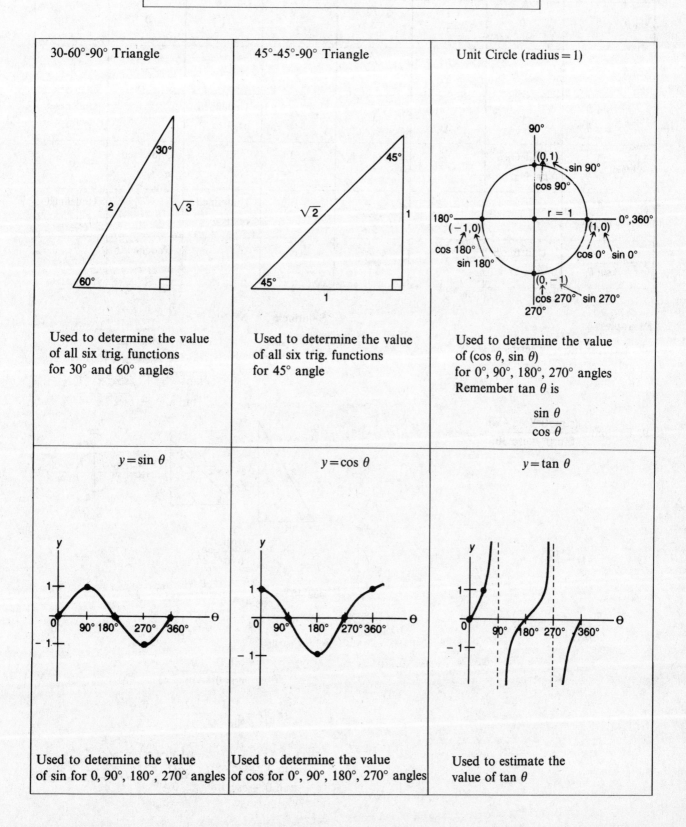

30-60°-90° Triangle

Used to determine the value
of all six trig. functions
for 30° and 60° angles

45°-45°-90° Triangle

Used to determine the value
of all six trig. functions
for 45° angle

Unit Circle (radius = 1)

Used to determine the value
of $(\cos \theta, \sin \theta)$
for 0°, 90°, 180°, 270° angles
Remember $\tan \theta$ is

$$\frac{\sin \theta}{\cos \theta}$$

$y = \sin \theta$

Used to determine the value
of sin for 0, 90°, 180°, 270° angles

$y = \cos \theta$

Used to determine the value
of cos for 0°, 90°, 180°, 270° angles

$y = \tan \theta$

Used to estimate the
value of $\tan \theta$

$\theta =$	$\theta = 0°$	30°	45°	60°	90°	180°	270°
Trigonometric Function	From: Unit circle	30°-60°-90°	45°-45°-90°	30°-60°-90°	Unit circle	Unit circle	Unit circle
$\sin \theta$	0	$\dfrac{1}{2}$	$\dfrac{\sqrt{2}}{2}$	$\dfrac{\sqrt{3}}{2}$	1	0	-1
$\cos \theta$	1	$\dfrac{\sqrt{3}}{2}$	$\dfrac{\sqrt{2}}{2}$	$\dfrac{1}{2}$	0	-1	0
$\tan \theta = \dfrac{\sin \theta}{\cos \theta}$	0	$\dfrac{\sqrt{3}}{3}$	1	$\sqrt{3}$	Undefined or $\pm\infty$	0	Undefined or $\pm\infty$
$\cot \theta = \dfrac{1}{\tan \theta}$	Undefined or $\pm\infty$	$\sqrt{3}$	1	$\dfrac{\sqrt{3}}{3}$	0	Undefined or $\pm\infty$	0
$\sec \theta = \dfrac{1}{\cos \theta}$	1	$\dfrac{2\sqrt{3}}{3}$	$\sqrt{2}$	2	Undefined or $\pm\infty$	-1	Undefined or $\pm\infty$
$\csc \theta = \dfrac{1}{\sin \theta}$	Undefined or $\pm\infty$	2	$\sqrt{2}$	$\dfrac{2\sqrt{3}}{3}$	1	Undefined or $\pm\infty$	-1

Examples:

8. Express $\tan x$ in terms of $\sec x$.

9. Evaluate: $\dfrac{(\sin 30°)(\sec 45°)}{\tan 0° + \csc 30°}$

Solutions:

8.
$$\tan^2 x + 1 = \sec^2 x$$
$$\tan^2 x = \sec^2 x - 1$$
$$\sqrt{\tan^2 x} = \pm\sqrt{\sec^2 x - 1}$$
$$\tan x = \pm\sqrt{\sec^2 x - 1}$$

9.

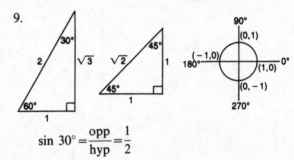

$$\sin 30° = \frac{\text{opp}}{\text{hyp}} = \frac{1}{2}$$

$$\sec 45° = \frac{1}{\cos 45°} = \frac{1}{\dfrac{1}{\sqrt{2}}} = \sqrt{2}$$

$$\tan 0° = \frac{\sin 0°}{\cos 0°} = \frac{0}{1} = 0$$

$$\csc 30° = \frac{1}{\sin 30°} = \frac{2}{1}$$

$$\frac{(\sin 30°)(\sec 45°)}{\tan 0° + \csc 30°} = \frac{\left(\dfrac{1}{2}\right)\left(\dfrac{\sqrt{2}}{1}\right)}{0 + 2} = \frac{\dfrac{\sqrt{2}}{2}}{2} = \frac{\sqrt{2}}{4}$$

10. Show: $\dfrac{\cos\theta+\cot\theta}{\cos\theta\cot\theta}=\tan\theta+\sec\theta$

10. Express all the trigonometric expressions in terms of $\sin\theta$ and $\cos\theta$ only.

$$\dfrac{\cos\theta+\cot\theta}{\cos\theta\cot\theta}\overset{?}{=}\tan\theta+\sec\theta$$

$$\dfrac{\left(\dfrac{\sin\theta}{\sin\theta}\right)\dfrac{\cos\theta}{1}+\dfrac{\cos\theta}{\sin\theta}}{\cos\theta\cdot\dfrac{\cos\theta}{\sin\theta}}\ \Bigg|\ \dfrac{\sin\theta}{\cos\theta}+\dfrac{1}{\cos\theta}$$

$$\dfrac{\dfrac{\sin\theta\cos\theta+\cos\theta}{\sin\theta}}{\dfrac{\cos^2\theta}{\sin\theta}}\ \Bigg|\ =\dfrac{\sin\theta+1}{\cos\theta}$$

$$\dfrac{(\sin\theta\cos\theta+\cos\theta)}{\sin\theta}\cdot\dfrac{\sin\theta}{\cos^2\theta}$$

$$\dfrac{\cos\theta(\sin\theta+1)}{\cos^2\theta}$$

$$\dfrac{\sin\theta+1}{\cos\theta}\ \Bigg|\ \dfrac{\sin\theta+1}{\cos\theta}$$

Q.E.D.

Quadrants

Quadrants are the divisions of the coordinate plane into four equal sections, usually denoted by roman numerals. Each quadrant is 90° in measure.

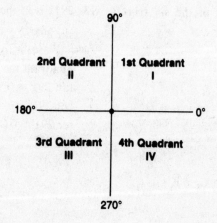

The six trigonometric functions have different signs (+ or −) in the different quadrants. The diagram below indicates which functions are positive (+) in which of the quadrants.

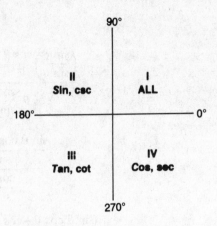

Many students remember the sequence using the sentence:
   ALL Students Take Calculus indicating in sequence I, II, III, IV... **All** sin tan cos.
The inverse functions follow the sign of their reciprocal trigonometric functions.

Examples:

11. If sin θ > 0 and tan θ < 0, in which quadrant is θ?

12. If csc θ is negative, is sin θ positive or negative?

13. If θ = 291°. What are the signs of the six trigonometric functions?

Solutions:

11. sin is positive in I and II , tan is negative in II and IV.

Sin (csc)	A
Tan (cot)	Cos (sec)

θ is in the second quadrant.

12. csc θ and sin θ are reciprocal functions.

$$\left(\sin\theta = \frac{1}{\csc\theta}\right) \quad \therefore \sin\theta < 0$$

13. 291° is in the fourth quadrant.

sin 291° < 0

cos 291° > 0 (cos is positive in the fourth quadrant.)

tan 291° < 0

cot 291° < 0

sec 291° > 0 (sec is positive in the fourth quadrant.)

csc 291° < 0

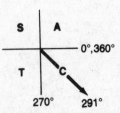

Evaluating Trigonometric Functions

There are **four categories** of angles to be analyzed:

- Those angles **less than 0°**, e.g., −30°, −240°
- Those angles **between 0° and 90°**, e.g., 45°, 38°
- Those angles **between 90° and 360°**, e.g., 225°, 300°
- Those angles **greater than 360°**, e.g., 420°, 2000°

I. **Angles between 0° and 90°.** These are first quadrant angles. For 30°, 45° and 60° use the 30°-60°-90° and 45°-45°-90° triangles. For other angles such as 28°, 86° use the trigonometric tables.

Examples:

14. Evaluate cos 60°.

15. Evaluate tan 17°.

Solutions:

14. $\cos 60° = \dfrac{\text{adj}}{\text{hyp}} = \dfrac{1}{2}$

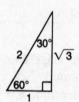

15. Refer to trigonometric tables:

Angle	sin	cos	**tan**
13°	.2250	.9744	.2309
14°	.2419	.9703	.2493
15°	.2588	.9659	.2679
16°	.2756	.9613	.2867
17°	.2924	.9563	**.3057**
18°	.3090	.9511	.3249
19°	.3256	.9455	.3443

tan 17° = .3057

☞ To change from angles between 0° and 45° to angles between 45° and 90°, and vice versa:

Use cofunctions.

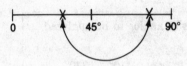

Examples:

16. Express csc 10° as a trigonometric function with an angle between 45° and 90°.

17. Express sin 79° as a trigonometric function of a positive acute angle less than 45°.

Solutions:

16. Using cofunctions: csc 10° = sec 80°.
    Ans: sec 80°.

17. Using cofunctions: sin 79° = cos 11°
    Ans: cos 11°.

II. **Angles between 90° and 360°.** These are second, third and fourth quadrant angles. The procedure is to convert these trigonometric functions into their equivalent in the first quadrant and utilize the procedures for 0° to 90° angles.

Procedure:

1. Draw a picture of the angle. Label which quadrants have positive trig. values. (Use: **A**LL **S**tudents **T**ake **C**alculus.)

2. The intent is to convert the given trig. function and angle into three (3) items.

+ or −	Trig. function	Reference angle
↑	↑	↑
The first item is + or − This is determined by comparing the trigonometric function involved with the quadrant rules associated with the angle location.	The second item is a repeat of the given trig. function.	The third item is the reference angle. It is the angle from the given angle to the nearest horizontal axis. It is always a positive value. Label the angle in the diagram.

3. Evaluate the first quadrant angle.

Evaluate sin 240°

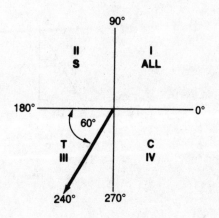

sin 240°

⇓

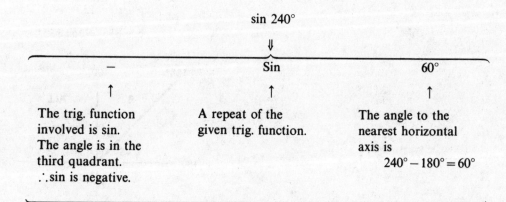

−	Sin	60°
↑	↑	↑
The trig. function involved is sin. The angle is in the third quadrant. ∴ sin is negative.	A repeat of the given trig. function.	The angle to the nearest horizontal axis is 240° − 180° = 60°

⇓

sin 240° = − sin 60°

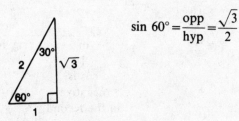

$$\sin 60° = \frac{\text{opp}}{\text{hyp}} = \frac{\sqrt{3}}{2}$$

$$\therefore \sin 240° = -\sin 60° = -\left(\frac{\sqrt{3}}{2}\right) = \frac{-\sqrt{3}}{2}$$

- − sin 60°: the angle is 60°. It is positive and acute. The " − " relates to the value not the angle.
- sin (− 60°): the angle is − 60°. It is negative and acute.

III. **Angles greater than 360°.** These are angles measuring more than one complete revolution. Subtract multiples of 360° until the angle is in the first, second, third or fourth quadrant. Then follow the rules for that quadrant.

Examples:

18. Evaluate cos 390°.

19. Evaluate cos 855°.

Solutions:

18.

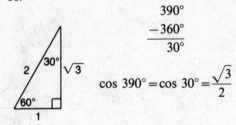

$$\begin{array}{r} 390° \\ -360° \\ \hline 30° \end{array}$$

$$\cos 390° = \cos 30° = \frac{\sqrt{3}}{2}$$

19.

$$\begin{array}{r} 855° \\ -360° \\ \hline 495° \\ -360° \\ \hline 135° \end{array}$$

$$\cos 855° = \cos 135°$$

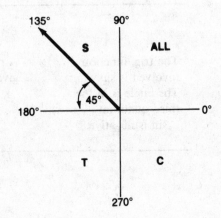

$-$	cos	45°
$+$ or $-$	Trig. funct.	Ref. angle
↑		↑
cos is negative in the second quadrant		$180° - 135° = 45°$ The angle to the nearest horizontal axis

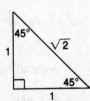

$$\cos 855° = \cos 135° = -\cos 45°$$

$$= -\left(\frac{\sqrt{2}}{2}\right) = -\frac{\sqrt{2}}{2}$$

IV. **Angles less than 0°.** These are negative angles. Add multiples of 360° until the angle is in the first, second, third or fourth equadrant. Then follow the rules for that quadrant.

Examples:

20. Evaluate sin (−330°).

21. Evaluate tan (−120°).

Solutions:

20.

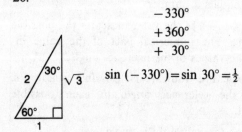

$$\begin{array}{r} -330° \\ +360° \\ \hline + \ 30° \end{array}$$

sin (−330°) = sin 30° = ½

21.

$$\begin{array}{r} -120° \\ +360° \\ \hline 240° \end{array}$$

tan (−120°) = tan 240°

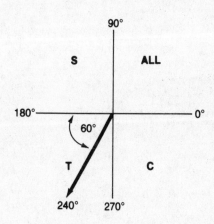

+	tan	60°
+ or −	Trig. function	Ref. angle

↑
tangent is
positive in the
third
quadrant

↑
240° − 180° = 60°
(the angle to the
nearest
horizontal axis)

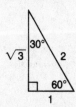

Solving for an Angle

Linear

Procedure:

1. Rearrange the equation so that the trigonometric function equals a number.

2. The sign $(+/-)$ of the value indicates the possible quadrants. The numerical part of the value indicates the values of the reference angle.

3. Draw a picture: indicate all possible quadrants. Include the reference angle in each possible quadrant.

4. Determine the value of the angle.

Quadratic

1. Solve the quadratic for the trigonometric function. Use factoring or the quadratic formula as required.

2. Determine the angles for each linear term.

Example: Solve $2 \sin \theta + 1 = 0$ for all values of θ, $0 \le \theta < 360$.

$$2 \sin \theta = -1$$
$$\sin \theta = -\tfrac{1}{2}$$

sin is "$-$" in the third and fourth quadrants.

Ref. $\angle = 30°$ $(\sin 30° = \tfrac{1}{2})$

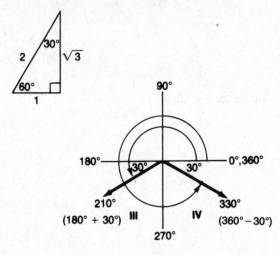

$$\theta = 210°, 330°$$

Solve $\tan \theta - 2\tan \theta \sin \theta = 0$ for all values of θ, $0 \le \theta < 360$.

$$\tan \theta (1 - 2 \sin \theta) = 0$$

$\tan \theta = 0$	$1 - 2 \sin \theta = 0$
$\tan \theta = \dfrac{\sin \theta}{\cos \theta}$	$\sin \theta = +\dfrac{1}{2}$
$\tan \theta$ is 0 when $\sin \theta$ is 0.	sin is "$+$" in the first and second quadrants. Ref. $\angle = 30°$

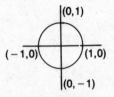

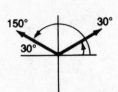

$\theta = 0°, 180°$ $\theta = 30°, 150°$

$$\theta = 0°, 30°, 150°, 180°$$

θ only has values if:

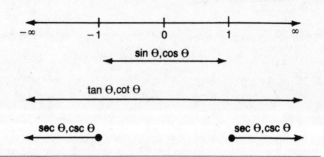

Example: Solve csc $\theta = \frac{1}{2}$, for θ. There are no solutions, csc θ is never $\frac{1}{2}$.
Solve sin $\theta = 2$, for θ. There are no solutions, sin θ is never 2.

Examples:

22. Solve $4 \csc^2 \theta - 1 = 0$ for all values $0 \leqslant \theta < 360$.

Solutions:

22.

$$4 \csc^2 \theta - 1 = 0$$
$$\sqrt{\csc^2 \theta} = \pm \sqrt{\frac{1}{4}}$$
$$\csc \theta = \pm \frac{1}{2}$$

csc $\theta = \frac{1}{2}$	csc $\theta = -\frac{1}{2}$
No solutions	No solutions
csc θ is never $\frac{1}{2}$.	csc θ is never $-\frac{1}{2}$.

Answer: { } or $\varnothing$ (null set).

Radians

Radians are a measure of angle size. The relationship between degrees and radians is:

$$2\pi \text{ radians} = 360°$$

	From	**To**	**Multiply By**
To Convert:	Degree measure	Radian measure	$\dfrac{\pi}{180}$
To Convert:	Radian measure	Degree measure	$\dfrac{180}{\pi}$

Examples:

23. Convert 210° to radians.

Solutions:

23.

$$210 \cdot \frac{\pi}{180} = \frac{210\pi}{180} = \frac{7\pi}{6} \text{ radians}$$

24. Convert $\frac{3\pi}{2}$ radians to degrees.

24.

$$\frac{3\pi}{2} \cdot \frac{180}{\pi} = \frac{3\pi(180)}{2\pi} = 270°$$

25. Evaluate $\csc \frac{5\pi}{4}$.

25.

$$\frac{5\pi}{4} \cdot \frac{180}{\pi} = 225°$$

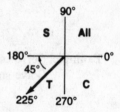

$$\csc \frac{5\pi}{4} = \csc 225°$$

$$= -\csc 45°$$

$$= -\sqrt{2}$$

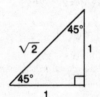

$$\sin 45° = \frac{1}{\sqrt{2}}$$

$$\csc 45° = \frac{\sqrt{2}}{1}$$

Practice Problems 17.1

1. $\cos 52°$ is equivalent to

A) $\sec 52°$ B) $\sin 52°$
C) $\sec 38°$ D) $\sin 38°$
E) $\csc 38°$

2. Express $\sin(-140°)$ as a function of a positive acute angle.

A) $-\sin 40°$ B) $\sin 40°$
C) $\sin 50°$ D) $-\sin 50°$
E) $\cos 50°$

3. If $\sec A = \frac{5}{4}$, find the value of $\cos A$.

A) $\frac{3}{5}$ B) $\frac{4}{5}$ C) $\frac{-3}{4}$

D) $\frac{-3}{5}$ E) $\frac{-4}{5}$

4. What is the smallest positive value of θ which satisfies the equation $2\cos^2 \theta - \cos \theta = 0$?

A) $30°$ B) $60°$ C) $90°$
D) $150°$ E) $180°$

5. For all values of A for which the expression is defined, which is an equivalent expression to $4 + \cos^2 A$?

A) $5 - \sec^2 A$ B) $5 - \sin^2 A$

C) $\frac{5}{\sec^2 A}$ D) $5 + \sin^2 A$ E) 5

6. Which values of x in the interval $0° \leqslant x < 360°$ satisfy the equation $2\cos x - 1 = 0$?

A) $30°$ and $330°$ B) $60°$ and $300°$
C) $30°$ and $150°$ D) $60°$ and $120°$
E) $60°$ and $240°$

7. Express $\frac{7\pi}{6}$ radians in degrees.

A) $120°$ B) $150°$ C) $210°$
D) $240°$ E) $420°$

8. Which value of x satisfies the equation $6\tan x - 2 = 4$?

A) $\frac{\pi}{6}$ B) $\frac{\pi}{4}$ C) $\frac{\pi}{3}$ D) $\frac{\pi}{2}$ E) π

9. Which value of x satisfies the equation $\sin(2x+30)° = \cos(3x+20)°$?

A) 8 B) 2 C) 10 D) 28 E) 30

10. A value of x for which the expression $\dfrac{1}{1-\cos x}$ is undefined is

A) 0° B) 30° C) 45° D) 60° E) 90°

Solutions on page 324

17.2 Graph of Trigonometric Functions

		Amplitude: One-half the range of the periodic function $=\frac{1}{2}$|maximum value $-$ minimum value|

Amplitude: One-half the range of the periodic function $=\frac{1}{2}$\|maximum value $-$ minimum value\|

Frequency: Number of times a periodic function repeats in an interval

Period: The interval for one complete cycle of the curve (wavelength)

	Domain	Range	Amplitude	Interval (for the basic curve)	Frequency	Period	Shift to left to rt.	
$y = a\sin(bx+c)$	All real	$-a \leqslant y \leqslant a$	$\|a\|$	360° or 2π	$\|b\|$	$\dfrac{360°}{\|b\|}$ or $\dfrac{2\pi}{\|b\|}$	$\dfrac{c}{b}$ $\dfrac{c}{b}>0$	$\dfrac{c}{b}<0$
$y = a\cos(bx+c)$	All real	$-a \leqslant y \leqslant a$	$\|a\|$	360° or 2π	$\|b\|$	$\dfrac{360°}{\|b\|}$ or $\dfrac{2\pi}{\|b\|}$	$\dfrac{c}{b}$ $\dfrac{c}{b}>0$	$\dfrac{c}{b}<0$
$y = a\tan(bx+c)$	All real except $[90(2n+1)]°$ n = integers	All real	Has none	180° or π	$\|b\|$	$\dfrac{180°}{\|b\|}$ or $\dfrac{\pi}{\|b\|}$	$\dfrac{c}{b}$ $\dfrac{c}{b}>0$	$\dfrac{c}{b}<0$
$y = a\cot(bx+c)$	All real except $(180n)°$ n = integers	All real	Has none	180° or π	$\|b\|$	$\dfrac{180°}{\|b\|}$ or $\dfrac{\pi}{\|b\|}$	$\dfrac{c}{b}$ $\dfrac{c}{b}>0$	$\dfrac{c}{b}<0$
$y = a\sec(bx+c)$	All real except $[90(2n+1)]°$ n = integers	$\|y\| \geqslant a$	Has none	360° or 2π	$\|b\|$	$\dfrac{360°}{\|b\|}$ or $\dfrac{2\pi}{\|b\|}$	$\dfrac{c}{b}$ $\dfrac{c}{b}>0$	$\dfrac{c}{b}<0$
$y = a\csc(bx+c)$	All real except $(180n)°$ n = integers	$\|y\| \geqslant a$	Has none	360° or 2π	$\|b\|$	$\dfrac{360°}{\|b\|}$ or $\dfrac{2\pi}{\|b\|}$	$\dfrac{c}{b}$ $\dfrac{c}{b}>0$	$\dfrac{c}{b}<0$

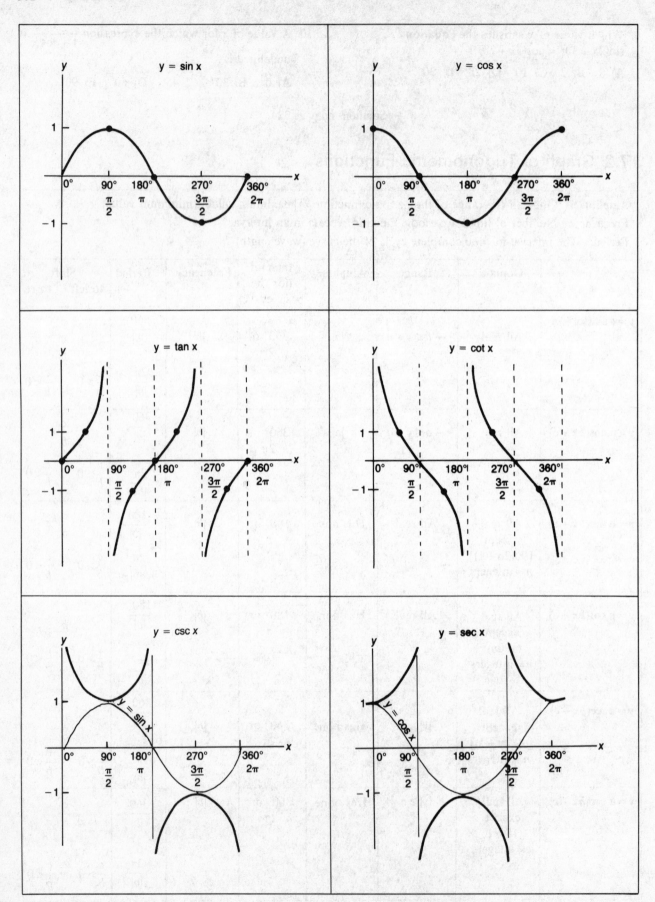

Practice Problems 17.2

1. The accompanying diagram represents the graph in the interval $-\pi \leqslant x \leqslant \pi$ of which function?

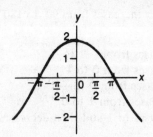

A) $y = \sin 2x$ B) $y = \cos 2x$
C) $y = \cos x$ D) $y = 2 \sin x$
E) $y = 2 \cos x$

2. What is the amplitude of $y = -3 \cos (2x - \pi)$?

A) 3 B) 2 C) $\dfrac{\pi}{2}$ D) 6 E) -3

3. What is the amplitude of the graph shown below?

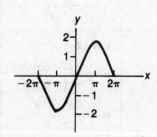

A) 2π B) 2 C) 1 D) -2 E) 4π

4. What is the period of $f(x) = 3 \sin \left(2x + \dfrac{\pi}{7} \right)$?

A) 3 B) 2 C) $\dfrac{\pi}{7}$ D) $\dfrac{\pi}{4}$ E) π

5. What is the phase shift if $y = 3 \sin \left(2x - \dfrac{\pi}{3} \right)$?

A) $\dfrac{\pi}{3}$ to the right B) $\dfrac{\pi}{3}$ to the left
C) $\dfrac{\pi}{6}$ to the right D) $\dfrac{\pi}{6}$ to the left
E) 3 to the left

6. What is the amplitude of $5 \tan(2x - 180)°$?

A) 5 B) -5 C) 2 D) 180 E) has none

7. What are the only value(s) all six basic trigonometric functions can have?

A) 0 only B) 1 only C) -1 only
D) 1 and -1 only E) 0 and 1 only

8. Find the period of $y = 4 \sec\left(\dfrac{3\pi}{4}x - \dfrac{\pi}{2}\right)$.

A) 4   B) $\dfrac{3\pi}{4}$ C) $\dfrac{-\pi}{2}$

D) $\dfrac{8}{3}$ E) $\dfrac{-2}{3}$

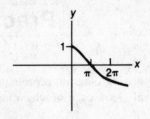

9. Which best expresses the equation of the graph shown?

A) $y = \sin \frac{1}{2}x$ B) $y = \cos \frac{1}{2}x$

C) $y = \frac{1}{2}\sin x$ D) $y = \frac{1}{2}\cos x$

E) $y = \frac{1}{2}$

10. As angle x increases from 90° to 180°, the value of $\sin x$:

A) increases from -1 to 0
B) increases from 0 to 1
C) decreases from 0 to -1
D) decreases from 1 to 0
E) increases to 1 and then decreases

Solutions on page 325

17.3 Trigonometric Identities for Two Angles, Double Angles and Half-Angles

	sin	cos	tan
Sum of Two Angles	$\sin(x+y) =$ $\sin x \cos y + \sin y \cos x$	$\cos(x+y) =$ $\cos x \cos y - \sin x \sin y$	$\tan(x+y) =$ $\dfrac{\tan x + \tan y}{1 - \tan x \tan y}$
Difference of Two Angles	$\sin(x-y) =$ $\sin x \cos y - \sin y \cos x$	$\cos(x-y) =$ $\cos x \cos y + \sin x \sin y$	$\tan(x-y) =$ $\dfrac{\tan x - \tan y}{1 + \tan x \tan y}$
Double-Angle	$\sin 2x =$ $2 \sin x \cos x$	$\cos 2x =$ • $\cos^2 x - \sin^2 x$ • $2\cos^2 x - 1$ • $1 - 2\sin^2 x$	$\tan 2x =$ $\dfrac{2\tan x}{1 - \tan^2 x}$
Half-Angle	$\sin \frac{1}{2}x =$ $\pm\sqrt{\dfrac{1 - \cos x}{2}}$	$\cos \frac{1}{2}x =$ $\pm\sqrt{\dfrac{1 + \cos x}{2}}$	$\tan \frac{1}{2}x =$ • $\pm\sqrt{\dfrac{1 - \cos x}{1 + \cos x}}$ • $\dfrac{\sin x}{1 + \cos x}$ • $\dfrac{1 - \cos x}{\sin x}$

Examples:

1. Evaluate cos 15° without use of a cosine table. (Use half-angle rules.)

2. Evaluate cos 15° without use of a cosine table. (Use the difference of two angles.)

Solutions:

1. 15° is half of 30°.

$$\cos \tfrac{1}{2}x = \pm\sqrt{\frac{1+\cos x}{2}}$$

Let $x = 30°$.

$$\cos \tfrac{1}{2}(30°) = \pm\sqrt{\frac{1+\cos 30°}{2}}$$

$$\cos 30° = \frac{\sqrt{3}}{2}$$

$$\cos 15° = \pm\sqrt{\frac{1+(\sqrt{3}/2)}{2}}$$

$$= \pm\sqrt{\frac{2+\sqrt{3}}{2}\cdot\frac{1}{2}}$$

$$= \pm\frac{\sqrt{2+\sqrt{3}}}{2}$$

Since 15° is in the first quadrant cos 15° is "+".

$$= \frac{1}{2}\sqrt{2+\sqrt{3}}$$

2. $15° = 45° - 30°$

$$\cos(x-y) = \cos x \cos y + \sin x \sin y$$

Let $x = 45$, $y = 30$.

$$\cos(45-30) = \cos 45 \cos 30 + \sin 45 \sin 30$$

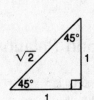 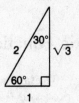

$$\cos 15 = \left(\frac{\sqrt{2}}{2}\right)\left(\frac{\sqrt{3}}{2}\right) + \left(\frac{\sqrt{2}}{2}\right)\left(\frac{1}{2}\right)$$

$$= \frac{\sqrt{6}}{4} + \frac{\sqrt{2}}{4} = \frac{\sqrt{6}+\sqrt{2}}{4}$$

☞ $\dfrac{\sqrt{6}+\sqrt{2}}{4}$ from example 2

$$= \tfrac{1}{2}\sqrt{2+\sqrt{3}} \text{ from example 1.}$$

3. If $\cos x = \frac{3}{5}$ and x is a positive acute angle, find $\tan \frac{1}{2}x$.

3. Since x is positive acute, x is in the first quadrant.

$\frac{1}{2}x$ is also in the first quadrant.

tan is '+' in the first quadrant.

$$\tan \tfrac{1}{2}x = \sqrt{\frac{1-\cos x}{1+\cos x}} = \sqrt{\frac{1-\frac{3}{5}}{1+\frac{3}{5}}}$$

$$= \sqrt{\frac{\frac{2}{5}}{\frac{8}{5}}} = \sqrt{\tfrac{1}{4}} = \tfrac{1}{2}$$

Practice Problems 17.3

1. If $\tan A = \frac{1}{3}$, find the value of $\tan 2A$.

A) $\frac{2}{3}$ B) $\frac{1}{9}$ C) $\frac{1}{2}$ D) $\frac{3}{4}$ E) $\frac{4}{3}$

2. If θ is a positive acute angle and $\cos \theta = \frac{7}{25}$, then $\sin \frac{1}{2}\theta$ is equal to

A) $\frac{4}{5}$ B) $-\frac{4}{5}$ C) $\frac{3}{5}$ D) $-\frac{3}{5}$ E) $\frac{3}{4}$

3. If $\cos \theta = \frac{3}{7}$ and $\tan \theta < 0$, find $\sin 2\theta$.

A) $\frac{8}{7}$ B) $\frac{6\sqrt{10}}{49}$ C) $\frac{-12\sqrt{10}}{58}$

D) $\frac{-7}{8}$ E) $\frac{-12\sqrt{10}}{49}$

4. If $\sin x = \frac{5}{6}$, what is the value of $\cos 2x$?

A) $\frac{-7}{18}$ B) $\frac{1}{6}$ C) $-\frac{2}{3}$

D) $-\frac{22}{3}$ E) $-\frac{1}{2}$

Solutions on page 326

17.4 Arc Functions and Relations

The inverse of $y = \sin x$ is $x = \sin y$. After swapping the x and y the next step in the inverse process is to solve for y.

$$\boxed{\text{If } x = \sin y \Rightarrow y = \text{arc sin } x \text{ or } y = \sin^{-1} x}$$

- The 'a' in arc and the 's' in $\sin^{-1}$ are lower case letters.
- The equations $y = \text{arc sin } x$ or $y = \sin^{-1} x$ are read: y represents the angles whose sine is x.

In this context, $\sin^{-1} x \neq \dfrac{1}{\sin x}$. This is a point of confusion.

Example:

For $0° \leqslant y < 360°$ solve $\sin y = \frac{1}{2}$

    or

For $0° \leqslant y < 360°$ solve $y = \text{arc} \sin \frac{1}{2}$

    or

For $0° \leqslant y < 360°$ solve $y = \sin^{-1} \frac{1}{2}$

Solution:

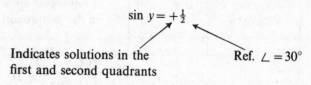

Indicates solutions in the first and second quadrants   Ref. $\angle = 30°$

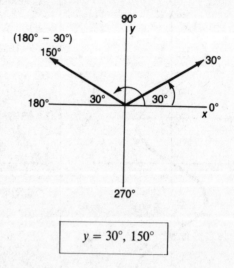

$$y = 30°, \ 150°$$

The relations $y = \text{arc} \sin x$ and $y = \sin^{-1} x$ are not functions. By restricting the range of the y values, an inverse trigonometric function can be defined:

$$y = \text{Arc} \sin x \quad \text{or} \quad y = \text{Sin}^{-1} x$$

☞ • The 'A' in Arc and the 'S' in Sin^{-1} are upper case letters.

The **principal value** of y is the angle defined within the restricted range of y.

	For Positive x ($x \geqslant 0$)		For Negative x ($x < 0$)	
	Restricted Range (Principal Value)	**Comment**	**Restricted Range (Principal Value)**	**Comment**
$\theta = \text{Arc} \sin x$ $= \text{Sin}^{-1} x$	$0° \leqslant \theta \leqslant 90°$ $0 \leqslant \theta \leqslant \dfrac{\pi}{2}$	θ represents an angle in the first quadrant	$-90° \leqslant \theta < 0°$ $-\dfrac{\pi}{2} \leqslant \theta < 0$	θ represents a **negative** angle in the fourth quadrant
$\theta = \text{Arc} \cos x$ $= \text{Cos}^{-1} x$	$0° \leqslant \theta \leqslant 90°$ $0 \leqslant \theta \leqslant \dfrac{\pi}{2}$	θ represents an angle in the first quadrant	$90° < \theta \leqslant 180°$ $\dfrac{\pi}{2} < \theta \leqslant \pi$	θ represents a **positive** angle in the second quadrant

$\theta = \text{Arc tan } x = \text{Tan}^{-1} x$	$0° \leqslant \theta \leqslant 90°$ $0 \leqslant \theta \leqslant \frac{\pi}{2}$	θ represents an angle in the first quadrant	$-90° \leqslant \theta < 0°$ $-\frac{\pi}{2} \leqslant \theta < 0$	θ represents a **negative** angle in the fourth quadrant

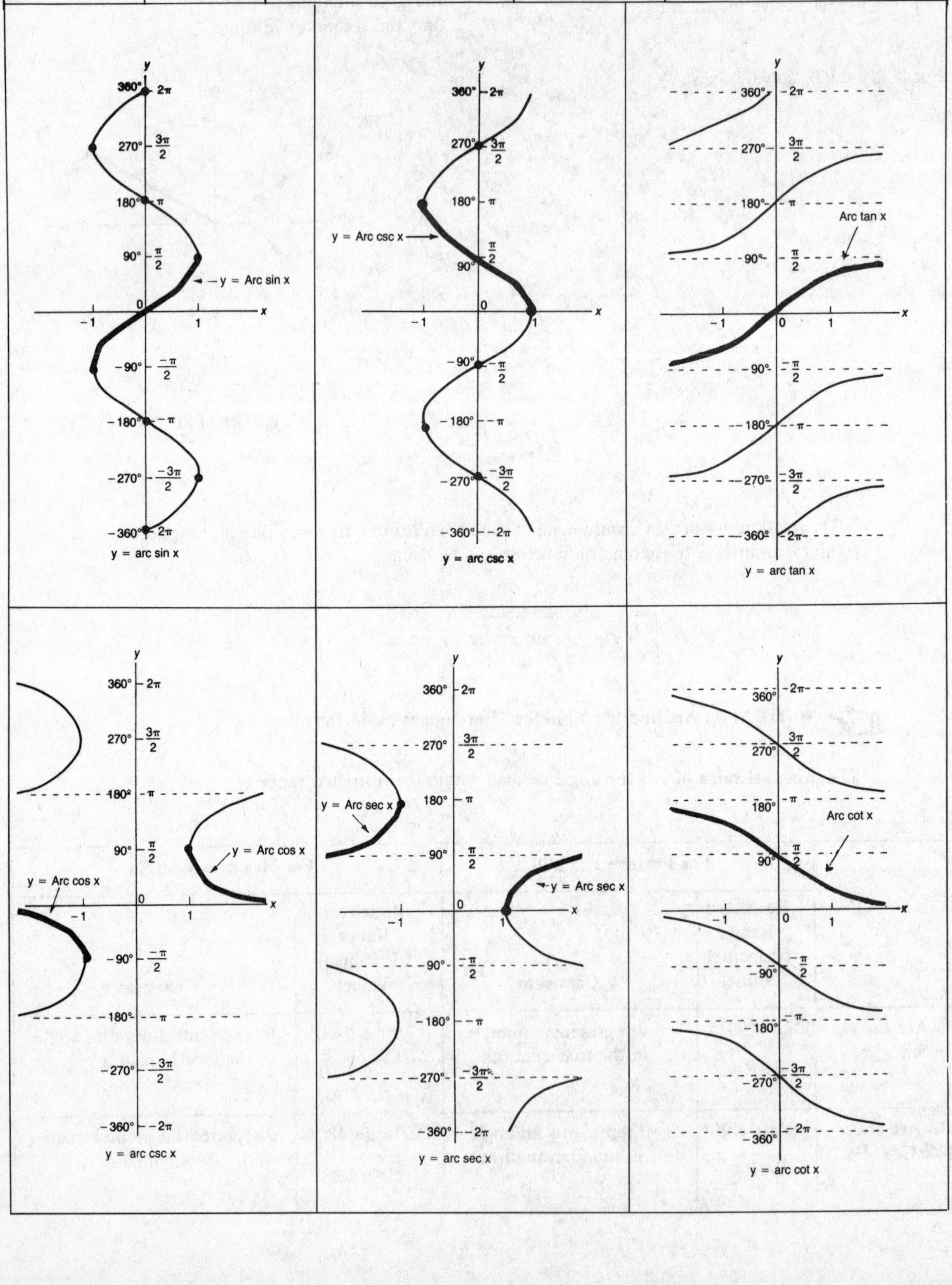
y = Arc sin x
y = arc sin x
y = Arc csc x
y = arc csc x
Arc tan x
y = arc tan x
y = Arc cos x
y = arc csc x
y = Arc sec x
y = arc sec x
Arc cot x
y = arc cot x

Examples:

1. Find θ, if $\theta = \text{Arc cos} \dfrac{\sqrt{3}}{2}$.

Solutions:

1.

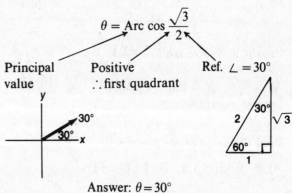

$$\theta = \text{Arc cos} \dfrac{\sqrt{3}}{2}$$

Principal value — Positive ∴ first quadrant — Ref. ∠ = 30°

Answer: $\theta = 30°$

2. Find θ, if $\theta = \text{Arc sin}\left(-\dfrac{\sqrt{3}}{2}\right)$.

2.

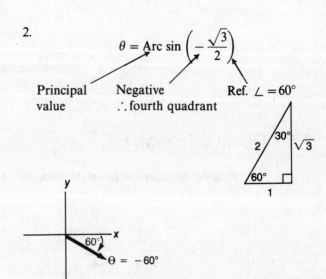

$$\theta = \text{Arc sin}\left(-\dfrac{\sqrt{3}}{2}\right)$$

Principal value — Negative ∴ fourth quadrant — Ref. ∠ = 60°

$\theta = -60°$

θ is a negative angle in the fourth quadrant.

Answer: $\theta = -60°$

☞ Negative angles are measured clockwise.

3. Evaluate $\sin\left(\text{Arc cos}\left(-\dfrac{3}{5}\right)\right)$.

3.

Let $\theta = \text{Arc cos}\left(-\dfrac{3}{5}\right)$

θ is a positive angle in the second quadrant.

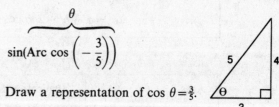

$$\sin\left(\text{Arc cos}\left(-\dfrac{3}{5}\right)\right)$$

Draw a representation of $\cos\theta = \dfrac{3}{5}$.

Evaluate $\sin\theta$: $\sin\theta = \dfrac{4}{5}$ and since θ is in the second quadrant

$$\sin\theta = +\dfrac{4}{5}$$

Practice Problems 17.4

1. Find θ, if $\theta = \text{Arc tan}\left(-\dfrac{\sqrt{3}}{3}\right)$.

   A) $-30°$ B) $-150°$ C) $30°$
   D) $150°$ E) $330°$

2. Find the value of $\cos(\text{Arc tan } \tfrac{3}{4})$.

   A) $\tfrac{4}{3}$ B) $\tfrac{3}{4}$ C) $\tfrac{3}{5}$ D) $\tfrac{4}{5}$ E) $-\tfrac{1}{4}$

3. The value of $\text{Arc sin}\left(\dfrac{\sqrt{3}}{2}\right) + \text{Arc tan } 1$ is

   A) $120°$ B) $105°$ C) $90°$
   D) $\dfrac{2+\sqrt{3}}{2}$ E) $75°$

4. Evaluate $\sin(2 \text{ Arc cos } \tfrac{1}{3})$.

   A) $\tfrac{2}{3}$ B) $17°$ C) 6
   D) $\dfrac{4\sqrt{2}}{9}$ E) $78°$

Solutions on page 327

17.5 Oblique Triangles

Oblique triangles are non-right triangles.

LAW OF SINES
$\dfrac{a}{\sin A} = \dfrac{b}{\sin B} = \dfrac{c}{\sin C}$

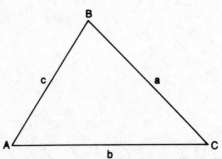

LAW OF COSINES
$a^2 = b^2 + c^2 - 2bc \cos A$ or $b^2 = a^2 + c^2 - 2ac \cos B$ or $c^2 = a^2 + b^2 - 2ab \cos C$

Area	Projection Equations
$\text{Area} = \tfrac{1}{2}ab \sin C$ $= \tfrac{1}{2}ac \sin B$ $= \tfrac{1}{2}bc \sin A$	$a = b \cos C + c \cos B$ $b = c \cos A + a \cos C$ $c = a \cos B + b \cos A$

Examples:

1. In triangle ABC, $a=6$, $b=8$ and $\cos C=\frac{3}{8}$. What is the length of side c?

2. In $\triangle NJL$, if $j=10$, $l=16$ and $m\angle N=30$, find the area of the triangle.

3. In $\triangle ABC$, $a=6$, $b=3$ and $\sin B=\frac{1}{4}$. What is the value of $\sin A$?

Solutions:

1. Use the law of cosines.

$$c^2=a^2+b^2-2ab\cos C$$
$$=6^2+8^2-2(6)(8)(\tfrac{3}{8})$$
$$=36+64-36$$
$$=64$$
$$c=\sqrt{64}=8$$

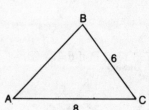

2. $A=\frac{1}{2}jl\sin N$
$=\frac{1}{2}(10)(16)\sin 30°$
$=\frac{1}{2}(10)(16)(\frac{1}{2})$
$=40$

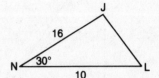

3.

$$\frac{a}{\sin A}=\frac{b}{\sin B}$$
$$\frac{6}{\sin A}=\frac{3}{\frac{1}{4}}$$
$$\frac{6}{\sin A}=\frac{12}{1}$$
$$\sin A=\frac{1}{2}$$

Practice Problems 17.5

1. In triangle NJL, $\sin N=8$ and $n=4$. The ratio $\frac{\sin L}{l}$ is

A) $\frac{1}{3}$ B) $\frac{1}{2}$ C) $\frac{2}{1}$ D) $\frac{5}{1}$ E) .6

2. In $\triangle ABC$, $a=5$, $b=8$ and $c=9$. The value of $\cos A$ is

A) $\frac{1}{12}$ B) $\frac{5}{6}$ C) $-\frac{5}{6}$ D) 25 E) $-\frac{1}{12}$

3. In $\triangle ABC$ if $c=20$, $m\angle A=35$, $m\angle B=68$, find a to the nearest integer.
($\cos 35°=.8192$, $\cos 68°=.3746$,
$\sin 35°=.5736$, $\sin 68°=.9272$,
$\sin 77°=.9563$.)

A) 11 B) 12 C) 13 D) 14 E) 15

4. In the figure below $\square ABCD$, $DB=8$, $AD=6$ and $m\angle ADB=38$. Find AB to the nearest integer. ($\cos 38°=.7880$.)

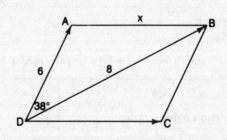

A) 2 B) 3 C) 4 D) 5 E) 6

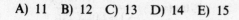

Solutions on page 328

17.6 Trigonometry Summary

Basics

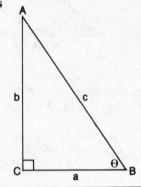

$$\sin \theta = \frac{\text{opp}}{\text{hyp}} = \frac{b}{c} \qquad \csc \theta = \frac{\text{hyp}}{\text{opp}} = \frac{c}{b}$$

$$\cos \theta = \frac{\text{adj}}{\text{hyp}} = \frac{a}{c} \qquad \sec \theta = \frac{\text{hyp}}{\text{adj}} = \frac{c}{a}$$

$$\tan \theta = \frac{\text{opp}}{\text{adj}} = \frac{b}{a} \qquad \cot \theta = \frac{\text{adj}}{\text{opp}} = \frac{a}{b}$$

$$\text{SOH} \qquad \text{CAH} \qquad \text{TOA}$$

Inverse (Reciprocal)

$$\sin \theta = \frac{1}{\csc \theta} \quad \Rightarrow \quad \csc \theta = \frac{1}{\sin \theta} \quad \Rightarrow \quad \sin \theta \cdot \csc \theta = 1$$

$$\cos \theta = \frac{1}{\sec \theta} \quad \Rightarrow \quad \sec \theta = \frac{1}{\cos \theta} \quad \Rightarrow \quad \cos \theta \cdot \sec \theta = 1$$

$$\tan \theta = \frac{1}{\cot \theta} \quad \Rightarrow \quad \cot \theta = \frac{1}{\tan \theta} \quad \Rightarrow \quad \tan \theta \cdot \cot \theta = 1$$

Cofunctions

$$\sin \theta = \cos(90 - \theta) \qquad \tan \theta = \cot(90 - \theta) \qquad \sec \theta = \csc(90 - \theta)$$

$$\cos \theta = \sin(90 - \theta) \qquad \cot \theta = \tan(90 - \theta) \qquad \csc \theta = \sec(90 - \theta)$$

Basics

$$\tan \theta = \frac{\sin \theta}{\cos \theta} \qquad \cot \theta = \frac{\cos \theta}{\sin \theta}$$

Pythagorean Equalities

$$\sin^2 \theta + \cos^2 \theta = 1 \qquad \tan^2 \theta + 1 = \sec^2 \theta \qquad \cot^2 \theta + 1 = \csc^2 \theta$$

sin	cos	tan
Sum of two angles:		
$\sin(x + y) = \sin x \cos y + \sin y \cos x$	$\cos(x + y) = \cos x \cos y - \sin x \sin y$	$\tan(x + y) = \dfrac{\tan x + \tan y}{1 - \tan x \tan y}$
Difference of two angles:		
$\sin(x - y) = \sin x \cos y - \sin y \cos x$	$\cos(x - y) = \cos x \cos y + \sin x \sin y$	$\tan(x - y) = \dfrac{\tan x - \tan y}{1 + \tan x \tan y}$

Double angle

$\sin 2x = 2 \sin x \cos x$

$\cos 2x = \begin{cases} \cos^2 x - \sin^2 x \\ 2 \cos^2 x - 1 \\ 1 - 2 \sin^2 x \end{cases}$

$\tan 2x = \dfrac{2 \tan x}{1 - \tan^2 x}$

Half angle

$\sin \frac{1}{2}x = \pm \sqrt{\dfrac{1 - \cos x}{2}}$

$\cos \frac{1}{2}x = \pm \sqrt{\dfrac{1 + \cos x}{2}}$

$\tan \frac{1}{2}x = \begin{cases} \dfrac{\sin x}{1 + \cos x}, \dfrac{1 - \cos x}{\sin x} \\ \pm \sqrt{\dfrac{1 - \cos x}{1 + \cos x}} \end{cases}$

Law of sines

$$\frac{a}{\sin A} = \frac{b}{\sin B} = \frac{c}{\sin C}$$

Law of cosines

$a^2 = b^2 + c^2 - 2bc \cos A$

or

$b^2 = a^2 + c^2 - 2ac \cos B$

or

$c^2 = a^2 + b^2 - 2ab \cos C$

Area (K)

$K = \frac{1}{2}ab \sin C$

or

$\frac{1}{2}bc \sin A$

or

$\frac{1}{2}ac \sin B$

18. Polar Coordinates

18.1 Essentials

Polar coordinates use an ordered pair (r, θ) to represent a point.

θ: The smallest angle of rotation from the polar axis
   to OP
   $\theta > 0$ counterclockwise
   $\theta < 0$ clockwise

r: The distance from the pole to point P

: The magnitude of OP
   $r > 0$ extends in the direction of θ
   $r < 0$ reflects through the pole (extends $180°$ from θ)

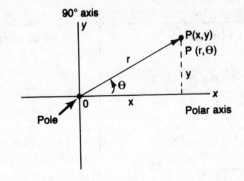

Example:

Plot (r, θ) A$(2, 30°)$
   B$(2, 170°)$
   C$(-2, 60°)$
   D$(2, 240°)$
   E$(2, 390°)$
   F$(3, -30°)$

 Note: In polar coordinates (r, θ), $(r, \theta + (360n°)$ and $(-r, \theta + 180°)$ all represent the same point (n is any integer).

Solution:

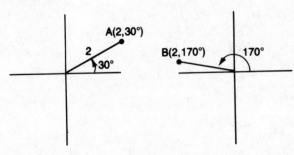

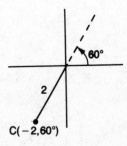 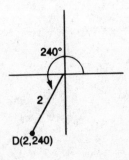

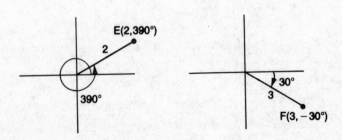

314

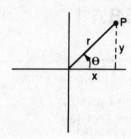

Conversion Table	
$x = r \cos \theta$	$r^2 = x^2 + y^2$
$y = r \sin \theta$	$\theta = \text{Arc} \tan \dfrac{y}{x}$
$x + yi = r(\cos \theta + i \sin \theta)$	

Examples:

1. Find the ordered pair of rectangular coordinates whose polar coordinates are $(4, 240°)$.

2. Convert the rectangular coordinates $(0, 2)$ into polar coordinates.

Solutions:

1.

$x = r \cos \theta = 4 \cos 240°$

$\quad = 4(-\tfrac{1}{2}) = -2$

$y = r \sin \theta = 4 \sin 240°$

$\quad = 4\left(-\dfrac{\sqrt{3}}{2}\right) = -2\sqrt{3}$

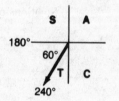

$\cos 240° = -\cos 60° = -\tfrac{1}{2}$

$\sin 240° = -\sin 60° = -\dfrac{\sqrt{3}}{2}$

$(-2, -2\sqrt{3})$

2. $r^2 = x^2 + y^2 = 0^2 + 2^2 = 4$

$\qquad r = \pm 2$

$\theta = \text{Tan}^{-1} \dfrac{y}{x} = \text{Tan}^{-1} \dfrac{2}{0} = \text{Tan}^{-1} \infty$

$\theta = 90°$ is the reference angle.
From the graph

$$\text{if } \theta = 90 \quad r = 2$$

we could also refer to the point as

$$\theta = 270 \quad r = -2$$

Solution: $(2, 90°)$ or $(-2, 270°)$

Practice Problems 18.1

1. Find the polar coordinates given by the rectangular coordinates $(-\sqrt{3}, 1)$.

   A) $(2, 30°)$ B) $(2, 60°)$ C) $(2, 120°)$
   D) $(2, 150°)$ E) $(-2, 315°)$

2. Transform the equation $r \sin \theta = -2$ into rectangular form.

   A) $x + y = -2$ B) $xy = 2$ C) $x = -2$
   D) $y = -2$ E) $\frac{x}{2} = 1$

3. Convert $1 + i\sqrt{3}$ to polar form.

   A) $2(\cos 60° + i \sin 60°)$
   B) $3(\cos 45° + i \sin 45°)$
   C) $(\cos 30° + i \sin 30°)$
   D) $4(\cos 60° + i \sin 60°)$
   E) $\sqrt{3}(\cos 30° + i \sin 30°)$

4. Convert $2\sqrt{6}(\cos 120° + i \sin 120°)$ into rectangular form.

   A) $-3\sqrt{2} - \sqrt{6}i$ B) $3\sqrt{2} + \sqrt{6}i$
   C) $\sqrt{6} + 3\sqrt{2}i$ D) $-\sqrt{6} - 3\sqrt{2}i$
   E) $-\sqrt{6} + 3\sqrt{2}i$

Solutions on page 328

19. DeMoivre's Theorem and Euler's Formula

19.1 Basics and Beyond

<table>
<tr><td colspan="2" align="center">

De Moivre's Theorem:

$$[r(\cos \theta + i \sin \theta)]^n = r^n(\cos n\theta + i \sin n\theta)$$

$$[r(\cos \theta + i \sin \theta)]^{1/n} = r^{1/n}\left(\cos \frac{\theta + 2\pi k}{n} + i \sin \frac{\theta + 2\pi k}{n}\right) \quad k = 0 \text{ to } n-1$$

$$= r^{1/n}\left(\cos \frac{\theta + 360k}{n} + i \sin \frac{\theta + 360k}{n}\right)$$

</td></tr>
<tr><td colspan="2" align="center">

Euler's Formula:

$$e^{i\theta} = \cos \theta + i \sin \theta$$

</td></tr>
<tr><td colspan="2" align="center">

Combined DeMoivre and Euler:

$$(e^{i\theta})^n = (\cos \theta + i \sin \theta)^n = \cos n\theta + i \sin n\theta$$

</td></tr>
<tr><td colspan="2" align="center">

Polar Multiplication:

$$[r_1(\cos \theta_1 + i \sin \theta_2] \cdot [r_2(\cos \theta_2 + i \sin \theta_2] = r_1 r_2[\cos(\theta_1 + \theta_2) + i \sin(\theta_1 + \theta_2)]$$

</td></tr>
<tr><td colspan="2" align="center">

Polar Division:

$$\frac{r_1(\cos \theta_1 + i \sin \theta_1)}{r_2(\cos \theta_2 + i \sin \theta_2)} = \frac{r_1}{r_2}(\cos(\theta_1 - \theta_2) + i \sin(\theta_1 - \theta_2))$$

</td></tr>
</table>

Examples:

1. Evaluate $e^{i\pi}$.

Solutions:

1. $e^{i\pi} = \cos \pi + i \sin \pi$

$\quad = -1 + i(0) = -1$

2. Use DeMoivre's theorem to find $(1+i)^4$.

2. $1+i=x+yi \quad x=1, \quad y=1$

$r=\sqrt{x^2+y^2}=\sqrt{1^2+1^2}=\sqrt{2}$

$\tan \theta=\dfrac{y}{x}=\dfrac{1}{1}=1 \quad \theta=\dfrac{\pi}{4}$

$1+i=\sqrt{2}\left(\cos \dfrac{\pi}{4}+i \sin \dfrac{\pi}{4}\right)$

$(1+i)^4=\left[\sqrt{2}\left(\cos \dfrac{\pi}{4}+i \sin \dfrac{\pi}{4}\right)\right]^4$

$=(\sqrt{2})^4\left(\cos 4\left(\dfrac{\pi}{4}\right)+i \sin 4\left(\dfrac{\pi}{4}\right)\right)$

$=4(\cos \pi+i \sin \pi)$

$=4(-1+i(0))=-4$

3. Find the three cube roots of 1.

3. Rewrite the cube root of 1 as $1^{1/3}$

$1=x+yi \quad x=1, y=0$

$r=\sqrt{x^2+y^2}=\sqrt{1^2+0^2}=1$

$\tan \theta=\dfrac{y}{x}=\dfrac{0}{1}=0. \quad \theta=0$

$1=1(\cos 0+i \sin 0)$

$1^{1/3}=[1(\cos(0+2\pi k)+i \sin(0+2\pi k)]^{1/3}$

$=1^{1/3}\left[\cos\left(\dfrac{0+2\pi k}{3}\right)+i \sin\left(\dfrac{0+2\pi k}{3}\right)\right]$

$k=0$

$=1^{1/3}(\cos 0+i \sin 0]$

$=1(1+i(0))=1$

$k=1$

$=1^{1/3}\left(\cos \dfrac{2\pi}{3}+i \sin \dfrac{2\pi}{3}\right)$

$=1\left(-\dfrac{1}{2}+\dfrac{i\sqrt{3}}{2}\right)=\dfrac{-1+i\sqrt{3}}{2}$

$k=2$

$=1^{1/3}\left(\cos \dfrac{4\pi}{3}+i \sin \dfrac{4\pi}{3}\right)$

$=1\left(-\dfrac{1}{2}-\dfrac{i\sqrt{3}}{2}\right)=\dfrac{-1-\sqrt{3}}{2}$

$\left\{1, \dfrac{-1+i\sqrt{3}}{2}, -\dfrac{1-i\sqrt{3}}{2}\right\}$

Practice Problems 19.1

1. Evaluate $e^{i(\pi/2)}$.

   A) 1 B) e C) i D) -1 E) $-i$

2. Use DeMoivre's theorem to evaluate $(\sqrt{3}-i)^5$.

   A) $32(\sqrt{3}-i)$ B) $-16(\sqrt{3}+i)$
   C) $16(-\sqrt{3}+i)$ D) $32(-\sqrt{3}+i)$
   E) $32(-\sqrt{3}-i)$

3. Find the two square roots of $2+2i\sqrt{3}$.

   A) $\sqrt{3}+1i$ and $\sqrt{3}-1i$
   B) $\sqrt{3}+1i$ and $-\sqrt{3}-1i$
   C) $\sqrt{3}-1i$ and $-\sqrt{3}-1i$
   D) $-\sqrt{3}-1i$ and $-\sqrt{3}+1i$
   E) $-\sqrt{3}-\sqrt{2}i$ and $\sqrt{3}+\sqrt{2}i$

4. Multiply $2(\cos 30° + i \sin 30°)$ by $7(\cos 10° + i \sin 10°)$.

   A) $14(\cos 300° + i \sin 300°)$
   B) $9(\cos 40° + i \sin 40°)$
   C) $14(\cos 40° + i \sin 40°)$
   D) $9(\cos 300° + i \sin 300°)$
   E) $2^7(\cos 3° + i \sin 3°)$

Solutions on page 329

20. Parametric Equations

20.1 Fundamental Facts

When a point (x, y) can be described in terms of a third variable t, the equations $x = f(t)$ and $y = g(t)$ are called **parametric equations** and the variable t is called the **parameter**.

Examples:

1. Express the equation $x = 2\cos t$ and $y = 3\sin t$ in terms of the rectangular coordinates x and y by eliminating the parameter t.

Solutions:

1. $\dfrac{x}{2} = \cos t \qquad \dfrac{y}{3} = \sin t$

$$\sin^2 t + \cos^2 t = 1 = \left(\frac{y}{3}\right)^3 + \left(\frac{x}{2}\right)^2$$

$$\frac{x^2}{4} + \frac{y^2}{9} = 1$$

This is an ellipse.

2. Eliminate the parameter t and express the equation in terms of the rectangular coordinates x and y.

$$x = 2 + t \qquad y = 2 + 3t$$

2.

$$t = x - 2 \quad t = \frac{y - 2}{3}$$

$$\therefore x - 2 = \frac{y - 2}{3}$$

$$3x - 6 = y - 2$$

$$y = 3x - 4$$

This is a line.

Practice Problems 20.1

1. Eliminate the parameter t.

$$x = t^2 - 2 \qquad y = t + 1$$

A) $y = x^2 - 2y + 1$ B) $x = y^2 - 2y - 1$
C) $x = y^2 - 2y + 1$ D) $x = y - 1$
E) $y = x^2 - 2x + 1$

2. Eliminate the parameter θ and identify the curve.

$$y = \tan \theta \qquad x = 2 \cot \theta$$

A) line   B) circle   C) ellipse
D) parabola   E) hyperbola

3. Eliminate the parameter t and identify the curve.

$$x = e^t \qquad y = e^{-t}$$

A) line   B) circle   C) ellipse
D) parabola   E) hyperbola

4. Eliminate the parameter ϕ and identify the curve.

$$y = 3 \sin \phi \qquad x = 3 \cos \phi$$

A) line   B) circle   C) ellipse
D) parabola   E) hyperbola

Solutions on page 330

Cumulative Review Test

1. Factor completely $ax + ay + a^2 + a$.

    A) $a(x+y)(a+1)$ B) $a(x+y+a)$
    C) $a(x+y+1)$ D) $a(x+y+a+1)$
    E) $a(x+y)+(a+1)$

2. $\dfrac{20(\cos 15° + i \sin 15°)}{5(\cos 3° + i \sin 3°)} =$

    A) 4   B) $15(\cos 5° + i \sin 5°)$
    C) $4(\cos 12° + i \sin 12°)$
    D) $4(\cos 5° + i \sin 5°)$
    E) $15(\cos 12° + i \sin 12°)$

3. $\tan 810° =$

    A) 1   B) 0   C) $\dfrac{\sqrt{3}}{3}$

    D) Undefined   E) $\cot 90°$

4. Simplify $b^{-3} \cdot b^{2/3}$.

    A) b^{-2} B) $(b^2)^{-2}$ C) $b^{-7/3}$
    D) $b^{11/3}$ E) b^2

5. The expression $\sec^2 x + \csc^2 x$ is equivalent to

    A) $\sin^2 x \cos^2 x$ B) $\dfrac{1}{\sin^2 x \cos^2 x}$
    C) $1 + \tan^2 x$ D) $1 - \tan^2 x$
    E) $\sin^2 x + \cos^2 x$

6. Express $-\sqrt{3} + i$ in polar notation.

    A) $2(\cos 150° + i \sin 150°)$
    B) $2(\cos 270° + i \sin 270°)$
    C) $2(\cos 30° + i \sin 30°)$
    D) $\sqrt{2}(\cos 150° + i \sin 150°)$
    E) $\sqrt{2}(\cos 30° + i \sin 30°)$

7. The expression $\cos 262°$ is equal to

    A) $\sin 8°$ B) $-\cos 8°$ C) $\sin 82°$
    D) $-\cos 82°$ C) $\dfrac{1}{\sec 82°}$

8. Which function is represented in the accompanying graph?

    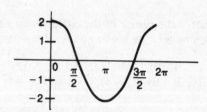

    A) $y = \sin 2x$ B) $y = 2 \sin x$
    C) $y = \cos 2x$ D) $y = 2 \cos x$
    E) $y = 2 \cos 2x$

9. The graph below represents which of the following?

    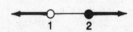

    A) $x < 1$ or $x > 2$ B) $x < 2$ or $x \geqslant 1$
    C) $x \leqslant 2$ or $x \geqslant 1$ D) $x \leqslant 1$ or $x > 2$
    E) $x < 1$ or $x \geqslant 2$

10. The parametric equations $x = 4 + 3\tan\theta$ and $y = -1 + 2\sec\theta$ represent the graph of a(n)

    A) line   B) parabola   C) circle
    D) ellipse   E) hyperbola

11. If the measure of an interior angle of a regular polygon is 108, how many sides does the polygon have?

    A) 4   B) 5   C) 6   D) 7   E) 11

12. If $A = \text{Arc} \tan \frac{3}{4}$, then $\tan \frac{1}{2}A$ is equal to

    A) $\sqrt{\frac{1}{7}}$ B) $\frac{1}{2}$ C) 3   D) $\frac{1}{3}$ E) $\frac{1}{9}$

13. In $\triangle JSL$, $j = 5$, $\sin J = .35$ and $\sin S = .21$. Find the measure of side s.

    A) .3   B) 3   C) $\frac{25}{3}$ D) .5   E) .03

14. $\left(\dfrac{\sqrt{3}+i}{2}\right)^{12} =$

   A) 3^6+2i B) -1 C) 1
   D) $2(\cos 30° + i \sin 30°)$
   E) $1(\cos 30° + i \sin 30°)$

15. $x = 2\tan\theta$, $y = \cot\theta$ is the parametric representation of a(n)

   A) line   B) circle   C) parabola
   D) ellipse   E) hyperbola

16. What is the center of the circle whose equation is given by $x^2 - 2x + y^2 + 6y + 4 = 0$?

   A) $(1, 3)$ B) $(-1, -3)$ C) $(-3, 1)$
   D) $(6, 2)$ E) $(1, -3)$

17. Find the value of $\tan\left(\text{Arc}\sin\dfrac{\sqrt{3}}{2}\right)$.

   A) $\sqrt{3}$ B) $-\dfrac{\sqrt{3}}{2}$ C) $-\sqrt{3}$
   D) $\frac{1}{2}$ E) $-\frac{1}{2}$

18. Find the length of side c in $\triangle ABC$ if $a = 3$, $b = 5$ and $\cos C = -\frac{1}{2}$.

   A) $\sqrt{19}$ B) $\sqrt{6}$ C) 7 D) $\sqrt{59}$ E) 8

19. The expression $\sin 35° \cos 22° + \cos 35° \sin 22°$ is equal to

   A) $\sin 13°$ B) $\sin 57°$ C) $\cos 13°$
   D) $\cos 57°$ E) $\tan 13°$

20. What is the solution set for the accompanying system of linear equations?

$$2x - y + z = 5$$
$$4x - 3y = 5$$
$$6x + 2y + 2z = 7$$

   A) $\{2, 1, 0\}$ B) $\{\frac{1}{2}, -1, 3\}$
   C) $\{1, -4, -1\}$ D) $\{2, 1, -1\}$
   E) $\{1, 0, 3\}$

Solutions on page 331

Diagnostic Chart for Cumulative Review Test

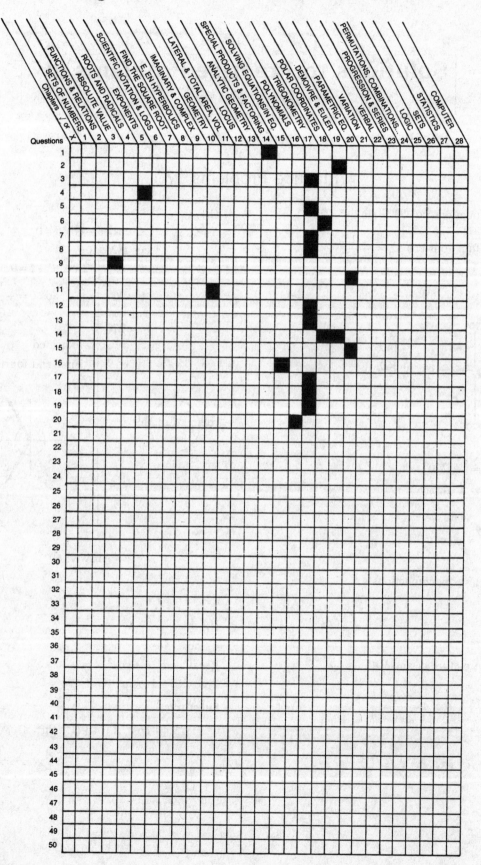

Solutions to Practice Problems in Unit Five

Practice Problems 17.1

1. **(D)** By cofunctions $\cos 52° = \sin 38°$.

   By inverse functions $\cos 52° = \dfrac{1}{\sec 52°}$.

2. **(A)**

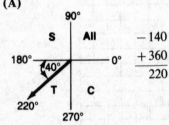

$$\sin(-140°) = \sin 220°$$
$$= -\sin 40°$$

or

$$= -\cos 50°$$
$$\uparrow$$
by cofunctions

3. **(B)** Draw a picture of angle A.

$$\sec A = \frac{\text{hyp}}{\text{adj}} = \frac{5}{4}$$

$$\cos A = \frac{1}{\sec} = \frac{+4}{5}$$

4. **(B)**

$$2 \cos^2 \theta - \cos \theta = 0$$
$$\cos \theta(2 \cos \theta - 1) = 0$$

$\cos \theta = 0$	$2 \cos \theta - 1 = 0$
$\theta = 90°$	$\cos \theta = +\dfrac{1}{2}$
$= 270°$	

indicated   Ref. $\angle = 60°$
First and fourth quadrants

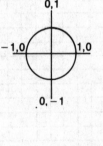

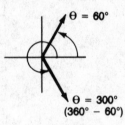

$\theta = 60°$

$\theta = 300°$
$(360° - 60°)$

Answer: $60°$

5. **(B)**

$4 + \cos^2 A$

Recall: $\sin^2 A + \cos^2 A = 1$

$\therefore \cos^2 A = 1 - \sin^2 A$

Substitute:

$$4 + \cos^2 A = 4 + (1 - \sin^2 A)$$
$$= 5 - \sin^2 A$$

6. (B)

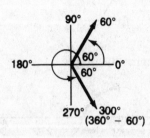

$$2 \cos x - 1 = 0$$

$$\cos x = \frac{+1}{2}$$

cos is "+" in
First and fourth quadrants

Ref. $\angle = 60°$

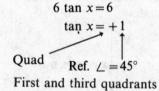

Answer: 60° and 300°

7. (C)

$$\frac{7\pi}{6} \times \frac{180}{\pi} = 210$$

8. (B)

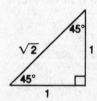

$$6 \tan x = 6$$

$$\tan x = +1$$

Quad   Ref. $\angle = 45°$

First and third quadrants

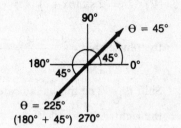

$$45° = \frac{\pi}{4}$$

$$225° = \frac{5\pi}{4}$$

9. (A) These are cofunctions.

$$(2x + 30) + (3x + 20) = 90$$

$$5x = 40$$

$$x = 8$$

10. (A) $\dfrac{1}{1 - \cos x}$ is undefined when

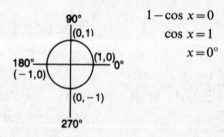

$$1 - \cos x = 0$$

$$\cos x = 1$$

$$x = 0°$$

Practice Problems 17.2

1. (E) At $x = 0$ y is a maximum
∴cos not sin.
Amplitude: 2
Period: $-\pi$ to $\pi = 2\pi$ ∴frequency: 1
∴$y = 2 \cos x$

2. (A) $y = a \cos(bx + c)$

Amplitude: $|a|$

$y = -3 \cos(2x - \pi)$

Amplitude: $|-3| = 3$

3. (B) Amplitude $= \frac{1}{2}|\text{max value} - \text{min value}|$

$$= \frac{1}{2}|2 - (-2)| = 2$$

4. (E) For $a \sin(bx + c)$

$$\text{the period} = \frac{2\pi}{|b|} = \frac{2\pi}{2} = \pi$$

5. **(C)** If $y = a \sin(bx + c)$

the phase shift $= \dfrac{c}{b} = \dfrac{-\dfrac{\pi}{3}}{2} = \dfrac{-\pi}{6}$.

Shift is $\dfrac{\pi}{6}$. The minus indicates that the shift is to the right.

6. **(E)** The tangent function has no amplitude.

7. **(D)** Examine the graphs. All can have the values 1 and -1.

8. **(D)** If $y = a \sec(bx + c)$

the period $= \dfrac{2\pi}{|b|} = \dfrac{2\pi}{\dfrac{3\pi}{4}} = \dfrac{8\pi}{3\pi} = \dfrac{8}{3}$.

9. **(B)** At $x = 0$ y is a max $\therefore \cos$ not sin.
Amplitude: 1

Half of a cycle is 2π; a complete cycle is 4π

$\therefore$ period $= 4\pi$, $\qquad P = \dfrac{2\pi}{\text{freq}}$ $\qquad \therefore$ freq $= \frac{1}{2}$

freq $= \frac{1}{2}$

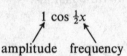

$1 \cos \frac{1}{2}x$

amplitude   frequency

10. **(D)** Examine the graph.

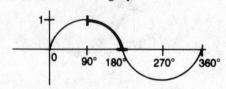

Decreases from 1 to 0

Practice Problems 17.3

1. **(D)**
$$\tan 2A = \dfrac{2 \tan A}{1 - \tan^2 A} = \dfrac{2(\frac{1}{3})}{1 - (\frac{1}{3})^2}$$

$$= \dfrac{\frac{2}{3}}{\frac{8}{9}}$$

$$= \dfrac{3}{4}$$

2. **(C)**
$$\text{Sin } \tfrac{1}{2}\theta = \pm\sqrt{\dfrac{1 - \cos \theta}{2}}$$

$$= \sqrt{\dfrac{1 - \frac{7}{25}}{2}}$$

$$= \sqrt{\dfrac{\frac{18}{25}}{2}} = \sqrt{\dfrac{18}{50}} = \sqrt{\dfrac{9}{25}}$$

$$= \dfrac{3}{5}$$

Since θ is positive acute (first quadrant), $\sin \frac{1}{2}\theta$ is '+'.

3. **(E)** Since $\cos \theta > 0$ and $\tan \theta < 0$, θ is in the fourth quadrant.

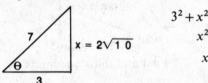

$$3^2 + x^2 = 7^2$$
$$x^2 = 40$$
$$x = 2\sqrt{10}$$

Sin $2\theta = 2 \sin \theta \cos \theta$

$$= 2\left(-\dfrac{2\sqrt{10}}{7}\right)\left(\dfrac{3}{7}\right) = \dfrac{-12\sqrt{10}}{49}$$

θ is in the fourth quadrant
$\sin \theta = $ "$-$" $\cos \theta = $ '+'

4. **(A)**
$$\cos 2x = 1 - 2 \sin^2 x$$

$$= 1 - 2\left(\dfrac{5}{6}\right)^2$$

$$= 1 - 2\left(\dfrac{25}{36}\right) = 1 - \dfrac{50}{36}$$

$$= \dfrac{-14}{36} = \dfrac{-7}{18}$$

Practice Problems 17.4

1. (A) $\theta = \text{Arc tan}\left(-\dfrac{\sqrt{3}}{3}\right)$

θ is a negative angle in the fourth quadrant.

Ref. $\angle = 30°$

$\theta = -30°$

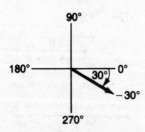

2. (D) Let $\theta = \text{Arc tan } \frac{3}{4}$. Find $\cos \theta$.
θ is a first quadrant angle.

Draw a diagram of $\tan \theta = \frac{3}{4}$

Find $\cos \theta$: $\cos \theta = \frac{4}{5}$.
Since the angle is in the first quadrant
$\underbrace{\cos(\text{Arc tan } \frac{3}{4})}_{\theta} = \cos \theta = +\frac{4}{5}$.

3. (B) $\text{Arc sin}\left(\dfrac{\sqrt{3}}{2}\right)$ is a first quadrant angle.

Ref. $\angle = 60°$

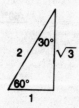

Arc tan 1 is a first quadrant angle.

Ref. $\angle = 45°$

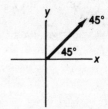

$\text{Arc sin } \dfrac{\sqrt{3}}{2} + \text{Arc tan } 1 = 60° + 45° = 105°$

4. (D) Let $\theta = \text{Arc cos } \frac{1}{3}$. This is a first quadrant angle.
Draw a diagram to represent $\cos \theta = \frac{1}{3}$.

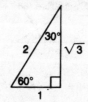

$$1^2 + x^2 = 3^2$$
$$x^2 = 8$$
$$x = 2\sqrt{2}$$

$\sin (2 \overbrace{\text{Arc cos } \frac{1}{3}}^{\theta}) = \sin 2\theta$

$= 2 \sin \theta \cos \theta$

$= 2\left(\dfrac{2\sqrt{2}}{3}\right)\left(\dfrac{1}{3}\right)$

$= \dfrac{4\sqrt{2}}{9}$

Practice Problems 17.5

1. (A) Use the law of sines.

$$\frac{n}{\sin N}=\frac{l}{\sin L} \quad \text{or} \quad \frac{\sin N}{n}=\frac{\sin L}{l}$$

$$\frac{.8}{4}=\frac{\sin L}{l}$$

$$\frac{\sin L}{l}=\frac{1}{5}$$

2. (B) Use the law of cosines.

$$a^2=b^2+c^2-2bc \cos A$$

$$5^2=8^2+9^2-2(8)(9) \cos A$$

$$25=64+81-144 \cos A$$

$$25=145-144 \cos A$$

 The 144 is multiplied by cos A.
You cannot subtract the 144 from the 145

$$\begin{array}{ll} 25= & 145-144 \cos A \\ -145 & -145 \\ \hline \dfrac{-120}{-144}= & \dfrac{-144 \cos A}{-144} \end{array}$$

$$\cos A=\frac{5}{6}$$

3. (B) Use the law of sines.

$$180°-(68°+35°)=77°$$

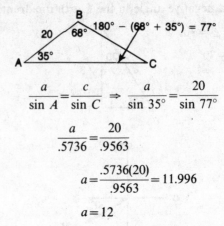

$$\frac{a}{\sin A}=\frac{c}{\sin C} \Rightarrow \frac{a}{\sin 35°}=\frac{20}{\sin 77°}$$

$$\frac{a}{.5736}=\frac{20}{.9563}$$

$$a=\frac{.5736(20)}{.9563}=11.996$$

$$a=12$$

4. (D)

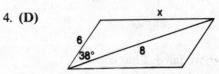

$$a^2=b^2+c^2-2bc \cos A$$

$$x^2=6^2+8^2-2(6)(8) \cos 38°$$

$$=36+64-96(.7880)$$

$$=100-75.65=24.65$$

$$x=\sqrt{24.65}=4.93$$

Answer: $x=5$

Practice Problems 18.1

1. (D) $(-\sqrt{3}, 1)$

Point is in the second quadrant.

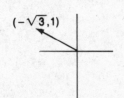

$(-\sqrt{3}, 1)$

$$r^2=(-\sqrt{3})^2+1^2=4$$

$$r=\pm 2$$

$$\theta=\text{Tan}^{-1}\frac{1}{-\sqrt{3}}=$$

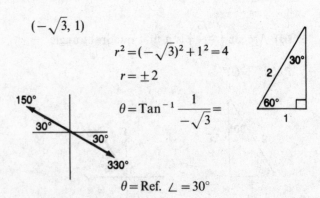

$$\theta=\text{Ref. } \angle=30°$$

From the graph, the point is in the second quadrant

$\therefore \theta=150°$ and the point is $(2, 150°)$. We could also refer to the point as $(-2, 330°)$.

2. **(D)** $r \sin \theta = -2$
From the conversion table

$r \sin \theta = y$

$\therefore y = -2$

3. **(A)** $x + yi = 1 + i\sqrt{3}$ $x = 1$

$(1, \sqrt{3})$ $y = \sqrt{3}$

$r^2 = 1^2 + (\sqrt{3})^2 = 4$ $r = \pm 2$

1st Quad

$\theta = \text{Tan}^{-1} \dfrac{\sqrt{3}}{1} = 60°$

$\theta = \text{reference angle} = 60°$

From the graph, the point is in the first quadrant
$\therefore \theta = 60°$ and the point is $(2, 60°)$. We could also refer to the point as $(-2, 240°)$

$$x + yi = r(\cos \theta + i \sin \theta)$$

$$1 + i\sqrt{3} = 2(\cos 60° + i \sin 60°)$$

or

$$= -2(\cos 240° + i \sin 240°)$$

4. **(E)**

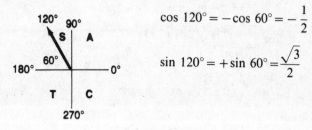

$$\cos 120° = -\cos 60° = -\frac{1}{2}$$

$$\sin 120° = +\sin 60° = \frac{\sqrt{3}}{2}$$

$$x = r \cos \theta = 2\sqrt{6}\left(-\frac{1}{2}\right)$$

$$= -\sqrt{6}$$

$$y = r \sin \theta = 2\sqrt{6}\left(\frac{\sqrt{3}}{2}\right)$$

$$= \sqrt{18} = 3\sqrt{2}$$

$$r(\cos \theta + i \sin \theta) = x + yi$$

$$= -\sqrt{6} + 3\sqrt{2}i$$

Practice Problems 19.1

1. **(C)**

$$e^{i(\pi/2)} = \cos \frac{\pi}{2} + i \sin \frac{\pi}{2}$$

$$= 0 + i(1) = i$$

2. **(B)**

$$\sqrt{3} - i = x + yi \quad x = \sqrt{3} \quad y = -1$$
$$r = \sqrt{x^2 + y^2} = \sqrt{3 + 1} = 2$$
$$\tan \theta = \frac{y}{x} = \frac{-1}{\sqrt{3}} \quad \theta = 330°$$

$$\sqrt{3} - i = 2(\cos 330° + i \sin 330°)$$

$$(\sqrt{3} - i)^5 = [2(\cos 330° + i \sin 330°)]^5$$

$$= 2^5(\cos 5 \cdot 330° + i \sin 5 \cdot 330°)$$

Note: $5(330°) = 1650°$
subtracting four multiples
of 360° yields: $1650°$
$\underline{-1440°}$
$210°$

$$= 2^5(\cos 210° + i \sin 210°)$$

$$= 32\left(\frac{-\sqrt{3}}{2} + i\left(-\frac{1}{2}\right)\right)$$

$$= 16(-\sqrt{3} - i)$$

$$= -16(\sqrt{3} + i)$$

$$\cos 210° = -\cos 30° = -\frac{\sqrt{3}}{2}$$

$$\sin 210° = -\sin 30° = -\frac{1}{2}$$

3. (B)

$2+2i\sqrt{3}=x+yi \quad x=2, \quad y=2\sqrt{3}$

$r=\sqrt{x^2+y^2}=\sqrt{2^2+(2\sqrt{3})^2}$

$\quad =\sqrt{4+12}=4$

$\tan\theta=\dfrac{y}{x}=\dfrac{2\sqrt{3}}{2}=\sqrt{3}, \quad \theta=60°$

$2+2i\sqrt{3}=4(\cos 60°+i\sin 60°)$

$(2+2i\sqrt{3})^{1/2}=[4(\cos 60°+i\sin 60°)]^{1/2}$

$k=0 \qquad =4^{1/2}\left(\cos\dfrac{60°}{2}+i\sin\dfrac{60°}{2}\right)$

$\qquad =2\left(\dfrac{\sqrt{3}}{2}+\dfrac{1}{2}\right)i=\sqrt{3}+1i$

$k=1 \qquad =4^{1/2}\left(\cos\dfrac{60°+360°}{2}+i\sin\dfrac{60°+360°}{2}\right)$

$\qquad =2\left(-\dfrac{\sqrt{3}}{2}-\dfrac{1}{2}\right)=-\sqrt{3}-1i$

4. (C)

$[2(\cos 30°+i\sin 30°)][7(\cos 10°+i\sin 10°)]$

$\quad =2\cdot7[\cos(30°+10°)+i\sin(30°+10°)]$

$\quad =14[\cos 40°+i\sin 40°]$

Practice Problems 20.1

1. (B)

$y=t+1$

$t=y-1$

$x=t^2-2$

$x=(y-1)^2-2$

$x+2=(y-1)^2$

$x+2=y^2-2y+1$

$x=y^2-2y-1$

2. (E)

$y=\tan\theta \qquad x=2\cot\theta$

$y=\dfrac{\sin\theta}{\cos\theta} \qquad x=\dfrac{2\cos\theta}{\sin\theta}$

$\dfrac{\cos\theta}{\sin\theta}=\dfrac{1}{y} \Rightarrow x=2\left(\dfrac{1}{y}\right)$

$x=\dfrac{2}{y}$

$xy=2$

This is a hyperbola.

3. (E)

$x=e^t \qquad y=e^{-t}=\dfrac{1}{e^t}$

$y=\dfrac{1}{e^t} \Rightarrow e^t=\dfrac{1}{y}$

$x=e^t=\dfrac{1}{y}$

$xy=1$

This is a hyperbola.

4. (B)

$y=3\sin\phi \qquad x=3\cos\phi$

$\sin\phi=\dfrac{y}{3} \qquad \cos\phi=\dfrac{x}{3}$

$\sin^2\phi+\cos^2\phi=1 \Rightarrow \left(\dfrac{y}{3}\right)^2+\left(\dfrac{x}{3}\right)^2=1$

$x^2+y^2=9$

This is a circle.

Cumulative Review Test

1. (D) GCF: a

$$\frac{ax+ay+a^2+a}{a}=x+y+a+1$$

$$ax+ay+a^2+a=a(x+y+a+1)$$

2. (C)

$$\frac{r_1(\cos\theta_1+i\sin\theta_1)}{r_2(\cos\theta_2+i\sin\theta_2)}$$

$$=\frac{r_1}{r_2}\left(\cos(\theta_1-\theta_2)+i\sin(\theta_1-\theta_2)\right)$$

$$=\frac{20}{5}\left(\cos(15°-3°)+i\sin(15°-3°)\right)$$

$$=4(\cos 12°+i\sin 12°)$$

3. (D)

$$\tan 810°=\tan 90°$$

$$=\frac{\sin 90°}{\cos 90°}=\frac{1}{0}=\text{Undefined}$$

$$\begin{array}{r} 810° \\ -360° \\ \hline 450° \\ -360° \\ \hline 90° \end{array}$$

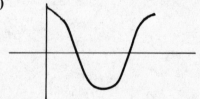

4. (C)

$$b^{-3}\cdot b^{2/3}=b^{-3+(2/3)}=b^{-(9/3)+(2/3)}$$

$$=b^{-7/3}$$

5. (B)

$$\sec^2 x+\csc^2 x=\frac{1}{\cos^2 x}+\frac{1}{\sin^2 x}$$

$$=\frac{\sin^2 x+\cos^2 x}{\sin^2 x\cos^2 x}$$

$$=\frac{1}{\sin^2 x\cos^2 x}$$

6. (A)

$$x+yi=-\sqrt{3}+1i \quad x=-\sqrt{3},\quad y=1$$

$$r=\sqrt{x^2+y^2}=\sqrt{3+1}=2$$

$$\tan\theta=\frac{y}{x}=\frac{1}{-\sqrt{3}}\quad \theta_{\text{ref}}=30°\quad \theta=150°$$

$$-\sqrt{3}+1i=2(\cos 150°+i\sin 150°)$$

7. (D)

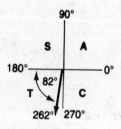

$$\cos 262°=-\cos 82°$$

$$=-\sin 8° \text{ (by cofunction rules)}$$

8. (D)

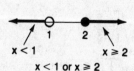

is a cos curve

with amplitude 2 and period 2π

$\therefore$ frequency $=1$.

$$y=2\cos x$$

9. (E)

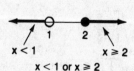

$$x<1 \text{ or } x\geq 2$$

10. (E)

$$x = 4 + 3\tan\theta \qquad y = -1 + 2\sec\theta$$

$$\tan\theta = \frac{x-4}{3} \qquad \sec\theta = \frac{y+1}{2}$$

$$\tan^2\theta + 1 = \sec^2\theta$$

$$\left(\frac{x-4}{3}\right)^2 + 1 = \left(\frac{y+1}{2}\right)^2$$

$$\left(\frac{x^2 - 8x + 16}{9}\right) + 1 = \frac{y^2 + 2y + 1}{4}$$

Multiply the entire equation by 36 to clear the fractions.

$$4x^2 - 32x + 64 + 36 = 9y^2 + 18y + 9$$

$$4x^2 - 32x - 9y^2 - 18y + 91 = 0$$

This is a hyperbola.

11. (B)

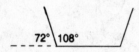

Each interior angle $= 108°$

Each exterior angle $= 180° - 108° = 72°$

Each exterior angle $= \dfrac{360}{n} = 72 \qquad n = 5$

5 sides

12. (D)

$$\tan\tfrac{1}{2}A = \sqrt{\frac{1 - \cos A}{1 + \cos A}}$$

$$= \sqrt{\frac{1 - \frac{4}{5}}{1 + \frac{4}{5}}}$$

$$= \sqrt{\frac{\frac{1}{5}}{\frac{9}{5}}} = \sqrt{\frac{1}{9}} = \frac{1}{3}$$

13. (B)

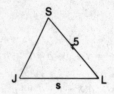

$$\frac{j}{\sin J} = \frac{s}{\sin S}$$

$$\frac{5}{.35} = \frac{s}{.21} \qquad s = \frac{.21(5)}{.35}$$

$$= 3$$

14. (C)

$$\frac{\sqrt{3} + i}{2} = \frac{\sqrt{3}}{2} + \frac{1}{2}i \quad x = \frac{\sqrt{3}}{2}, \ y = \frac{1}{2}$$

$$r = \sqrt{x^2 + y^2} = \sqrt{\tfrac{3}{4} + \tfrac{1}{4}} = 1$$

$$\tan\theta = \frac{y}{x} = \frac{1}{\sqrt{3}}$$

$$\theta_{\text{ref}} = 30°, \ \theta = 30°$$

$$\frac{\sqrt{3} + i}{2} = 1(\cos 30° + i \sin 30°)$$

$$\left(\frac{\sqrt{3} + i}{2}\right)^{12} = 1^{12}(\cos 12 \cdot 30° + i \sin 12 \cdot 30°)$$

$$= 1(\cos 360° + i \sin 360°)$$

$$= 1(1 + i(0)) = 1$$

15. (E)

$$\tan\theta = \frac{x}{2} \qquad \cot\theta = y$$

$$\tan\theta = \frac{1}{\cot\theta} \qquad \therefore \frac{x}{2} = \frac{1}{y}$$

$$xy = 2$$

This is a hyperbola.

16. (E)

$$x^2 - 2x \qquad + y^2 + 6y \qquad = -4$$

$$\underbrace{x^2 - 2x + (-1)^2}_{} + \underbrace{y^2 + 6y + (+3)^2}_{} = -4 + (-1)^2$$

$$+ (3)^2$$

$$(x-1)^2 \qquad + (y+3)^2 \qquad = 6$$

center: $(1, -3)$ \qquad radius: $\sqrt{6}$

17. **(A)**   First quadrant

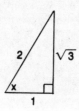

$$\tan x = \frac{\sqrt{3}}{1} = \sqrt{3}$$

18. **(C)**

$$c^2 = a^2 + b^2 - 2ab \cos c$$
$$= 9 + 25 - 30(-\tfrac{1}{2})$$
$$= 34 + 15 = 49$$
$$c = 7$$

19. **(B)** The identity

$\sin(x + y) = \sin x \cos y + \sin y \cos x$,
where $x = 35°$ and $y = 22°$.

$\sin 35° \cos 22° + \sin 22° \cos 35°$
$$= \sin(35° + 22°) = \sin 57°$$

20. **(B)**

Using equations 1 with 3 and equation 2 alone eliminate z

$2x - y + z = 5$
$6x + 2y + 2z = 7$
Multiply the first equation by -2 and add it to the second equation to eliminate z.
$-4x + 2y - 2z = -10$
$\underline{6x + 2y + 2z = 7}$

$\boxed{2x + 4y = -3}$

$4x - 3y = 5$

The variable z is already removed

$\boxed{4x - 3y = 5}$

$2x + 4y = -3$
$4x - 3y = 5$

Multiply first equation by -2 and add to the second to eliminate x.

$-4x - 8y = 6$
$\underline{4x - 3y = 5}$
$-11y = 11$

$\boxed{y = -1}$

Substitute $y = -1$ into one of the 2 variable equations

$$2x + 4y = -3$$
$$2x + 4(-1) = -3$$
$$2x - 4 = -3$$
$$2x = 1$$

$\boxed{x = \tfrac{1}{2}}$

Substitute $y = -1$ and $x = \tfrac{1}{2}$ into one of the original 3 variable equations.

$$2x - y + z = 5$$
$$2(\tfrac{1}{2}) - (-1) + z = 5$$
$$1 + 1 + z = 5$$
$$2 + z = 5$$

$\boxed{z = 3}$

UNIT SIX

21. Variation

21.1 Basics and Beyond

Types of Variation

Type	Equation of Variation	Using Ratios	Alternate Phrase	Sentence
Direct	$x = Ky$	$\dfrac{x_1}{y_1} = \dfrac{x_2}{y_2} \cdots = K$	proportional	x varies directly as y
Direct (Square)	$x = Ky^2$	$\dfrac{x_1}{y_1^2} = \dfrac{x_2}{y_2^2} \cdots = K$	proportional	x varies directly as the square of y
Inverse	$x = \dfrac{K}{y}$	$x_1 y_1 = x_2 y_2 \cdots = K$	indirectly	x varies inversely as y
Inverse (Square)	$x = \dfrac{K}{y^2}$	$x_1 y_1^2 = x_2 y_2^2 = \cdots = K$	indirectly	x varies inversely as the square of y
Joint	$x = Kyz$	$\dfrac{x_1}{y_1 z_1} = \dfrac{x_2}{y_2 z_2} \cdots = K$		x varies jointly as y and z
Combined	$x = \dfrac{Kyz}{w}$	$\dfrac{x_1 w_1}{y_1 z_1} = \dfrac{x_2 w_2}{y_2 z_2} \cdots = K$ note: K is a constant		x varies jointly with y and z and inversely as w

Examples:

1. Determine if the table expresses a direct variation, indirect variation or neither.

x	y
4	8
$\frac{1}{2}$	1
1	2
$\frac{1}{3}$	$\frac{2}{3}$
50	100

Solutions:

1. Test $x_1 y_1 = x_2 y_1 = K$

   $4(8) \neq \frac{1}{2}(1)$ Not inverse

   Test $\dfrac{x_1}{y_1} = \dfrac{x_2}{y_2} \cdots = K$

   $\dfrac{4}{8} = \dfrac{\frac{1}{2}}{1} = \dfrac{1}{2} = \dfrac{\frac{1}{3}}{\frac{2}{3}} = \dfrac{50}{100} = \dfrac{1}{2}$

   Variation is direct

337

2. If x varies jointly as y and the $\sqrt{z}$, and if $x=3$ when $y=\frac{1}{4}$ and $z=16$, find x when $y=2$ and $z=25$.

2.
$$\frac{x_1}{y_1\sqrt{z_1}}=\frac{x_2}{y_1\sqrt{z_2}}$$

$$\frac{3}{\frac{1}{4}\sqrt{16}}=\frac{x}{2\sqrt{25}}$$

$$\frac{3}{1}=\frac{x}{2\sqrt{25}}$$

$$\frac{3}{1}=\frac{x}{10}$$

$$x=30$$

Practice Problems 21.1

1. If x varies inversely as the square of y, and if $x=3$ when $y=4$, find x when $y=2$.

   A) 96 B) 48 C) 24 D) 12 E) $\sqrt{12}$

2. If y varies inversely as x, when x is multiplied by 2, y would

   A) be multiplied by 2
   B) be divided by 2
   C) be multiplied by 4
   D) be multiplied by $\frac{1}{2}$
   E) remain unchanged

3. If y varies directly as x and $y=\frac{1}{4}$ when $x=\frac{1}{2}$, find y when $x=4$.

   A) $\frac{1}{2}$ B) 2 C) $\frac{1}{4}$ D) 4 E) $\frac{1}{8}$

4. If x varies jointly as y and z and inversely as the square of w, and if $x=3$ when $z=6$, $y=8$ and $w=4$, find x when $z=10$, $y=1$ and $w=2$.

   A) $\frac{5}{4}$ B) 43.2 C) 28.8 D) $\frac{10}{4}$ E) 5

Solutions on page 359

22. Verbal Problems

22.1 Fundamentals

This section is relegated to a brief review of the basic verbal problems.

Type	Problem	Solution
Number	The greater of two numbers is 5 more than twice the smaller. If the sum of the numbers is 17, find the greater number.	Let $x=$ greater number $y=$ smaller number $x+y=17$ $x=2y+5$ $x+y=17$ $(2y+5)+y=17$ $3y=12$ $y-4$ $x=\boxed{13}$
Consecutive Integer	The sum of three consecutive integers is 57. Find the integers.	Let $x=$ first integer $x+1=$ second integer $x+2=$ third integer $x+(x+1)+(x+2)=57$ $3x+3=57$ $3x=54$ $x=\boxed{18}$ $x+1=\boxed{19}$ $x+2=\boxed{20}$
Consecutive Integers (odd/even)	Find the second of three consecutive odd integers if the sum of the first and third is 38.	Let $x=$ first odd integer $x+2=$ second odd integer $x+4=$ third odd integer $x+(x+4)=38$ $2x+4=38$ $2x=34$ $x=17$ $x+2=\boxed{19}$

Coin	Joan has 45 coins which are worth $3.50. If the coins are nickels and dimes only, how many dimes does she have?	Let x = number of nickels $45 - x$ = number of dimes Unit Type Qty × (Value) = Worth

Type	Qty	Unit (Value)	Worth
Nickels	x	5¢	$5x$¢
Dimes	$45 - x$	10¢	$(45 - x)10$¢

$$\text{Total worth} = 5x + (45 - x)10 = 350 \text{ (in cents)}$$
$$5x - 10x + 450 = 350$$
$$-5x = -100$$
$$x = 20 \text{ nickels}$$

$$45 - x = \boxed{25 \text{ dimes}}$$

Age	Mr. Unger is 24 years older than his son Jerry. In 8 years, Mr. Unger will be twice as old as Jerry will be then. How old is Mr. Unger now?	Let x = Jerry's age now $x + 24$ = Mr. Unger's age now

	Age now	Age in 8 yrs
Mr. Unger	$x + 24$	$x + 24 + 8$
Jerry	x	$x + 8$

$$\text{In 8 yrs } (x + 24 + 8) = 2(x + 8)$$
$$x + 32 = 2x + 16$$
$$x = 16, \; x + 24 = \boxed{40}$$

Investment	Joan invests $7200, part at 4% and the rest at 5%. If the annual income from both investments was the same, find her total income from each investment.	Let x = amount invested at 4% $7200 - x$ = amount invested at 5%

Principal	× Rate	= Income
x	.04	$.04x$
$7200 - x$	.05	$.05(7200 - x)$

$$.04x = .05(7200 - x)$$
$$9x = 5(7200)$$
$$x = 4000$$
$$\text{Investment income} = .04(4000) = \boxed{160}$$

Mixture	Shirley mixes nuts worth 36¢ a pound with nuts worth 52¢ a pound to make a 300 pound mixture worth 40¢ a pound. How many pounds of the 52¢ nuts did she use?	Let x = amt of the 52¢ nuts $300 - x$ = amt of the 36¢ nuts (¢ per pound) × (# of pounds) = Cost $36x + 52(300 - x) = 40(300)$ $36x + 52(300) - 52x = 40(300)$ $52(300) - 40(300) = 52x - 36x$ $$\frac{12(300)}{16} = \frac{16x}{16}$$ $$\frac{900}{4} = x$$ $225 = x$ $300 - x = \boxed{75}$

Table (cost):

(¢ per pound)	(# of pounds)	Cost
36¢	x	$36x$
52¢	$300 - x$	$52(300 - x)$
40¢	300	$40(300)$

Motion (same and opposite direction)	At 10 a.m. two cars started traveling toward each other from towns 287 miles apart. They passed each other at 1:30 p.m. If the rate of the faster car exceeds the rate of the slower car by 6 m.p.h., find the rate in miles per hour of the faster car.	Let x = rate of slower car $x + 6$ = rate of faster car

Rate × Time = Distance

Rate	Time	Distance
x	$3\frac{1}{2}$ hrs	$(3\frac{1}{2})(x)$
$x + 6$	$3\frac{1}{2}$ hrs	$(3\frac{1}{2})(x + 6)$

10 a.m. to 1:30 p.m. was $3\frac{1}{2}$ hrs

$$\frac{7}{2}(x) + \frac{7}{2}(x + 6) = 287$$
$$\frac{7}{2}x + \frac{7}{2}x + \frac{7}{2}(6) = 287$$
$$7x + 21 = 287$$
$$7x = 266$$
$$x = 38$$
$$x + 6 = \boxed{44}$$

Motion (overtake)	Susie left her home at 11 a.m. traveling at 30 m.p.h. At 1 p.m. Norm left home and started after her on the same road at 45 m.p.h. At what time did Norm catch up to Susie?	Let t = time Susie traveled $t-2$ = time Norm traveled Rate $\times$ Time = Distance Susie | 30 | t | $30t$ Norm | 45 | $t-2$ | $45(t-2)$ To catch up means to have gone the same distance. $$30t = 45(t-2)$$ $$90 = 15t$$ $$6 = t$$ $\therefore$ 11 a.m. + 6 hrs = 5 p.m.
Work	Joan can paint a house in 6 hours. Her younger sister Karen can do the same job in 9 hours. In how many hours can they do the job if they work together?	Joan's rate = $\frac{1}{6}$ Karen's rate = $\frac{1}{9}$ Let x = amount of hours they worked together Rate $\times$ Time = Did Joan | $\frac{1}{6}$ | x | $\frac{x}{6}$ Karen | $\frac{1}{9}$ | x | $\frac{x}{9}$ $$\frac{x}{6}+\frac{x}{9}=1 \quad \text{(one job)}$$ $$\frac{3x+2x}{18}=1$$ $$5x=18$$ $$\boxed{x=\tfrac{18}{5}\ \text{hrs}}$$
Percent	If 25% of a number is 80, find the number.	Let x = the number Note: % means division by 100, "of" implies multiplication $$\frac{25}{100}\cdot x = 80$$ $$x = \frac{8000}{25} = \boxed{320}$$

Percent (cont.)	Of 140 seniors, 35 were on the wrestling team. What percent of the seniors were on the wrestling team?	$\% = \dfrac{\text{part}}{\text{whole}} \times 100$ $= \dfrac{35}{140} \times 100 = \dfrac{1}{4} \times 100 = \boxed{25\%}$
	6 is 20% of what number?	Let $x =$ the number $6 = \frac{20}{100} \cdot x$ $\frac{600}{20} = x$ $\boxed{30 = x}$
Perimeter	If the sides of a pentagon are x, x, $x+1$, $2x+1$ and $5x-7$, and the perimeter is 45, find the largest side.	$(x)+(x)+(2x+1)+(x+1)+(5x-7)=45$ $10x - 5 = 45$ $x = 5$ The largest side is $5x-7 = 25-7 = \boxed{18}$
Area	The length of a rectangle is 7 less than 3 times the width. If the area is 6, find the width of the rectangle.	Let $x =$ width of rectangle $3x - 7 =$ length of rectangle $A = lw = (3x-7)(x) = 6$ $3x^2 - 7x - 6 = 0$ $(3x+2)(x-3) = 0$ $3x+2 = 0 \quad \mid \quad x-3 = 0$ $x = -\frac{2}{3} \quad \mid \quad \boxed{x=3}$ Reject
Digit	The sum of the digits of a two-digit number is 9. If the number with the digits reversed is subtracted from 3 times the original number the difference is 9. Find the number.	Let $t =$ ten digits $u =$ units digit $t + u =$ sum of the digits $10t + u =$ original number $10u + t =$ number with the digits reversed ――――――― $t + u = 9$ $3(10t + u) - (10u + t) = 9$ ――――――― $30t + 3u - 10u - t = 9$ $29t - 7u = 9 \qquad t + u = 9$ $29(9 - u) - 7u = 9 \qquad \therefore t = 9 - u$ $261 - 29u - 7u = 9$ $-36u = -252$ $u = 7$ $t = 2$ Number $= \boxed{27}$

Ratio	An angle which measures 150° is divided into three parts whose ratios are $2:5:8$. Find the measure of the three parts.	Let $2x =$ the measure of the part of ratio 2 $5x =$ the measure of the part of ratio 5 $8x =$ the measure of the part of ratio 8 $$2x + 5x + 8x = 150$$ $$15x = 150$$ $$x = 10$$ $$2x = 20$$ $$5x = 50$$ $$8x = 80$$
Current	A boat can travel 12 m.p.h. in still water. If it can travel 22 miles down a stream in the same time that it can travel 11 miles up the stream, what is the rate of the stream?	$s =$ rate of the stream Distance ÷ Rate = Time Up: 11, $12-s$, $\dfrac{11}{12-s}$ Down: 22, $12+s$, $\dfrac{22}{12+s}$ $$D = RT \qquad \frac{D}{R} = T$$ $$\frac{11}{12-s} = \frac{22}{12+s}$$ $$11(12+s) = 22(12-s)$$ $$11(12) + 11s = 22(12) - 22s$$ $$33s = 22(12) - 11(12) = 11(12)$$ $$33s = 11(12)$$ $$s = \frac{11(12)}{33} \qquad S = \boxed{4}$$
Percent Mixture	If a 40% acid solution is mixed with a 60% acid solution how much of each is necessary to yield 60 gallons of a 50% acid solution?	$\left(\begin{array}{c}\text{Amount}\\\text{of solution}\end{array}\right) \times (\%\text{ acid}) = \left(\begin{array}{c}\text{Amount}\\\text{of acid}\end{array}\right)$ x, $.40$, $.40x$ $60-x$, $.60$, $.60(60-x)$ 60, $.50$, $.50(60)$ Let $x =$ amount of 40% solution $60 - x =$ amount of 60% solution $$.40x + .60(60-x) = .50(60)$$ Multiply equation by 100 to clear the decimals. $$40x + 60(60-x) = 50(60)$$ $$4x + 360 - 6x = 300$$ $$-2x = -60$$ $$x = 30$$ $$60 - x = 30$$ 30 gallons of each

Practice Problems 22.1

1. The tens digit of a two-digit number is three times the units digit. The number obtained by reversing the digits is 18 less than the original number. Find the original number.

   A) 84   B) 31   C) 26   D) 13   E) 93

2. How many ounces of pure acid must be added to 30 ounces of a 20% solution of acid to make it a 50% solution?

   A) 18   B) 30   C) 28   D) 13   E) 180

3. A new machine can package in 3 hours the whole day's production. An older machine takes 5 hours to do the same job. If both machines are run concurrently, how many hours will it take for them to package a day's output?

   A) $\frac{5}{3}$ B) $\frac{3}{5}$ C) $\frac{8}{15}$ D) $\frac{15}{8}$ E) 8

4. A car leaves town A at 1:00 p.m. traveling at 45 m.p.h. A second car leaves town A at 2:00 p.m. traveling the same route at 55 m.p.h. How many hours will it take the second car to overtake the first car?

   A) 1   B) $5\frac{1}{2}$ C) 10   D) $4\frac{1}{2}$ E) $3\frac{3}{4}$

5. A man invests $6000 for one year, part in a certificate which pays 16%, and the remainder in bonds which pay 9%. If the annual income from both investments is the same, find the amount invested at 16%.

   A) 2160   B) 3840   C) 2750
   D) 3250   E) 3000

6. Find the largest of three positive consecutive odd integers such that the square of the smallest is 9 more than the sum of the other two.

   A) 3   B) 5   C) 7   D) 9   E) 11

Solutions on page 360

23. Arithmetic and Geometric Progressions and Series

23.1 Essentials

Arithmetic Progression: A progression of terms such that the difference between any term and the term immediately preceeding it is a constant.

Arithmetic Series: The sum of the terms of an arithmetic progression.

Geometric Progression: A progression of terms such that the ratio between any term and the term immediately preceeding it is a constant.

Geometric Series: The sum of the terms of a geometric progression.

$$\underset{\text{1st term}}{a_1 = a}, \underset{\text{2nd term}}{a_2}, \underset{\text{3rd term}}{a_3}, \underset{\text{4th term}}{a_4}, \ldots \underset{k\text{th term}}{a_k}, \underset{k+1 \text{ term}}{a_{k+1}}, \ldots \underset{n-1 \text{ term}}{a_{n-1}}, \underset{n\text{th term}}{a_n = l}$$

The a's represent the values of the respective terms.

a (without a subscript) usually represents the value of a_1 the first term.

l usually represents the value of a_n or the last term.

n represents the number of terms involved.

	Arithmetic	Geometric
Common Difference	$d = a_{k+1} - a_k$	—
Common Ratio	—	$r = \dfrac{a_{k+1}}{a_k}$
Mean	Given two real numbers a and b **Arithmetic Mean** $= \dfrac{a+b}{2}$	Given two real numbers a and b **Geometric Mean** $= \pm\sqrt{ab}$ $\sqrt{ab}$ if a and b are positive $-\sqrt{ab}$ if a and b are negative
Last Term (a_n)	$l = a_n = a + (n-1)d$	$l = a_n = a \cdot r^{n-1} \quad (n \geq 1)$
Sum of the First n Terms	$S = \dfrac{n}{2}(a+l)$ $= \dfrac{n}{a}(2a + (n-1)d)$	$S = \dfrac{a - rl}{1-r} \quad (r \neq 1)$ $= \dfrac{a(1-r^n)}{1-r}$

	Arithmetic	**Geometric**
Sum of an Infinite Series	—	$S_\infty = \dfrac{a}{1-r}$ $\quad(-1 < r < 1)$
Sample: Using Letters	$a,\ a+d,\ a+2d,\ a+3d,\ \ldots\ a+(n-1)d$ $\displaystyle\sum_{k=1}^{n} a+(k-1)d$	$a,\ ar,\ ar^2,\ ar^3\ \cdots\ ar^{n-1}$ $\displaystyle\sum_{k=1}^{n} ar^{k-1}$

$\sum$ is the summation symbol. Σ is the capital greek letter sigma.	$\prod$ is the product symbol. Π is the capital greek letter pi.

$\displaystyle\sum_{k=a}^{b} (3k+7)$ means: Substitute the integers consecutively from a to b inclusive for the value of k. Once all the terms have been established—add them.

$\displaystyle\sum_{k=1}^{5} (2k-1)$

$= [2(1)-1] + [2(2)-1] + [2(3)-1] + [2(4)-1] + [2(5)-1]$
$= 1+3+5+7+9 = 25$

(Note: This is an arithmetic progression.)

$\displaystyle\prod_{k=a}^{b} (3k+7)$ means: Substitute the integers consecutively from a to b inclusive for the value of k. Once all the terms have been established—multiply them.

$\displaystyle\prod_{k=1}^{4} (2k-1)$

$= [2(1)-1] \cdot [2(2)-1] \cdot [2(3)-1] \cdot [2(4)-1]$

$= 1 \cdot 3 \cdot 5 \cdot 7 = 105$

To convert a never ending repeating decimal to $\dfrac{a}{b}$ form.

This process can be achieved using an infinite geometric series, but there is an easier procedure.

Step	Description	Example Convert $.3333\overline{3}$ to $\dfrac{a}{b}$ form
1	Let $x =$ the repeating decimal	$x = .333\overline{3}$
2	Represent $10x$	$10x = 3.333\overline{3}$
3	The goal is to find any two 10^n multiples of x, i.e., $10x$, $100x$, $1000x$, etc., which repeats exactly the same endless decimal (n is an integer)	$10x$ has decimal $.333\overline{3}$ x has decimal $.333\overline{3}$ These are the same!
4	Subtract these 10^n multiples of x to eliminate the decimals	$\begin{aligned}10x &= 3.333\overline{3} \\ x &= .333\overline{3} \\ \hline 9x &= 3\end{aligned}$
5	Solve for x and reduce	$x = \tfrac{3}{9} = \tfrac{1}{3}$

Examples:

1. Find the 21st term of the progression

$$1, 4, 7, 10 \ldots$$

2. Find the sum of the first 20 terms of the series

$$1+4+7+10+ \ldots$$

3. Find the two geometric means between 3 and 24.

4. Express $.1666\overline{6}$ in $\dfrac{a}{b}$ form.

Solutions:

1. $d=4-1=3, \quad a=1, \quad n=21$

$$\begin{aligned}
a_n=a_{21}=l &=a+(n-1)d \\
l &=1+(21-1)3 \\
&=1+20(3) \\
&=61
\end{aligned}$$

2. $S_{20}=\dfrac{n}{2}(a+l), \quad d=4-1=3$

$$l=a+(n-1)d$$
$$l=1+(20-1)3$$
$$=1+19(3)$$
$$=58$$
$$S_{20}=\dfrac{20}{2}(1+58)=590$$

3. $\underline{3} \; \_\_ \; \_\_ \; \underline{24}, \quad n=4$

$$\begin{aligned}
a_4=l &= ar^{n-1} \\
24 &= 3r^3 \\
r^3 &= 8 \\
r &= 2
\end{aligned}$$

$$\therefore \quad \underline{3} \quad \underline{3\cdot 2} \quad \underline{3\cdot 2^2} \quad \underline{24}$$

$$\text{6 and 12}$$

4. Method 1

$$x= \quad .1666\overline{6}$$
$$\left.\begin{array}{l} 10x= \; 1.6666\overline{6} \\ 100x= 16.6666\overline{6} \end{array}\right\} \quad \begin{array}{l} 100x=16.6666\overline{6} \\ \underline{10x= \; 1.6666\overline{6}} \\ 90x=15 \end{array}$$

$$x=\dfrac{15}{90}=\dfrac{1}{6}$$

Method 2

$$x=.1+\underbrace{.06+.006+.0006 \ldots}$$
$$\text{This is a geometric series.}$$

$$r=\dfrac{.006}{.06}=.1$$

$$S_\infty=\dfrac{a}{1-r}=\dfrac{.06}{1-.1}=\dfrac{.06}{.9}=\dfrac{6}{90}=\dfrac{1}{15}$$

$$x=.1+\underbrace{\dfrac{1}{15}}_{S_\infty}=\dfrac{1}{10}+\dfrac{1}{15}=\dfrac{25}{150}=\dfrac{1}{6}$$

Practice Problems 23.1

1. Find the 18th term of the arithmetic progression

   $$2, 7, 12, 17 \ldots$$

   A) 90   B) 98   C) 78   D) 87   E) 92

2. Find the 8th term of the geometric progression

   $$5, 1, \tfrac{1}{5}, \tfrac{1}{25} \ldots$$

   A) $\frac{1}{15625}$ B) $\frac{1}{625}$ C) $\frac{1}{125}$ D) 6.4
   E) $\frac{1}{3125}$

3. Find the sum of the first 20 terms of the series

   $$2 + 6 + 10 + 14 + \ldots$$

   A) 810   B) 850   C) 860
   D) 840   E) 800

4. Find the two arithmetic means between 3 and 18.

   A) 10 and 11   B) 8 and 13   C) $10\frac{1}{2}$ only
   D) 6 and 12   E) 8 and -8

5. Find the sum of the first 10 terms of the geometric series

   $$2 + 4 + 8 + 16 + \ldots$$

   A) -2048 B) -2046 C) 2048
   D) 2046   E) 1023

6. Write $.1111\bar{1}$ in $\frac{a}{b}$ form.

   A) $\frac{11}{100}$ B) $\frac{111}{1000}$ C) $\frac{1}{9}$ D) $\frac{1}{90}$ E) $\frac{10}{9}$

7. Find the sum of the infinite geometric series

   $$\tfrac{1}{2} + \tfrac{1}{4} + \tfrac{1}{8} + \tfrac{1}{16} + \ldots$$

   A) 1   B) 2   C) 3   D) $\frac{19}{20}$ E) $\frac{153}{154}$

8. $\displaystyle\sum_{k=1}^{99}(k+3) =$

   A) 396   B) 5247   C) 1030
   D) 5427   E) 10197

9. Find the result of the infinite product

   $17^{1/2} \cdot 17^{1/4} \cdot 17^{1/8} \cdot 17^{1/16} \ldots 17^{1/2n} \ldots$

   A) 17   B) $17^{\frac{111}{112}}$ C) 17171717
   D) 17^{17} E) 1

Solutions on page 361

24. Permutations, Combinations, Probability and the Binomial Theorem

24.1 Permutations

Permutation: An *arrangement* of items in some *specified order.*

Fundamental Counting Principle: If one activity can be done in any of g ways and once completed a second independent activity can be done in h ways, then both activities can be done in the stated order in $g \cdot h$ ways.

Permutation of n things taken n at a time	$_nP_n = n! = n(n-1)(n-2)\ldots(3)(2)(1)$
Permutation of n things taken r at a time	$_nP_r = P(n, r) = {}^nP_r = P_r^n = P_{n,r} =$ $\underbrace{n(n-1)(n-2)\cdots}_{r \text{ factors}} = \dfrac{n!}{(n-r)!}$
The number of distinguishable permutations of n objects of which a are alike	$\dfrac{n!}{a!}$
The number of circular permutations of n things	$(n-1)!$
Math Notes	$0! = 1$ $\qquad$ $n! = n \cdot (n-1)!$

Examples:

1. A diner has 3 choices of appetizer, 2 choices of soup, 4 choices of entrée and 10 choices of dessert. How many different four-course meals are available?

2. How many different 7-letter permutations can be formed from the word "college"?

Solutions:

1. By the fundamental counting principle:

$$\underset{\text{appetizer}}{3} \cdot \underset{\text{soup}}{2} \cdot \underset{\text{entrée}}{4} \cdot \underset{\text{dessert}}{10} = 240$$

2. College has 7 letters
   2 l's
   2 e's

Number of distinguishable permutations

$$= \frac{7!}{2!2!} = 1260$$

for the   for the
2-l's   2-e's

3. In how many arrangements can 8 people be seated around a circular table?

4. How many different 6-letter permutations are possible from the letters in the word "waiter" with a "w" or "e" as the first letter?

3. Circular permutations $= (n-1)! = 7! = 5040$

4.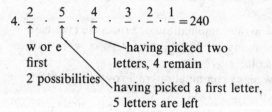

$$\frac{2}{\uparrow} \cdot \frac{5}{\uparrow} \cdot \frac{4}{\uparrow} \cdot \frac{3}{\_} \cdot \frac{2}{\_} \cdot \frac{1}{\_} = 240$$

w or e first
2 possibilities

having picked two letters, 4 remain

having picked a first letter, 5 letters are left

Practice Problems 24.1

1. How many different 4-digit permutations are there from the numeral 1981?

   A) 10   B) 19   C) 12   D) 24   E) 48

2. There are 5 boys and 3 girls in a math class. In how many ways can the 8 students be arranged in a line?

   A) 40320   B) 1024   C) 56
   D) 20   E) 8

3. There are 5 boys and 3 girls in a math class. In how many different ways may the 8 students be ar-

ranged in a line if all 3 girls are to be in line ahead of the 5 boys.

   A) 40320   B) 720   C) 620
   D) 500   E) 210

4. There are 4 different roads from town A to town B, 6 different roads from town B to town C and 2 different roads from town C to town D. How many different direct routes are available to travel from town A to town D, through towns B and C?

   A) 6   B) 12   C) 24   D) 36   E) 48

Solutions on page 362

24.2 Combinations

Combination: A selection (picking) of items regardless of order.

Combinations of n things taken r at a time	$\begin{aligned} {}_nC_r &= \binom{n}{r} = {}^nC_r = C(n, r) \\ &= C_{n,r} \\ &= \frac{{}_nP_r}{r!} = \frac{n!}{r!(n-r)!} \qquad r \geqslant n > 0 \end{aligned}$
Math Notes	${}_nC_r = {}_nC_{n-r}$ $\qquad$ ${}_nC_n = {}_nC_0 = 1$ • **At least:** means equal to or greater than • **At most:** means less than or equal to

Example:

A vase contains 4 yellow roses, 3 pink roses and 5 red roses.

a) How many combinations of 3 roses are possible?

b) How many combinations of 3 roses will be of one color only?

c) How many combinations of 3 roses will contain exactly one rose of each color?

d) How many combinations of 3 roses will contain at least 2 yellow roses?

Solutions:

1. a) This is selecting any 3 of the 12 $(4+3+5)$ roses

$$_{12}C_3 = \frac{12 \cdot 11 \cdot 10}{3 \cdot 2} = \boxed{220}$$

b) $\quad _4C_3 \quad$ or $\quad _3C_3 \quad$ or $\quad _5C_3$

selecting 3 $\quad$ selecting 3 $\quad$ selecting 3

yellow $\qquad$ pink $\qquad$ red

out of 4 $\qquad$ out of 3 $\qquad$ out of 5

☞ "or" indicates + (adding)

"and" indicates · (multiplication)

$_4C_3 + {_3C_3} + {_5C_3} = 4+1+10 = \boxed{15}$

c) $\quad _4C_1 \quad$ and $\quad _3C_1 \quad$ and $\quad _5C_1$

selecting 1 $\qquad$ 1 pink $\qquad$ 1 red

yellow $\qquad$ out of 3 $\qquad$ out of 5

out of 4

$= {_4C_1} \cdot {_3C_1} \cdot {_5C_1}$

$= 4 \cdot 3 \cdot 5 = \boxed{60}$

d) At least 2 yellow roses:

2 yellow and 1 non-yellow

$= \quad _4C_2 \quad \cdot \quad _8C_1 \qquad = 6 \cdot 8 = 48$

2 yellow $\quad$ 1 non-yellow

out of 4 $\qquad$ out of 8

$\qquad\qquad$ non yellow

or

3 yellow: $\quad _4C_3 \qquad = 4$

3 yellow

out of 4

$({_4C_2} \cdot {_8C_1}) + {_4C_3} = 48 + 4 = 52$

Practice Problems 24.2

1. There are 5 boys and 3 girls in a class. How many 6-member committees can be formed consisting of exactly 4 boys and 2 girls?

A) 720 B) 48 C) 24 D) 15 E) 8

2. A vase contains 6 yellow, 4 red and 2 pink roses. How many selections of 4 roses will have 2 yellow, 1 red and 1 pink?

A) 2 B) 4 C) 48 D) 100 E) 120

3. Evaluate $_{30}C_{28}$.

A) 5040 B) 840 C) 435
D) 345 E) 58

4. From a regular deck of 52 cards, in how many ways can 5 cards be selected containing all four aces?

A) 0 B) 1 C) 4 D) 48 E) 120

Solutions on page 362

24.3 Probability

Probability of an event:	$\dfrac{\text{Number of favorable or successful outcomes}}{\text{Total number of all alternative ways}}$

Probability of *A or B* (**inclusive** events)	$P(A) + P(B) - P(A \text{ and } B)$
Probability of *A or B* (**exclusive** events)	$P(A) + P(B)$
Probability of *A and B* (**independent** events)	$P(A) \cdot P(B)$
Probability of *A given B* (**dependent** events)	$P(A \text{ given } B) = \dfrac{P(A \text{ and } B)}{P(B)}$

Exclusive: not more than one of the events can occur in a single trial

Inclusive: more than one event can occur in a single trial

Independent: if one event does not affect the other event

Probability of *r* successes out of *n* attempts p = probability of success in a single trial q = probability of failure $(1-p)$	$_nC_r p^r q^{n-r}$
Probability (**Certainty**) = 1 Probability (**Never or Impossibility**) = 0	$P(A \cup B)$ = probability (A **or** B) $P(A \cap B)$ = probability (A **and** B) $P(A/B)$ = probability (A **given** B)

Examples:

1. With one fair die, find the probability of throwing 3 fours in 5 attempts.

2. If three cards are selected from an ordinary deck, find the probability that all three are black.

3. Given a parallelogram, a rhombus, a rectangle and a square, if one of these quadrilaterals is selected at random, what is the probability that its diagonals bisect each other?

Solutions:

1. Probability of a four $= \frac{1}{6}$.

$$p = p(\text{success}) = \tfrac{1}{6} \quad q = p(\text{failure}) = \tfrac{5}{6}$$
$$n = 5 \text{ attempts}, \quad r = 3 \text{ successes}$$
$$_5C_3 (\tfrac{1}{6})^3 (\tfrac{5}{6})^2 = (10)(\tfrac{1}{216})(\tfrac{25}{36}) = (\tfrac{125}{3888})$$

2. Number of ways to succeed $= {}_{26}C_3$, selecting 3 black out of 26.

Total number of ways $= {}_{52}C_3$, selecting 3 cards out of 52.

$$P = \frac{{}_{26}C_3}{{}_{52}C_3} = \frac{26 \cdot 25 \cdot 24}{3 \cdot 2 \cdot 1} \cdot \frac{3 \cdot 2 \cdot 1}{52 \cdot 51 \cdot 50} = \frac{\cancel{26} \cdot \cancel{25} \cdot 24}{\cancel{3 \cdot 2 \cdot 1}} \cdot \frac{\cancel{3 \cdot 2 \cdot 1}}{\cancel{52} \cdot 51 \cdot \cancel{50}}$$

$$= \frac{24}{4(51)} = \frac{2}{17}$$

3. All parallelograms have diagonals that bisect each other.

∴ All four are successes.

$P(\text{diagonals bisecting each other}) = \text{certainty} = 1$

4. What is the probability of 6 heads out of 10 tosses of a fair coin?

4. $P(\text{head}) = \frac{1}{2}$ $P(\text{tail}) = \frac{1}{2}$

$p = p(\text{success}) = p(\text{head}) = \frac{1}{2}$
$q = p(\text{failure}) = \frac{1}{2}$

$r = 6$ success
$n = 10$ attempts

$$_{10}C_6(\tfrac{1}{2})^6(\tfrac{1}{2})^4 = (210)(\tfrac{1}{64})(\tfrac{1}{16}) = \tfrac{105}{512}$$

Practice Problems 24.3

1. From a standard deck of 52 cards, two cards are drawn at random without replacement. What is the probability that both cards are aces?

   A) $\frac{6}{2652}$ B) $\frac{4}{2652}$ C) $\frac{12}{2652}$ D) $\frac{4}{52}$
   E) $\frac{4}{48}$

2. If a card is randomly selected from a standard deck of 52 cards, what is the probability of picking a black king or a club?

   A) $\frac{1}{52}$ B) $\frac{13}{2652}$ C) $\frac{14}{52}$ D) $\frac{15}{52}$
   E) $\frac{12}{2652}$

3. On a loaded coin the $P(\text{head}) = \frac{1}{3}$. Find the probability of 2 heads on 7 throws of this coin.

   A) $_7C_2(\tfrac{1}{3})^2(\tfrac{2}{3})^5$ B) $_7C_2(\tfrac{1}{3})^5(\tfrac{2}{3})^2$ C) $_7C_2(\tfrac{1}{3})^7(\tfrac{2}{3})^2$
   D) $_7C_2(\tfrac{1}{3})^7$ D) $_7C_2(\tfrac{2}{3})^7$

4. From a group of 3 boys and 4 girls, what is the probability that a randomly selected subgroup of 5 will consist of 3 boys and 2 girls?

   A) $\frac{5}{7}$ B) $\frac{2}{7}$ C) $\frac{1}{3}$ D) $\frac{5}{8}$ E) $\frac{2}{9}$

Solutions on page 362

24.4 Binomial Theorem

Binomial theorem provides the **expansion of $(a+b)^n$**.

$$(a+b)^n = a^n + na^{n-1}b + \frac{n(n-1)}{2 \cdot 1} a^{n-2}b^2 + \frac{n(n-1)(n-2)}{3 \cdot 2 \cdot 1} a^{n-3}b^3 + \cdots + b^n$$

(good for all real n)

or

$$= \sum_{k=1}^{n+1} {}_nC_{k-1}(a)^{n-(k-1)}(b)^{k-1} \quad [n \text{ is a positive integer}]$$

kth term of $(a+b)^n = {}_nC_{k-1}(a)^{n-(k-1)}(b)^{k-1}$

Math Notes	$(a+b)^n$: • has $n+1$ terms • in each term the sum of the exponents is n • the exponent of b is one less than the number of the term • ${}_nC_{k-1} = {}_nC_{n-(k-1)}$

Examples:

1. Find the 3rd term of $(5x+y)^4$.

2. Find the 4th term of $(\sqrt{x}-2\sqrt[3]{y})^4$.

Solutions:

1. kth term $= {}_nC_{k-1}(a)^{n-(k-1)}(b)^{k-1}$

$$k=3, \qquad k-1=2 \qquad (\underbrace{5x}_{a}+\underbrace{y}_{b})^{4-n}$$

$$= {}_4C_2(5x)^{4-2}(y)^2 = \frac{4\cdot3}{2\cdot1}(5x)^2(y)^2$$

$$= 6(25x^2)(y^2)$$

$$= 150x^2y^2$$

2. kth term $= {}_nC_{k-1}(a)^{n-(k-1)}(b)^{k-1}$

$$k=4 \qquad k-1=3 \qquad (\underbrace{\sqrt{x}}_{a}-\underbrace{2\sqrt[3]{y}}_{b})^{4-n}$$

$$= {}_4C_3(\sqrt{x})^{4-3}(-2\sqrt[3]{y})^3$$

$$= \frac{4\cdot3\cdot2}{3\cdot2\cdot1}(\sqrt{x})^1(-2\sqrt[3]{y})^3 = 4(\sqrt{x})(-8y)$$

$$= -32y\sqrt{x}$$

Practice Problems 24.4

1. What is the last term in the expansion of $(2x-3y)^4$?

   A) $-81y^4$ B) $81y^4$ C) $3y^4$
   D) $-54xy^3$ E) $324y^4$

2. Find the middle term of $(2x^2-y)^4$.

   A) ${}_4C_2(2x^2)^2(-y)^2$ B) ${}_4C_2(2x)^2(y)^2$
   C) ${}_4C_3(2x^2)^3(-y)$ D) ${}_4C_3(2x)^2(-y)^2$
   E) ${}_4C_2(2x^2)^4$

3. Find the 4th term of the expansion $(x+y)^6$.

   A) $30x^3y^4$ B) $15x^4y^2$ C) $20x^3y^3$
   D) $15x^2y^3$ E) $30x^3y^3$

4. What is the middle term of $(x+\frac{1}{x})^6$?

   A) $\dfrac{15}{x^2}$ B) $20x^3$ C) $20x$ D) 20

   E) $\dfrac{20}{x}$

Solutions on page 363

Cumulative Review Test

1. The solution set of the equation

    $2x^3 - 3x^2 - 11x + 6 = 0$

    is

    A) $\{-\frac{1}{2}, -2, 5\}$ B) $\{\frac{1}{2}, -2, 3\}$ C) $\{-1, 2, 3\}$
    D) $\{1, -2, \frac{1}{2}\}$ E) $\{-3, -2, -1\}$

2. If x varies directly as the square of y and $x = 100$ when $y = 5$, find the value of x when $y = 4$.

    A) 64   B) 75   C) 80   D) 210   E) 125

3. Find the sum of the infinite series

    $4 + 2 + 1 + \frac{1}{2} + \frac{1}{4} + \dots$

    A) 7.79   B) 7.93   C) 7.9   D) 8   E) $8\frac{1}{2}$

4. The sum of the digits of a two-digit number is 10. If the digits are reversed and the resulting number is divided by the original number, the quotient is 1 and the remainder is 36. Find the original number.

    A) 28   B) 37   C) 73   D) 19   E) 91

5. The last term in the expansion $(2x + 3y)^4$ is

    A) $-81y^4$ B) $81y^4$ C) $3y^4$
    D) $-54xy^3$ E) $8x^4$

6. Evaluate $\dfrac{|a-3|}{4} + a^0 + a^{1/2}$ when $a = 4$.

    A) $3\frac{1}{4}$ B) $2\frac{1}{4}$ C) $2\frac{3}{4}$ D) $3\frac{3}{4}$ E) $1\frac{3}{4}$

7. What is the numerical value of $\sin\dfrac{\pi}{6} + \cos\dfrac{\pi}{2}$?

    A) $\dfrac{1}{2}$ B) 1   C) $\dfrac{3}{2}$ D) $\dfrac{\sqrt{3}}{2}$ E) $\dfrac{2+\sqrt{3}}{2}$

8. How many liters of water must be evaporated from 84 liters of a 20% salt solution to raise it to a 35% salt solution?

    A) 14   B) 16   C) 24   D) 36   E) 34

9. Figure I is a rectangle, II is a square, III is a right triangle, IV is an acute scalene triangle and V is a parallelogram. The lengths of particular segments of the figures are indicated. Which figure has the greatest area?

    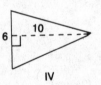 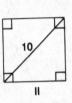 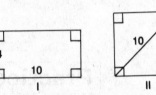 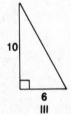

    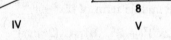

    A) I   B) II   C) III   D) IV   E) V

10. Which value of x does not satisfy the equation $\tan^2 x = 1$?

    A) 45°   B) 135°   C) 225°
    D) 300°   E) 315°

11. If x varies directly as y and $x = 16$ when $y = 12$, find x, when $y = 21$.

    A) 28   B) 24   C) $\frac{64}{7}$ D) $\frac{63}{4}$ E) 62

12. Find the third term of the expansion $(2x - y)^4$.

    A) $12x^2y^2$ B) $-12x^2y^2$
    C) $24x^2y^2$ D) $4x^3y$
    E) $4xy^3$

356

13. Find the common ratio of the infinite geometric progression whose first term is 3 and whose sum is 6.

    A) $\frac{1}{2}$ B) 2 C) $-\frac{1}{2}$ D) -2 E) $-\frac{3}{2}$

14. In how many different ways can the letters of the word "cosine" be arranged?

    A) 24 B) 6! C) 6^6 D) 6^2 E) 2^6

15. Find the 20th term of the arithmetic progression 20, 16, 12, 8 . . . .

    A) -57 B) -56 C) -60
    D) -64 E) -68

16. Which graph does not represent a function?

A) 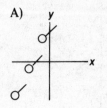 B) 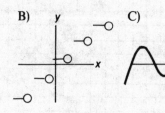 C)

D) 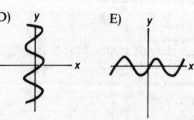 E)

17. A committee of 4 is to be chosen from 3 boys and 4 girls. How many different ways are there to choose committees which will contain exactly 2 boys and 2 girls?

    A) 18 B) 35 C) 12 D) 7 E) 6

18. The product of two numbers is 54 and their arithmetic mean is $\frac{15}{2}$. Find the larger number.

    A) 5 B) 6 C) 7 D) 8 E) 9

19. The area of a rectangle is 30m². If the length of the rectangle is increased by 4 inches and the width is decreased by 1 inch, the area of the rectangle becomes 33 square inches. Find the width of the original rectangle.

    A) 8 B) $\frac{15}{4}$ C) 3 D) 10 E) $\frac{11}{4}$

20. The first term of an arithmetic progression is 8 and the fifth term is 2. Find the common difference.

    A) $-\frac{3}{2}$ B) -2 C) -1 D) $-\frac{1}{2}$ E) $\frac{1}{2}$

Solutions on page 364

Diagnostic Chart for Cumulative Review Test

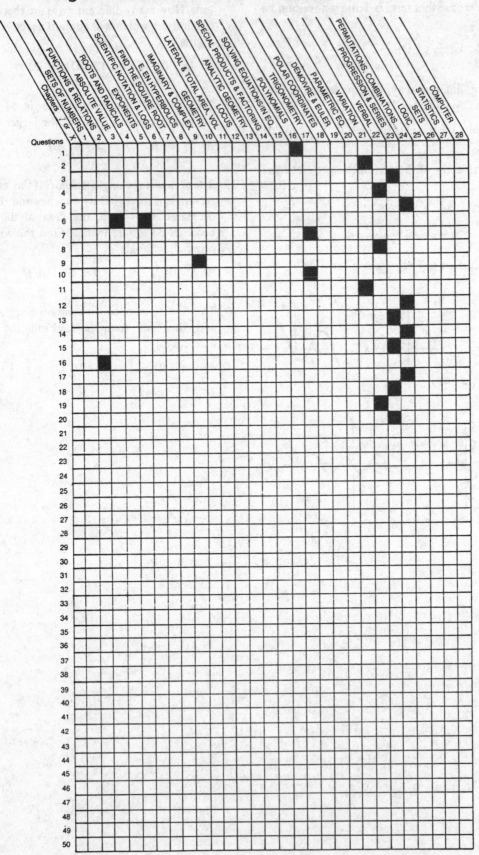

Solutions to Practice Problems in Unit Six

Practice Problems 21.1

1. **(D)** $x_1 y_1^2 = x_2 y_2^2 = k$

   $3(4)^2 = x_2(2)^2 = k$

   $48 = 4x$

   $12 = x$

2. **(B)** $y = \dfrac{k}{x}$

   When $x = 1$ $y = k$

   When $x = 2$ (doubled)

   $y = \dfrac{k}{2}$

   If x is doubled, y is halved y is divided by 2.

3. **(B)** $\dfrac{x_1}{y_1} = \dfrac{x_2}{y_2} = K$

   $\dfrac{\frac{1}{2}}{\frac{1}{4}} = \dfrac{4}{y}$

   $\dfrac{1}{2} \cdot \dfrac{4}{1} = \dfrac{4}{y}$

   $2 = \dfrac{4}{y}$

   $2y = 4$

   $y = 2$

4. **(D)** $\dfrac{x_1 w_1^2}{y_1 z_1} = \dfrac{x_2 w_2^2}{y_2 z_2}$

   $\dfrac{3(4)^2}{8(6)} = \dfrac{x(2)^2}{1(10)}$

   $\dfrac{3(16)}{8(6)} = \dfrac{48}{48} = \dfrac{4x}{10}$

   $1 = \dfrac{4x}{10}$

   $x = \dfrac{10}{4}$

Practice Problems 22.1

1. (B) Let $t=$ tens digit

$\quad\quad u=$ units digit

$$t=3u$$

$$(10u+t)+18=10t+u$$

$$10u+t+18=10t+u$$

$$9t-9u=18$$

$$t-u=2$$

$$(3u)-u=2$$

$$2u=2$$

$$u=1$$

$$t=3$$

$$\boxed{31}$$

2. (A)

Let $x=$ amount of pure acid

$\binom{\text{Amount}}{\text{of solution}} \times \ (\% \ \text{acid}) \ = \ \text{Amount acid}$

x	1.00	$1.00x$
30	.20	6
$30+x$	.50	$.50(30+x)$

$$1.00x+6=.50(30+x)$$

Multiply by 100 to clear the decimals.

$$100x+600=50(30+x)$$

$$100x+600=1500+50x$$

$$50x=900$$

$$x=18$$

3. (D)

Let $t=$ time working together

	Rate $\times$	Time $=$	Work accomplished
New	$\dfrac{1}{3}$	t	$\dfrac{t}{3}$
Old	$\dfrac{1}{5}$	t	$\dfrac{t}{5}$

New rate $=\frac{1}{3}$　　Old rate $=\frac{1}{5}$

$$\frac{t}{3}+\frac{t}{5}=1 \quad \text{(did 1 job)}$$

$$5t+3t=15$$

$$8t=15$$

$$t=\frac{15}{8}$$

4. (D)

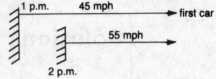

Let $t=$ time traveled by first car

$t-1=$ time traveled by second car

	Rate $\times$	Time $=$	Distance
First car	45	t	$45t$
Second car	55	$t-1$	$55(t-1)$

$$45t=55(t-1)$$

$$10t=55$$

$$t=5.5 \text{ hrs}$$

It will take the second car $t-1=4.5$ hrs to overtake the first car.

5. (A)

Let $x=$ amount invested at 16%

Principal $\times$	Rate $=$	Interest
x	.16	$.16x$
$6000-x$	.09	$.09(6000-x)$

$$.16x=.09(6000-x)$$

Multiply by 100 to clear the decimals.

$$16x=9(6000-x)$$

$$25x=54000$$

$$x=2160$$

6. (D)

Let $x=$ first positive odd integer

$x+2=$ second odd integer

$x+4=$ third odd integer

$$(x)^2=(x+2)+(x+4)+9$$

$$x^2=2x+15$$

$$x^2-2x-15=0$$

$$(x-5)(x+3)=0$$

$$x=5 \quad | \quad x=-3$$

$$x+2=7 \quad | \quad \text{Reject (not positive)}$$

$$x+4=9 \quad |$$

Answer: The largest is 9.

Practice Problems 23.1

1. **(D)**

$$a_{18} = l = a + (n-1)d$$
$$a = 2, \quad d = 7 - 2 = 5, \quad n = 18$$
$$l = 2 + (18-1)5 = 2 + 85 = 87$$

2. **(A)**

$$a_8 = l = a \cdot r^{n-1}$$
$$a = 5 \quad r = \tfrac{1}{5} \quad n = 8$$
$$l = 5\left(\frac{1}{5}\right)^{8-1} = 5\left(\frac{1}{5}\right)^7 = \frac{1}{5^6} = \frac{1}{15625}$$

3. **(E)**

$$S_{20} = \frac{n}{2}(a+l)$$

$$a = 2, \quad d = 6 - 2 = 4, \quad l = a + (n-1)d$$
$$n = 20 \qquad\qquad = 2 + (20-1)4 = 78$$

$$S = \frac{20}{2}(2+78) = 800$$

4. **(B)** $\dfrac{3}{-} \;\; - \;\; - \;\; \dfrac{18}{-}$

$$a = 3, \quad l = 18, \quad n = 4$$
$$l = a + (n-1)d$$
$$18 = 3 + (4-1)d$$
$$15 = 3d$$
$$5 = d$$

$$3,\; 3+5 = 8, \quad 3 + 2(5) = 13,\; 18$$

The two arithmetic means are 8 and 13.

$$\frac{3}{-} \quad \frac{8}{-} \quad \frac{13}{-} \quad \frac{18}{-}$$

5. **(D)**

$$S_{10} = \frac{a(1-r^n)}{1-r}$$

$$a = 2, \quad r = \tfrac{4}{2} = 2, \quad n = 10$$

$$S_{10} = \frac{2(1-2^{10})}{1-2} = \frac{2(1-1024)}{-1}$$

$$= \frac{2(-1023)}{-1} = 2046$$

6. **(C)**

$$\left.\begin{array}{l} x = .11111\overline{1} \\ 10x = 1.11111\overline{1} \end{array}\right\}$$

$$10x = 1.1111\overline{1}$$
$$\underline{x = .1111\overline{1}}$$
$$9x = 1$$
$$x = \tfrac{1}{9}$$

7. **(A)**

$$S_\infty = \frac{a}{1-r} \qquad a = \tfrac{1}{2}$$

$$r = \frac{\tfrac{1}{4}}{\tfrac{1}{2}} = \tfrac{1}{2}$$

$$= \frac{\tfrac{1}{2}}{1 - \tfrac{1}{2}} = \frac{\tfrac{1}{2}}{\tfrac{1}{2}} = 1$$

8. **(B)**

$$\sum_{k=1}^{99} (k+3) = [1+3] + [2+3] + [3+3] \ldots [99+3]$$

$$= 4 + 5 + 6 \ldots 102$$

there are 99 terms.

$$S_{99} = \frac{n}{2}(a+l)$$

$$= \frac{99}{2}(4+102) = 5247$$

9. **(A)** $x^a \cdot x^b = x^{a+b}$

$$17^{1/2} \cdot 17^{1/4} \cdot 17^{1/8} \ldots$$
$$= 17^{\tfrac{1}{2} + \tfrac{1}{4} + \tfrac{1}{8} + \tfrac{1}{16} \ldots}$$

$$\frac{1}{2} + \frac{1}{4} + \frac{1}{8} + \frac{1}{16} \ldots$$

$$S_\infty = \frac{a}{1-r} = \frac{\tfrac{1}{2}}{1 - \tfrac{1}{2}} = 1$$

$$= 17^1 = 17$$

Practice Problems 24.1

1. **(C)** Distinguishable permutations of n items of which a are alike $= \dfrac{n!}{a!}$.

   1981 $n = 4$ numbers
   There are 2-1's.

   $$= \frac{4!}{2!} = \frac{24}{2} = 12$$

2. **(A)**

   $$\underline{8 \cdot 7 \cdot 6 \cdot 5 \cdot 4 \cdot 3 \cdot 2 \cdot 1} = 40320$$

   $8! = 40320$

3. $\underset{\uparrow}{\dfrac{3 \cdot 2 \cdot 1 \cdot 5 \cdot 4 \cdot 3 \cdot 2 \cdot 1}{\quad}} = 720$

   any of the 5 remaining boys
   the remaining girl
   any of   one girl having been seated,
   3 girls   any of 2 remaining girls.

4. **(E)**
   $4 \cdot 6 \cdot 2 = 48$ different routes

Practice Problems 24.2

1. **(D)** 5 boys and 3 girls

   $$\underset{\substack{\text{select 4} \\ \text{boys out of 5}}}{{}_5C_4} \cdot \underset{\substack{\text{select 2} \\ \text{girls out of 3}}}{{}_3C_2} = 5 \cdot 3 = 15$$

2. **(E)** 6 yellow, 4 red, 2 pink

   $$\underset{\substack{\text{select 2} \\ \text{yellow out} \\ \text{of 6}}}{{}_6C_2} \cdot \underset{\substack{\text{select 1} \\ \text{red out} \\ \text{of 4}}}{{}_4C_1} \cdot \underset{\substack{\text{select 1} \\ \text{pink out} \\ \text{of 2}}}{{}_2C_1} = 15 \cdot 4 \cdot 2 = 120$$

   $${}_6C_2 = \frac{6 \cdot 5}{2 \cdot 1} = 15 \quad {}_4C_1 = 4 \quad {}_2C_1 = 2$$

3. **(C)**

   $${}_{30}C_{28} = {}_{30}C_2 = \frac{30 \cdot 29}{2 \cdot 1} = 435$$

4. **(D)** 4 aces,   48 non-aces

   $$\underset{\substack{\text{select 4} \\ \text{aces out} \\ \text{of 4}}}{{}_4C_4} \cdot \underset{\substack{\text{select 1} \\ \text{non-ace} \\ \text{out of} \\ \text{48 non-aces}}}{{}_{48}C_1} = 1 \cdot 48 = 48$$

Practice Problems 24.3

1. **(C)**

   $P(\text{1st ace}) = \frac{4}{52}$

   $P(\text{2nd ace}) = \dfrac{3}{51}$ ← only 3 aces left
   ← only 51 cards left

   $P(\text{ace and ace}) = P(\text{1st ace}) \cdot P(\text{2nd ace})$

   $$= \frac{4}{52} \cdot \frac{3}{51} = \frac{12}{2652}$$

2. **(C)**

   $P(\text{black king}) = \frac{2}{52}$

   $P(\text{club}) = \frac{13}{52}$

   $P(\text{king of clubs}) = \frac{1}{52}$

   $P(A \text{ or } B) = P(A) + P(B) - P(A \text{ and } B)$

   $$= \frac{2}{52} + \frac{13}{52} - \frac{1}{52} = \frac{14}{52}$$

3. **(A)**

$p = p(\text{success}) = \frac{1}{3}, \quad n = 7 \text{ attempts}$

$q = p(\text{failure}) = \frac{2}{3}, \quad r = 2 \text{ successes}$

$_7C_2(\frac{1}{3})^2(\frac{2}{3})^5$

4. **(B)**

$\text{Success} = \quad _3C_3 \qquad _4C_2 \qquad = 1 \cdot 6 = 6$

$\qquad\qquad\qquad \uparrow \qquad\qquad \uparrow$

$\qquad\qquad \text{select 3} \quad \text{select 2}$

$\qquad \text{boys out of 3} \quad \text{girls out of 4}$

All possible ways to pick 5

$= \quad _7C_5 \quad = _7C_2 = \frac{7 \cdot 6}{2 \cdot 1} = 21$

$\qquad \uparrow$

$\qquad \text{pick 5 out}$
$\qquad \text{of 7}$

$P\left(\begin{array}{c} \text{3 boys and 2 girls} \\ \text{out of a subgroup of 5} \end{array}\right) = \frac{6}{21} = \frac{2}{7}$

Practice Problems 24.4

1. **(B)** The last term is the 5th term

$\text{5th term} = _nC_{k-1}(a)^{n-(k-1)}(b)^{k-1}$

$k = 5,$

$\therefore k - 1 = 4 \qquad \underset{a}{(2x} \underset{b}{- 3y)}^{4 \leftarrow n}$

$\qquad = _4C_4(2x)^{4-(4)}(-3y)^4$

$\qquad = 1 \cdot 1 \cdot 81y^4$

$\qquad = 81y^4$

2. **(A)** There are 5 terms; the 3rd term is in the middle.

$\text{3rd term} = _nC_{k-1}(a)^{n-(k-1)}(b)^{k-1}$

$k = 3,$

$\therefore k - 1 = 2 \qquad \underset{a}{(2x^2} \underset{b}{- y)}^{4 \leftarrow n}$

$\qquad = _4C_2(2x^2)^2(-y)^2$

3. **(C)**

$\text{4th term} = _nC_{k-1}(a)^{n-(k-1)}(b)^{k-1}$

$k = 4,$

$k - 1 = 3 \qquad = _6C_3(x)^{6-3}(y)^3$

$\qquad = \frac{6 \cdot 5 \cdot 4}{3 \cdot 2 \cdot 1} x^3 y^3$

$\qquad = 20x^3y^3$

4. **(D)** There are 7 terms; the 4th term is in the middle

$\text{4th term} = _nC_{k-1}(a)^{n-(k-1)}(b)^{k-1}$

$k = 4,$

$\therefore k - 1 = 3 \qquad \left(\underset{a}{x} + \underset{b}{\frac{1}{x}}\right)^{6 \leftarrow n}$

$\qquad = _6C_3(x)^{6-3}\left(\frac{1}{x}\right)^3$

$\qquad = \frac{6 \cdot 5 \cdot 4}{3 \cdot 2 \cdot 1} x^3\left(\frac{1}{x}\right)^3 = 20$

Cumulative Review 1-24

1. **(B)** $P(x) = 2x^3 - 3x^2 - 11x + 6$

$$\frac{p}{q} = \frac{\pm 6, \ \pm 3, \ \pm 2, \ \pm 1}{\pm 2}$$

$P(x) = 2x^3 - 3x^2 - 11x + 6$ has   2   sign changes

$\therefore 2$ or 0 positive roots.

$$P(-x) = -2x^3 - 3x^2 + 11x + 6 \text{ has 1 sign change}$$

$\therefore 1$ negative root.

$P(-1) = -2 - 3 + 11 + 6 \neq 0$

$P(-2) = -16 - 12 + 22 + 6 = 0 \ \sqrt{}$

$x = -2$ is a root, $x + 2$ is a factor

$$\begin{array}{r|rrrr} -2 & 2 & -3 & -11 & 6 \\ & & -4 & 14 & -6 \\ \hline & 2 & -7 & 3 & \boxed{0} \end{array}$$

$P(x) = 2x^3 - 3x^2 - 11x + 6$
$\quad = (x+2)(x^2 - 7x + 3)$
$\qquad\qquad (2x-1)(x-3)$
$\quad = (x+2)(2x-1)(x-3) = 0$
$\qquad x = -2 | x = \frac{1}{2} | x = 3$

2. **(A)**

$$\frac{x_1}{y_1^2} = \frac{x_2}{y_2^2} \quad \frac{100}{5^2} = \frac{x}{4^2}$$

$$x = \frac{(100)(16)}{25} = 64$$

3. **(D)** $S_\infty = \dfrac{a}{1-r} \quad r = \dfrac{2}{4} = \dfrac{1}{2}$

$$= \frac{4}{1 - (\frac{1}{2})} = 8$$

4. **(B)**

$t + u = 10$

$$\frac{10u + t}{10t + u} = 1 \text{ remainder } 36$$

$\therefore$ by cross multiplication $10u + t = (10t + u) + 36$
rearrange the terms and divide by 9

$9u - 9t = 36$

$u - t = 4 \Rightarrow -t + u = 4$

combining the two equations

$$\begin{array}{r} t + u = 10 \\ -t + u = \ \ 4 \\ \hline 2u = 14 \\ u = \ \ 7 \end{array}$$

Since $t + u = 10$ and $u = 7$

$$t + 7 = 10$$
$$t = \ \ 3$$

The number is 37.

5. **(B)**   The last term is the 5th term

$$5\text{th term} = {}_nC_{k-1}(a)^{n-(k-1)}(b)^{k-1}$$

$k = 5, \quad \therefore k - 1 = 4$

$$(\underbrace{2x}_{a} + \underbrace{3y}_{b})^4$$

$5\text{th term} = {}_4C_4(2x)^{4-4}(3y)^4 = 81y^4$

6. **(A)**

$$\frac{|4-3|}{4} + 4^0 + 4^{1/2} = \frac{1}{4} + 1 + 2 = 3\frac{1}{4}$$

7. **(A)**

$$\sin \frac{\pi}{6} + \cos \frac{\pi}{2} = \frac{1}{2} + 0 = \frac{1}{2}$$

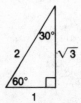

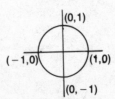

Note: $\sin \dfrac{\pi}{6} = \sin 30° = \dfrac{1}{2}$

$$\cos \frac{\pi}{2} = \cos 90° = 0$$

8. **(D)** Let $x =$ the amount of water evaporated.

$$\begin{pmatrix} \text{Amount} \\ \text{of solution} \end{pmatrix} \times \quad (\% \text{ salt}) \quad = \quad \begin{matrix} \text{Amount} \\ \text{of salt} \end{matrix}$$

84	.20	16.8
x	0	0
$84 - x$	.35	$.35(84 - x)$

$$16.8 = .35(84 - x)$$
$$16.8 = 29.4 - .35x$$

Multiply by 100 to clear the decimals.

$$1680 = 2940 - 35x$$
$$x = 36$$

9. **(B)**

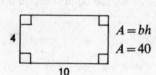

$A = bh$
$A = 40$

$A = \dfrac{d^2}{2}$
$= \boxed{50}$

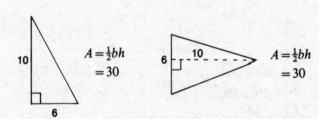

$A = \frac{1}{2}bh$
$= 30$

$A = \frac{1}{2}bh$
$= 30$

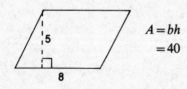

$A = bh$
$= 40$

10. **(D)**

$$\tan^2 x - 1 = 0$$
$$(\tan x + 1)(\tan x - 1) = 0$$

$\tan x = 1$	$\tan x = -1$
$x = 45°$	$x = 135°$
$x = 225°$	$x = 315°$

11. **(A)**

$$\frac{x_1}{y_1} = \frac{x_2}{y_2}$$
$$\frac{16}{12} = \frac{x}{21}$$
$$x = 28$$

12. **(C)**

$$k\text{th term} = {}_nC_{k-1}(a)^{n-(k-1)}(b)^{k-1}$$

$k = 3$ $\qquad \underbrace{(2x}_{a} \underbrace{- y)^{4n}}_{b}$

$\therefore k - 1 = 2$

$$= {}_4C_2(2x)^{4-2}(-y)^2$$
$$= \frac{4 \cdot 3}{2 \cdot 1}(4x^2)(y^2)$$
$$= 24x^2 y^2$$

13. **(A)**

$$S_\infty = \frac{a}{1 - r} \quad \Rightarrow \quad 6 = \frac{3}{1 - r}$$
$$6 - 6r = 3$$
$$-6r = -3$$
$$r = \tfrac{1}{2}$$

14. **(B)**

$$6 \cdot 5 \cdot 4 \cdot 3 \cdot 2 \cdot 1 = 6!$$

15. **(B)** This is an arithmetic progression.

$d = -4$
$l = a + (n - 1)d$
$l = 20 + (20 - 1)(-4)$
$\ = 20 + (19)(-4)$
$\ = -56$

16. **(D)** Use the vertical line test.

17. **(A)**

${}_3C_2$: pick 2 boys out of 3

${}_4C_2$: pick 2 girls out of 4

$${}_3C_2 \cdot {}_4C_2 = \frac{3 \cdot 2}{2 \cdot 1} \cdot \frac{4 \cdot 3}{2 \cdot 1}$$
$$= 3 \cdot 6 = 18$$

18. (E)

Let x = first number

y = second number

$xy = 54$

$$\frac{x+y}{2} = \frac{15}{2}$$

$$x + y = 15$$

$$xy = 54 \quad \therefore x = \frac{54}{y}$$

Substitute $\frac{54}{y}$ for x in $x + y = 15$

$$x + y = 15$$

$$\frac{54}{y} + y = 15$$

$$54 + y^2 = 15y \Rightarrow y^2 - 15y + 54 = 0$$

$$(y-9)(y-6) = 0$$

$$\begin{array}{c|c} y = 9 & y = 6 \\ x = 6 & x = 9 \end{array}$$

19. (B)

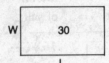

$$wl = 30 \Rightarrow w = \frac{30}{l}$$

$$(w-1)(l+4) = 33$$

$$wl + 4w - l - 4 = 33$$

Substitute $\frac{30}{l}$ for w

$$\left(\frac{30}{l}\right)(l) + 4\left(\frac{30}{l}\right) - l = 37$$

$$30 + \frac{120}{l} - l = 37$$

Multiply each term by l and set the equation equal to zero.

$$30l + 120 - l^2 = 37l$$

$$l^2 + 7l - 120 = 0$$

$$(l+15)(l-8) = 0$$

$$\begin{array}{c|c} l = -15 & l = 8 \therefore w = \frac{15}{4} \\ \text{reject} & \end{array}$$

20. (A)

$$l = a + (n-1)d$$

$$2 = 8 + (5-1)d$$

$$2 = 8 + 4d$$

$$-6 = 4d$$

$$-\tfrac{3}{2} = d$$

UNIT SEVEN

25. Logic

25.1 Basics and Beyonds

Symbols—Summary

	Name	p: It is blue; q: It is heavy
$\sim$	**Negation** (not)	$\sim p$: It is **not** blue
$\wedge$ ($\cdot$)	**Conjunction** (and)	$p \wedge q$: It is blue **and** it is heavy
$\vee$	**Disjunction** (or)	$p \vee q$: It is blue **or** it is heavy
$\rightarrow$ ($\supset$)	**Conditional** If —, then — (implies)	$p \rightarrow q$: **If** it is blue, **then** it is heavy or It is blue **implies** it is heavy
$\leftrightarrow$	**Biconditional** If and only if (iff)	$p \leftrightarrow q$: It is blue **if and only if** it is heavy or **If** it is blue, **then** it is heavy **AND** **If** it is heavy, **then** it is blue
	Tautology	A logical sentence whose truth value is always **True**
	Contradiction	A logical sentence whose truth value is always **False**
	Converse (reverse, backwards)	Statement $p \rightarrow q$: If it is blue, then it is heavy Converse $q \rightarrow p$: If it is heavy, then it is blue
	Inverse*	Statement $p \rightarrow q$: If it is blue, then it is heavy Inverse $\sim p \rightarrow \sim q$: If it is **not** blue, then it is **not** heavy
	Contrapositive (Inverse of the converse) or (converse of the inverse)	Statement $p \rightarrow q$: If it is blue, then it is heavy Contrapositive $\sim q \rightarrow \sim p$: If it is not heavy, then it is not blue

*The inverse is obtained by negating the hypothesis and negating the conclusion.

Truth Table-Summary

p	q	~p	~q	p∧q	p∨q	statement p→q	p↔q	converse q→p	inverse ~p→~q	contrapositive ~q→~p	tautology	contradiction
		negation (not)		conjunction (and)	disjunction (or)	conditional (if_, then_)	biconditional (iff)	converse	inverse	contrapositive		
T	T	F	F	T	T	T	T	T	T	T	T	F
T	F	F	T	F	T	F	F	T	T	F	T	F
F	T	T	F	F	T	T	F	F	F	T	T	F
F	F	T	T	F	F	T	T	T	T	T	T	F

Given a statement
- The contrapositive is ALWAYS logically equivalent.
- The truth or falsity of the inverse and converse vary with the particular statement.

Note: The converse and inverse are logically equivalent.
(One is the contrapositive of the other.)

Laws of Inference

p→q p ——— q	p implies q Given: p ——— Conclusion: q	**Law of Detachment** or **Modus Ponens** (M.P.)
p→q ——— ~q→~p	p implies q ——— Conclusion: not q implies not p	**Law of Contrapositive**
~(~p) ——— p	not(not p) ——— Conclusion: p	**Law of Double Negation**
p→q ~q ——— ~p	p implies q Given: not q ——— Conclusion: not p	**Law of Modus Tollens** (M.T.)
p→q q→r ——— p→r	p implies q q implies r ——— Conclusion: p implies r	**Law of Hypothetical Syllogism** (chain rule)

$p \lor q$ $\sim q$ ___ p	$p \lor q$ $\sim p$ ___ q	p or q Given: not q ___ Concl: p	p or q Given: not p ___ Concl: q	**Law of Disjunctive Inference** (Disjunctive Syllogism)
$p \land q$ ___ p	$p \land q$ ___ q	p and q ___ Concl: p	p and q ___ Concl: q	**Law of Simplification**
$\sim(p \land q)$ ___ $\sim p \lor \sim q$	$\sim(p \lor q)$ ___ $\sim p \land \sim q$	not (p and q) ___ Concl: not p or not q	not (p or q) ___ Concl: not p and not q	**De Morgan's Laws**
p q ___ $p \cdot q$		Given: p Given: q ___ Concl: p and q		**Law of Conjunction**
p ___ $p \lor q$		Given: p ___ Concl: p or q		**Law of Addition**
$p \to q$ ___ $p \to (p \land q)$		p implies q ___ Concl: p implies (p and q)		**Law of Absorption**

Quantifiers

$\forall_x$:	Universal Quantifier	All, No For all x Each For each x Every, None For every x
$\exists_x$:	Existential Quantifier	Some For some At least one There is at least one

The negation of $\forall_x$: p is $\exists_x$: $\sim p$ The negation of "for all p" is "for some not p"

The negation of $\exists_x$: p is $\forall_x$: $\sim p$ The negation of "for some p" is "for all not p"

Examples:

1. Using the laws of inference:

    Given: $S \rightarrow W$
    $\quad\quad\quad P \rightarrow \sim W$
    $\quad\quad\quad S \vee \sim H$
    $\quad\quad\quad H \vee \sim A$
    $\quad\quad\quad A$

    prove $\sim P$.

Solutions:

1.

Statements	Reasons
1) $S \rightarrow W$	Given
2) $P \rightarrow \sim W$	Given
3) $S \vee \sim H$	Given
4) $H \vee \sim A$	Given
5) A	Given
6) H	4,5 Disjunctive inference
7) S	3,6 Disjunctive inference
8) W	1,7 Modus ponens
9) $\sim P$	2,8 Modus tollens

2. Is $\sim(p \vee q) \leftrightarrow (\sim p \wedge \sim q)$ a tautology?

2.

p	q	$p \vee q$	$\sim(p \vee q)$	$\sim p$	$\sim q$	$\sim p \wedge \sim q$	$\sim(p \vee q)$ $\leftrightarrow$ $(\sim p \wedge \sim q)$
T	T	T	F	F	F	F	T
T	F	T	F	F	T	F	T
F	T	T	F	T	F	F	T
F	F	F	T	T	T	T	T

Yes it is a tautology. Its truth value is always true.

3. What is the negation of "All trees have green leaves"?

3. The negation of $\forall_x : p$ is $\exists_x : \sim p$.

   $\forall_x : p =$ All trees have green leaves

   $\exists_x : \sim p =$ Some trees do not have green leaves

4. By DeMorgan's Law what is logically equivalent to $\sim(p \vee \sim q)$?

4. $\sim(p \vee q) \rightarrow \sim p \wedge \sim q$ $\quad \therefore \sim(p \vee \sim q) \rightarrow \boxed{\sim p \wedge q}$

   Proof:

p	q	$\sim q$	$p \vee \sim q$	$\sim(p \vee \sim q)$	$\sim p$	$\sim p \wedge q$
T	T	F	T	F	F	F
T	F	T	T	F	F	F
F	T	F	F	T	T	T
F	F	T	T	F	T	F

$\quad\quad\quad\quad\quad\quad\quad\quad\quad\quad\quad\quad\quad\quad\quad\quad\quad \uparrow \quad\quad\quad\quad\quad \uparrow$
logically equivalent

Practice Problems 25.1

1. If x is an integer, which statement is true?

   A) $\forall_x : x^2 + 2x = 0$
   B) $\exists_x : x^2 + 2x = 0$
   C) $\exists_x : x^2 < 0$ D) $\forall_x x^2 \leqslant 0$
   E) $\forall_x : x^2 > 3$

2. Which statement is false when p is false and q is false?

   A) $p \wedge q$ B) $p \rightarrow q$ C) $\sim p \rightarrow \sim q$
   D) $p \leftrightarrow q$ E) $q \rightarrow p$

3. Which is the converse of the statement $p \rightarrow \sim q$?

   A) $p \rightarrow \sim q$ B) $p \rightarrow q$
   C) $\sim p \rightarrow \sim q$ D) $q \rightarrow \sim p$
   E) $\sim q \rightarrow p$

4. The inverse of $\sim p \rightarrow q$ is

   A) $p \rightarrow \sim q$ B) $\sim p \rightarrow \sim q$
   C) $\sim q \rightarrow p$ D) $q \rightarrow \sim p$
   E) $p \rightarrow q$

5. Let $p = x$ is a prime number
   Let $q = x < 10$
   If $x = 11$ which statement is true?

   A) $\sim p \vee q$ B) $p \wedge q$ C) $\sim p \rightarrow q$
   D) $p \rightarrow q$ E) $p \leftrightarrow q$

6. Which is logically equivalent to the statement "If I live, then I love"?

   A) If I don't live, then I don't love
   B) If I love, then I live
   C) If I don't love, then I don't live
   D) If I don't love, then I live
   E) If I don't live, then I love

7. What is the negation of the statement 'Some students do not do their homework'?

   A) Some students do their homework
   B) All students do not do their homework
   C) All students do their homework
   D) No student does his homework
   E) At least one student does his homework

8. The negation of $r \wedge \sim t$ is

   A) $\sim r \vee t$ B) $\sim r \wedge t$
   C) $\sim r \vee \sim t$ D) $\sim r \wedge \sim t$
   E) $r \rightarrow t$

Solutions on page 390

26. Sets

26.1 Fundamentals

We often consider a collection, or SET, of items, numbers or things.

{ }	Bracket notation. Used to indicate a set. Each number or item in a set is called a **MEMBER** or an **ELEMENT** of the set.
$\in$	Used to indicate set membership. If $x = \{0, 1, 3, 7\}$, then $3 \in x$, that is, 3 is a member of set x. $5 \notin x$ means 5 is not a member of set x.
{ } **or** $\emptyset$	The empty or null set.
$\subset$	Indicates a subset. If all the elements of a set B are included in set D, then B is called a subset of D, written $B \subset D$. For example, if $R = \{-4, 0, 2\}$ and $S = \{0, 2\}$, then $S \subset R$. The symbol $\not\subset$ means is not a subset.
$\cup$	Indicates a union. The union of two sets is a new set, each of whose elements are in either or both of the original sets. For example, if $M = \{1, 2, 3, 4\}$ and $N = \{2, 4, 6\}$ then $M \cup N = \{1, 2, 3, 4, 6\}$.
$\cap$	Indicates intersection. The intersection of two sets is a new set, whose elements are only those elements shared by the original sets. For example, if $M = \{1, 2, 3, 4\}$ and $N = \{2, 4, 6\}$ then $M \cap N = \{2, 4\}$.

Introduction to Venn Diagrams

The circle represents set X; the square represents the **universe of available data**, and the space outside the circle but within the square are elements not contained in set X but in the universe of the data.

Sets X and Y have no elements in common. They are disjoint.

The area indicated by b is the intersection of sets X and Y ($X \cap Y$). b represents the common elements of sets X and Y.

The entire shaded area represents $X \cup Y$, the set of elements of X together with the elements of Y.

Since set Y is wholly contained within set X, set Y is a subset of X. $Y \subset X$.

The elements of a set may be indicated using SET-BUILDER notation.
$\{x/x+5=7\}$ means the set of all numbers x such that $x+5=7$. This set has a single element 2.

$\{x/x>3\}$ means the set of all numbers x which are greater than 3.

If $A=\{x/x^2=16\}$, then the elements of A are all solutions to the equation

$$x^2=16, A=\{4, -4\}$$

If $B=\{x/|x|=-3\}$, the elements of B are all the solutions to $|x|=-3$. Since there are no solutions, $B=\{\ \}$ or $B=\varnothing$.

Examples:

1. If $Q=\{$set of rational numbers$\}$ and $H=\{$set of irrational numbers$\}$, then $Q\cup H$ represents what?

2. If $x=\{$odd integers$\}$ and $y=\{$prime numbers$\}$, find the element which is NOT in the set $X\cap Y$.

Solutions:

1. $Q\cup H$ is the union of the set of rational numbers with the set of irrational numbers; this yields the set of real numbers indicated by R.

$$Q\cup H = R$$

2. The set y is the numbers 2, 3, 5, 7, 11, 13, 17, 19, 23..., all of which are odd with the exception of 2. The only element of y which is not contained in x is 2.

Practice Problems 26.1

1. If $R=\{5, 10, 15\}$ and $S=\{5\}$, which statement is true?

   A) $10\in S$ B) $S\notin R$ C) $R\subset S$
   D) $S\subset R$ E) $R\cup S = S$

2. Which is not a subset of $\{-2, -1, 1, 2\}$?

   A) $\{\ \}$ B) $\{0\}$ C) $\{1\}$
   D) $\{-2, -1, 1, 2\}$ E) $\{-2, -1\}$

3. If $P=\{2, 4\}, Q=\{1, 2, 3, 4, 5\}, R=\{1, 3, 5\}$, which is NOT true?

   A) $P\cup R=Q$ B) $Q\cap R=R$ C) $R\subset Q$
   D) $P\subset R$ E) $P\cap R=\varnothing$

4. If $A=\{x/|x-2|=5\}$, then $A=$

   A) $\{\ \}$ B) $\{5\}$ C) $\{5, -5\}$
   D) $\{-3, 7\}$ E) $\{3, -7\}$

5. If $M=\{x/x>0\}$ and $N=\{x/x^2+5x-24=0\}$, the $M\cap N=$

   A) $\{3, -8\}$ B) $\{-3, 8\}$ C) $\{8\}$
   D) $\{3\}$ E) $\{\ \}$

6. If $A=\{\pi, \sqrt{2}, e, \sqrt{17}\}$ and $B=\{$all rational numbers$\}$ then $A\cap B$ is

   A) $\{\pi, e\}$ B) $\{\sqrt{2}, \sqrt{17}\}$
   C) $\{$whole numbers$\}$
   D) $\{$real numbers$\}$ E) $\{\ \}$

Solutions on page 390

26.2 Applications of Venn Diagrams

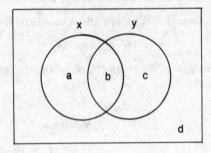

$\odot X$ represents the set of all x
$\odot Y$ represents the set of all y

    a represents the set of all x that is not also y
    c represents the set of all y that is not also x
    b represents the set of items that are both x and y
    d represents the set of all items that are neither x nor y

The set of all items is $a+b+c+d$.

Examples:

1. At a dog show there are 23 dogs. 6 have black spots, 14 have brown spots and 7 have neither brown nor black spots. How many dogs have both brown and black spots?

Solutions:

1.

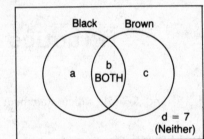

$$a+b=6$$
$$c+b=14$$

$$\text{Total} = \underset{6}{\underbrace{a+b}}+c+\underset{7}{\underbrace{d}}=23$$

$$\therefore c=10$$

Since $c+b=14$ and $c=10$

$$\therefore b=4$$

4 dogs have both black and brown spots.

2. Of 100 cats, 82 had long hair and 40 had black stripes. If 23 had both long hair and black stripes, how many had neither?

2.

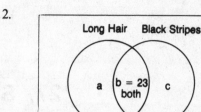

$$a+b=82$$
$$b+c=40$$
$$b=23$$

$$b+c=40$$
$$\underline{b=23}$$
$$\therefore c=17$$

$$\text{Total} = \underbrace{a+b}_{82} + \underbrace{c}_{17} + d = 100$$

$$82+17+d=100$$
$$99+d=100$$
$$d=1$$

Practice Problems 26.2

1. In a pantry there are 28 cans of vegetables. Eight of these have labels with white lettering, 18 have labels with green lettering and 8 have labels with neither white nor green lettering. How many cans have both white and green lettering?

   A) 2 B) 10 C) 8 D) 6 E) 26

2. In a class of 175 students everyone takes math and/or English. If 125 take math and 117 take English. How many take both math and English?

   A) 50 B) 58 C) 67 D) 123 E) 76

3. In a kennel of 100 dogs 40 dogs have only brown spots. Fifty dogs have neither brown nor black spots. How many have black spots (with or without brown spots)?

   A) 10 B) 25 C) 50 D) 60 E) 90

4. In a class of 40 students, twice as many students take English as math. If 12 students take both and 10 take neither, how many students take math?

   A) 6 B) 8 C) 10 D) 12 E) 14

Solutions on page 391

27. Statistics

27.1 Measures of Central Tendency

	Given sample values $x_1, x_2, x_3, \ldots, x_n$	Given sample values $x_1, x_2, x_3, \ldots, x_n$ with respective frequencies $f_1, f_2, f_3, \ldots, f_n$
The **Mean**	$\bar{x} = \dfrac{\text{sum of items}}{\text{quantity of items}} = \dfrac{\sum\limits_{i=1}^{n} x_i}{n}$	$\bar{x} = \dfrac{\text{sum of items}}{\text{quantity of items}} = \dfrac{\sum\limits_{i=1}^{n} f_i x_i}{\sum\limits_{i=1}^{n} f_i}$
	Sometimes called the Average, sometimes called the Arithmetic Mean.	
The **Median**	The **middle** value when the items are sequenced in numerical order. If there are two middle values, use their arithmetic mean.	
The **Mode**	The **most frequent value**. If more than one value occurs with the maximum frequency, then each of them is a mode.	

Examples:

1. Given the ages:

    15, 15, 15, 16, 17, 18, 20

    find the mean, mode and median.

2. Given the ages:

    14, 15, 15, 15, 16, 16, 16, 17

    find the mean, mode and median.

Solutions:

1. $\text{mean} = \bar{x} = \dfrac{15 + 15 + 15 + 16 + 17 + 18 + 20}{7}$

    $= \dfrac{116}{7} = 16\frac{4}{7}$

    median: the middle value is 16

    mode = 15 the most frequent value

2. $\text{mean} = \dfrac{14 + 15 + 15 + 15 + 16 + 16 + 16 + 17}{8}$

    $= \dfrac{124}{8} = 15\frac{1}{2}$

    median: there are two middle values, 15 and 16

    $\text{median} = \dfrac{15 + 16}{2} = 15\frac{1}{2}$

    mode = 15 and 16

Practice Problems 27.1

1. Given the following table, which score is the mode?

Score	Frequency
98	2
95	3
92	2
87	1
84	2

   A) 98   B) 95   C) 92   D) 87   E) 84

2. If the mean of four scores is 85, what is the sum of the four scores?

   A) $21\frac{1}{4}$ B) 85   C) 170   D) 340
   E) 500

3. The scores on a test were 75, 75, 85, 90 and 100, which statement about these scores is true?

   A) the mean and the median are the same.
   B) the mode is greater than the median.
   C) the mode is greater than the mean.
   D) the median is greater than the mean.
   E) the mode and median are the same.

4. Express the mean of $(2x+1)$, $(x+1)$ and $(3x-8)$ in terms of x.

   A) $6x-6$ B) $2x-2$ C) $3x-3$
   D) $4x-4$ E) $5x-5$

Solutions on page 392

27.2 Measures of Dispersion

	Given sample values $x_1, x_2, x_3, \ldots, x_n$	Given sample values $x_1, x_2, x_3, \ldots, x_n$ with respective frequencies $f_1, f_2, f_3, \ldots, f_n$
The **Variance**	The average of the squares of the deviations from the mean	
	$v = \sigma^2 = \dfrac{\sum\limits_{i=1}^{n}(x_i - \bar{x})^2}{n} = \dfrac{1}{n}\sum\limits_{i=1}^{n}(x_i - \bar{x})^2$	$v = \sigma^2 = \dfrac{1}{n}\sum f_i(x_i - \bar{x})^2$
The **Standard Deviation**	The square root of the variance	
	$\sigma = \text{S.D.} = \sqrt{\dfrac{1}{n}\sum\limits_{i=1}^{n}(x_i - \bar{x})^2}$	$\sigma = \text{S.D.} = \sqrt{\dfrac{1}{n}\sum f_i(x_i - \bar{x})^2}$
The **Range**	The difference between the highest and lowest values of x_i	
	$R = x_{\max} - x_{\min}$	

Examples:

1. Using the accompanying data

find: the mean
the median
the mode
the variance
the standard deviation

Score	Frequency
70	3
75	4
80	7
85	6

Solutions:

1. The mean $= \dfrac{\sum f_i x_i}{n}$

$$= \frac{70(3) + 75(4) + 80(7) + 85(6)}{20}$$

$$= \frac{210 + 300 + 560 + 510}{20}$$

$$\bar{x} = \frac{1580}{20} = 79$$

There are 20 items, the median is between the 10th and 11th in sequence. Both values are 80.

∴ The median $= 80$.

The most frequent value is the mode.

The mode $= 80$.

x_i (score)	f_i (freq)	$\bar{x}$	$x_i - \bar{x}$	$(x_i - \bar{x})^2$	$f_i(x_i - \bar{x})^2$
70	3	79	−9	81	243
75	4	79	−4	16	64
80	7	79	1	1	7
85	6	79	6	36	216
Total	20				530

$$v = \frac{1}{n} \sum f_i(x_i - \bar{x})^2 = \frac{1}{20}(530) = 26.5$$

$$\sigma = \sqrt{v} = \sqrt{26.5} \approx 5.1$$

Practice Problems 27.2

1. Find the standard deviation for the following set of data. (Answer may be left in radical form.)

Value	Frequency
60	1
75	4
80	3
90	2

A) 78   B) 660   C) $\sqrt{660}$

D) $\sqrt{66}$ E) $\sqrt{17}$

2. The ages of ten teachers at Jamaica High School are 33, 23, 36, 29, 36, 36, 33, 29, 36 and 29. Find the standard deviation of these ages.

A) $\sqrt{174}$ B) $\sqrt{29}$ C) 27
D) $\sqrt{27}$ E) $\sqrt{17.4}$

3. Find the range for the following data

75, 99, 88, 91, 95, 92, 83, 82, 81.

A) 99   B) 75   C) 83   D) 24   E) 18

4. Find the variance for the following set of data.

Score	Frequency
50	4
58	4
62	3
64	6
65	2
68	1

A) 68   B) 64   C) 60
D) 31.9   E) $\sqrt{31.9}$

Solutions on page 392

27.3 Normal Distribution and σ-Limits

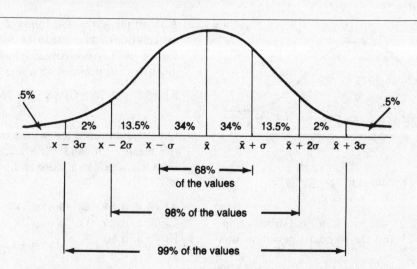

The data indicated in the diagram are accurate but approximate.

Examples:

1. On a standardized test with a normal distribution, the mean was 75 and the standard deviation was 4.0. Approximately what percentage of scores would fall between the range 71 to 79?

Solutions:

1.

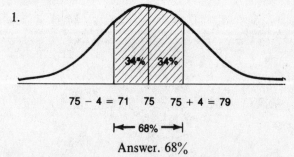

Answer. 68%

2. What approximate percentage of scores on a normal distribution would be expected to fall within two standard deviations of the mean?

2.

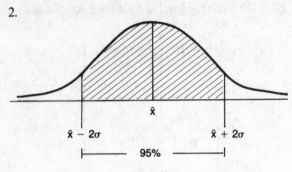

Answer: 95%

Practice Problems 27.3

1. In the accompanying diagram, the shaded area represents approximately 95% of the scores on a standardized test with a normal distribution. If the scores range from 52 to 96, which could be the standard deviation?

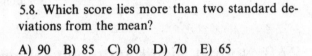

A) 11   B) 22   C) 44   D) 74   E) 76

2. For a standardized test with a normal distribution, the mean was 75 and the standard deviation was

5.8. Which score lies more than two standard deviations from the mean?

A) 90   B) 85   C) 80   D) 70   E) 65

3. A set of measures that follow a bell curve (normal distribution) has a mean of 50 and standard deviation of 5. Approximately what percent of the measures fall between 45 and 55?

A) 13.5   B) 34   C) 68   D) 95   E) 99

4. On a standardized test with a normal distribution, the mean is 76 and the standard deviation is 4. Between which two scores will ~68% of the scores fall?

A) 68 and 84   B) 74 and 84
C) 72 and 80   D) 74 and 78
E) 76 and 80

Solutions on page 393

28. Computer Programs

28.1 Fundamental Facts

Computer programs are detailed step by step instructions. They are followed in sequence unless they direct to do otherwise from within the program.

As an illustration, the program below will be analyzed in detail.

1. **LET** A = 9
2. **LET** B = −1
3. **IF** A = B², **GO TO** INSTRUCTION 6
4. **INCREASE** B by 2
5. **GO TO** INSTRUCTION 3
6. **PRINT** THE VALUE OF B
7. **STOP**

Instructions 1 and 2 provide us with starting values for A and B.
Instruction 3 asks us to test if A = B².
Since A = 9 and B = −1, $9 \neq (-1)^2$.
They are not equal; go on to the next instruction.
Instruction 4 tells us to replace the existing value of B (which is −1) with +1 (−1 + 2 = +1).
Instruction 5 tells us to go back to instruction 3 and test if A = B².
Since A = 9 and B = +1, $9 \neq (+1)^2$.
They are not equal; go on to the next instruction.
Instruction 4 tells us to replace the existing value of B (which is +1) with +3 (+1 + 2 = +3).
Instruction 5 tells us to go back to instruction 3 and test A = B².
Since A = 9 and B = 3, $9 = 3^2$.
They are equal. We are directed to instruction 6, which tells us to print the current value of B.

$$B = +3$$

Practice Problems 28.1

1. What is the final value of X that will be printed by instruction 6?

   1. LET X = 0
   2. LET F = X² + 8X − 20
   3. IF F = 0, GO TO INSTRUCTION 6
   4. INCREASE X BY 1
   5. GO BACK TO INSTRUCTION 2
   6. PRINT THE VALUE OF X
   7. PRINT THE SENTENCE "THIS IS A ROOT OF THE EQUATION"
   8. STOP

   A) −10   B) −20   C) 8   D) 2   E) 0

2. What is the final value of A that will be printed by instruction 7?

   1. LET A = 10
   2. LET B = 10
   3. IF $\frac{A}{B} = \frac{2}{3}$ GO TO 7
   4. LET A BECOME A − 1
   5. LET B BECOME B + 1
   6. GO BACK TO INSTRUCTION 3
   7. PRINT THE VALUE OF A
   8. STOP

   A) 11   B) 12   C) 8   D) 9   E) 10

3. What will this program print?

    1. LET A = 5
    2. LET B = A + 7
    3. LET C = B + 2
    4. IF $A^2 + B^2 = C^2$, GO TO INSTRUCTION 7
    5. PRINT "THE VALUES OF A, B, AND C DO NOT REPRESENT THE LENGTHS OF THE SIDES OF A RIGHT TRIANGLE"
    6. STOP
    7. PRINT "THE VALUES OF A, B, AND C REPRESENT THE LENGTHS OF THE SIDES OF A RIGHT TRIANGLE"
    8. STOP
    9. PRINT THE VALUE OF A
    10. PRINT THE VALUE OF B
    11. PRINT THE VALUE OF C
    12. STOP

A) THE VALUE OF A
B) THE VALUE OF B
C) THE VALUE OF C
D) THE VALUES A, B. AND C DO NOT REPRESENT THE LENGTHS OF THE SIDES OF A RIGHT TRIANGLE
E) THE VALUE OF A, B, AND C REPRESENT THE LENGTHS OF THE SIDES OF A RIGHT TRIANGLE

4. A, B and C represent the coefficients of a quadratic equation $Ax^2 + Bx + C = 0$. What will this program print?

    1. LET A = -2
    2. LET B = -4
    3. LET C = -2
    4. IF $B^2 - 4AC > 0$ GO TO INSTRUCTION 7
    5. IF $B^2 - 4AC = 0$ GO TO INSTRUCTION 9
    6. IF $B^2 - 4AC < 0$ GO TO INSTRUCTION 11
    7. PRINT "ROOTS ARE REAL AND UNEQUAL"
    8. STOP
    9. PRINT "ROOTS ARE REAL AND EQUAL"
    10. STOP
    11. PRINT "ROOTS ARE IMAGINARY"
    12. STOP
    13. PRINT "ROOTS ARE RATIONAL"
    14. STOP
    15. PRINT "ROOTS ARE IRRATIONAL"
    16. STOP

A) ROOTS ARE REAL AND UNEQUAL
B) ROOTS ARE REAL AND EQUAL
C) ROOTS ARE IMAGINARY
D) ROOTS ARE RATIONAL
E) ROOTS ARE IRRATIONAL

Solutions on page 394

Cumulative Review Post-Test

1. The expression $\log r + \log r^2$ is equal to

   A) $\log(r+r^2)$ B) $3 \log r$ C) $\log r$
   D) r^3 E) r

2. In a plane, the locus of points a distance 6 units from a given line is

   A) a line   B) a circle
   C) two intersecting lines
   D) two parallel lines
   E) 3 points

3. In the accompanying figure, $DE \parallel AB$. If $CD = 6$, $CA = 18$ and $DE = 4$, what is the length of AB?

   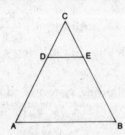

   A) 16   B) 14   C) 12   D) 6   E) 8

4. What is the length of the line which connects the points $J(5,6)$ and $L(2,-1)$?

   A) $\sqrt{34}$ B) $\sqrt{58}$ C) 8   D) $\sqrt{20}$ E) 14

5. If $\log_3 9^{2x+1} = 4$, solve for x.

   A) 2   B) 1   C) $\frac{1}{2}$ D) $-\frac{1}{2}$ E) -1

6. Find the reciprocal of $1 + \sqrt{2}$.

   A) $1 + \sqrt{2}$ B) $\sqrt{2} - 1$ C) $1 - \sqrt{2}$
   D) $\frac{\sqrt{2}-1}{3}$ E) $1 + \frac{\sqrt{2}}{2}$

7. The sum of the interior angles of a polygon is 1,980°. How many sides does the polygon have?

   A) 13   B) 14   C) 15   D) 16   E) 17

8. What is the solution set of the equation $|5 - 2x| = 7$?

   A) $\{6, -1\}$ B) $\{6\}$ C) $\{1\}$
   D) $\{-1\}$ E) $\{ \}$

9. The expression $x^{-1} + y^{-1}$ is equal to

   A) $(x+y)^{-1}$ B) $\frac{1}{x+y}$ C) $x+y$
   D) $\frac{x+y}{xy}$ E) $\frac{xy}{x+y}$

10. The graph of $x = \sin \theta$, $y = 3\csc \theta$ is a(n)

    A) line   B) circle   C) ellipse
    D) parabola   E) hyperbola

11. If A is a positive acute angle, find the smallest value of A that satisfies the equation

    $$\sin(A + 40)° = \cos 10°.$$

    A) -30 B) -20 C) 0   D) 20   E) 40

12. If $f(x) = x + \frac{1}{x}$, then $f\left(\frac{1}{x}\right) =$

    A) $x^2 + 1$ B) $f(x)$ C) $x^2 + x$
    D) $\frac{2}{x}$ E) $\frac{x^2 + 1}{x^2}$

13. If the roots of a quadratic equation are 3 and 1, find the equation.

    A) $x^2 + 4x + 3 = 0$ B) $x^2 - 4x - 3 = 0$
    C) $x + 4x - 3 = 0$ D) $x^2 + 2x + 3 = 0$
    E) $x^2 - 4x + 3 = 0$

14. In a group of 40 students, 25 applied to **NJL** College and 30 applied to Joshua University. If 3 students applied to neither NJL or Joshua how many applied to both schools?

    A) 6   B) 12   C) 18   D) 19   E) 11

15. In the accompanying figure *AB* and *AC* are tangents to circle *O* from point *A*. If major arc *BC* = 210°, what is the measure of angle *A*?

    A) 30   B) 45   C) 90   D) 150   E) 75

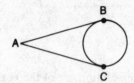

16. The graph shown represents which equation?

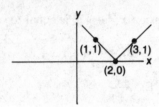

    A) $y = |x + 2|$ B) $y = |x| + 2$
    C) $y = |x - 2|$ D) $y = |x| - 2$
    E) $y + 2 = |x|$

17. Which is not an identity?

    A) $\sin^2 2\theta + \cos^2 2\theta = 1$
    B) $\dfrac{\tan 7x}{\sin 7x} = \sec 7x$
    C) $\tan 3t + \cot 3t = 1$
    D) $\dfrac{1}{\sin 4r} = \csc 4r$
    E) $\tan^2 m^2 + 1 = \sec^2 m^2$

18. Simplify the product $(1 + 2\sqrt{7}i)^2$.

    A) 29   B) $-27 + 4\sqrt{7}i$ C) -13
    D) $29 + 4\sqrt{7}i$ E) $15i$

19. The equation $\dfrac{x}{3} + \dfrac{y}{2} = 1$ represents a(n)

    A) line   B) parabola   C) circle
    D) ellipse   E) hyperbola

20. What is the remainder when $x^5 - 1$ is divided by $x - 1$?

    A) 0   B) -2 C) 2   D) -1 E) 1

21. Multiply $(x + 2)(x^2 - 2x + 4)$.

    A) $x^3 - 6x^2 + 2x + 8$ B) $x^3 + 8$
    C) $x^3 + 6x^2 - 3x + 4$ D) $x^3 - 8$
    E) $x^3 + 2x^2 - 4x + 8$

22. In the accompanying figure, the area of circle *O* is 36π and the area of a sector of the circle is 4π. Find the number of degrees in the measure of the central angle of the sector.

    A) 40   B) 4   C) 36   D) 144   E) 20

23. What is the negation of $\sim p \vee \sim q$?

    A) $p \wedge q$ B) $\sim p \wedge \sim q$ C) $p \vee q$
    D) $p \wedge \sim q$ E) $p \vee \sim q$

24. If $\log 2 = .3010$, $\log 2y$ is equal to

    A) $.3010y$ B) $.3010 + y$ C) $\dfrac{.3010}{y}$
    D) $.3010 + \log y$ E) $.3010 \log y$

25. Solve the following system of equations:

    $$x^2 - 3y^2 = 13$$
    $$x + 3y = 1$$

    A) (4,1) and (5,−2)
    B) (−5,2) and (4,−1)
    C) (3,−1) and (0,$\frac{1}{3}$)
    D) (6,−1) and (5,−2)
    E) (4,2) and (−5,−1)

26. Express $\sin(-42°)$ as a function of a positive acute angle.

    A) $\sin 42°$ B) $\cos 42°$
    C) $-\sin 48°$ D) $-\cos 42°$
    E) $-\cos 48°$

27. $2(\cos 30° + i \sin 30°) =$

    A) $1 + 2i$ B) $\sqrt{3} + i$ C) $2 + \dfrac{\sqrt{3}}{2}$
    D) $\dfrac{1 + 2\sqrt{3}}{2}$ E) $1 + \sqrt{3}i$

28. Find the average of $\frac{1}{4}$, 25% and .25.

    A) .21   B) .22   C) .24   D) .25   E) .26

29. Find the $\sqrt{3}$ to the nearest tenth.

    A) 1.5   B) 1.57   C) 1.6   D) 1.7   E) 1.8

30. Find the reciprocal of $2i$.

    A) $\frac{1}{2}i$ B) $\dfrac{2}{i}$ C) $-\frac{1}{2}i$ D) -4 E) $\frac{1}{2}$

31. In the accompanying diagram BDC is a straight line and $DE \perp AEC$. If $m\angle ADB = 80$ and $m\angle EDC = 50$, find the $m\angle DAC$.

    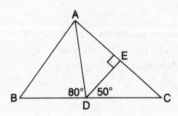

    A) 45   B) 60   C) 50   D) 40   E) 25

32. The graph of the equation $x^2 + 2x + \dfrac{y^2}{2} = 18$ is a(n)

    A) line   B) parabola
    C) circle   D) ellipse
    E) hyperbola

33. $[\frac{1}{2}(\cos 10° + i \sin 10°)]^3 =$

    A) $\frac{1}{2}(\cos 1000° + i \sin 1000°)$
    B) $\frac{1}{8}(\cos 10° + i \sin 10°)$
    C) $\frac{1}{8}(\cos 30° + i \sin 30°)$
    D) $\frac{1}{2}(\cos 30° + i \sin 30°)$
    E) $\frac{1}{6}(\cos 13° + i \sin 13°)$

34. The number of points which are at a given distance d from a line and also equally distant from two given points on the given line is

    A) 0   B) 1   C) 2   D) 3   E) 4

35. In the accompanying diagram AB, BC, CD and DA are tangent to the circle at R, V, T and S,

respectively. If $VC = 8$ and $SD = 10$, find CD.

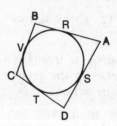

A) 21   B) 18   C) 14   D) 12   E) 9

36. The expression $\cos 2A$ is equivalent to

    A) $2 \sin^2 A - 1$ B) $\sin^2 A - \cos^2 A$
    C) $1 - 2 \cos^2 A$
    D) $(\cos A + \sin A)(\cos A - \sin A)$
    E) $1 - \sin^2 2A$

37. The period of the graph $y = 3 \cos \frac{1}{2}x$ is

    A) 1   B) 180°   C) 3   D) 720°   E) $\frac{1}{2}$

38. Find the sum of the infinite series

    $2, \frac{4}{3}, \frac{8}{9}, \ldots$.

    A) $8\frac{3}{4}$ B) $7\frac{1}{5}$ C) 6   D) $5\frac{3}{8}$ E) $5\frac{1}{4}$

39. The sum of the digits of a two-digit number is 11. The number obtained by interchanging the digits exceeds twice the original number by 34. Find the original number.

    A) 47   B) 38   C) 29   D) 56   E) 65

40. If x varies inversely as t and if $x = 6$ when $t = 2$, find x when $t = 4$.

    A) $1\frac{1}{2}$ B) 3   C) 1   D) 12   E) 36

41. If the volume of a right circular cylinder, whose height $= 10$, is 90π, find the lateral area.

    A) 3   B) 60   C) 60π D) 78π E) 180π

42. How many irrational roots does the following equation have?

    $$x^4 - 7x^3 + 9x^2 + 13x - 4 = 0$$

    A) 4   B) 3   C) 2   D) 1   E) 0

43. The measure, in degrees, of an angle of 1 radian is

    A) 180   B) $\dfrac{180}{\pi}$ C) 360

    D) $\dfrac{360}{\pi}$ E) π^2

44. Evaluate $\sum_{k=3}^{5} (k^2 - 1)$.

   A) 24   B) 12   C) 48   D) 49   E) 47

45. If the probability of throwing a four on a loaded die is $\frac{1}{5}$, what is the probability of throwing exactly 2 fours out of 5 trials.

   A) $_5C_2(\frac{1}{5})(\frac{4}{5})$ B) $_5C_2(\frac{1}{5})^3$
   C) $_5C_2(\frac{1}{5})^2(\frac{4}{5})^3$ D) $_5C_3(\frac{1}{5})^3(\frac{4}{5})^2$
   E) $_5C_3(\frac{1}{5})(\frac{4}{5})^4$

46. Which is the negation of the statement "All squares are parallelograms"?

   A) All squares are not parallelograms.
   B) No squares are parallelograms.
   C) Some squares are parallelograms.
   D) Some squares are not parallelograms.
   E) If it is a square, then it is a parallelogram.

47. The average of $2P$, $3Q$ and a third number is A. In terms of A, P and Q represent the third number.

   A) $A - 2P - 3Q$ B) $A - 2P + 3Q$
   C) $3A - 2P + 3Q$
   D) $3A - 2P - 3Q$ E) $A + 2P - 3Q$

48. What is the value of C that will be printed?

   1. LET A = 1
   2. LET B = 1
   3. LET F = A^2 − B
   4. IF F ≥ 0 GO TO INSTRUCTION 6
   5. IF F < 0 GO TO INSTRUCTION 8
   6. PRINT C = 2
   7. STOP
   8. PRINT C = 7
   9. STOP

   A) 1   B) 2   C) 0   D) 7   E) 2

49. If $\ln x + \ln(x + 2) = \ln 3$, solve for x.

   A) 1   B) 0   C) 2   D) e E) e^2

50. The numbers 6, $-\frac{3}{4}$, $.175$, $.17\overline{1717}$ are *all* elements of which of the following?

   A) whole numbers   B) integers
   C) fractions   D) rational
   E) irrational

Solutions on page 394

Diagnostic Chart Post-Test

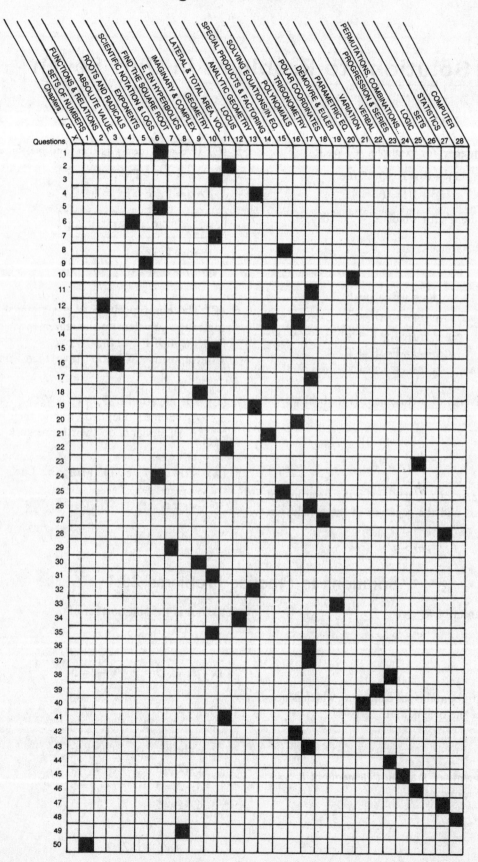

Solutions to Problems in Unit Seven

Practice Problems 25.1

1. **(B)** x is an integer, for example, -2 or 1.

$$(-2)^2 + 2(-2) = 0 \quad \text{True}$$
$$(1)^2 + 2(1) \neq 0$$

There are some x's for which $x^2 + 2x = 0$

$$\therefore \exists_x : x^2 + 2x = 0.$$

2. **(A)**

$$p \wedge q \Rightarrow F \wedge F \Rightarrow F$$
$$p \rightarrow q \Rightarrow F \rightarrow F \Rightarrow T$$
$$\sim p \rightarrow \sim q \Rightarrow T \rightarrow T \Rightarrow T$$
$$p \leftrightarrow q \Rightarrow F \leftrightarrow F \Rightarrow T$$
$$q \rightarrow p \Rightarrow F \rightarrow F \Rightarrow T$$

3. **(E)** The converse of a statement is its backwards statement:

$$p \rightarrow \sim q$$

Converse: $\qquad \sim q \rightarrow p$

4. **(A)** The inverse of a statement is the negation of the hypothesis and the negation of the conclusion
statement: $\qquad \sim p \rightarrow q$
Inverse: $\qquad p \rightarrow \sim q$

5. **(C)** If $x = 11$, then p is true and q is false.

$$\sim p \vee q \Rightarrow \sim T \vee F \Rightarrow F \vee F \Rightarrow F$$
$$p \wedge q \Rightarrow T \wedge F \Rightarrow F$$

$$\boxed{\sim p \rightarrow q \Rightarrow \sim T \rightarrow F \Rightarrow F \rightarrow F \Rightarrow T}$$

$$p \rightarrow q \Rightarrow T \rightarrow F \Rightarrow F$$
$$p \leftrightarrow q \Rightarrow T \leftrightarrow F \Rightarrow F$$

6. **(C)** The contrapositive of a statement is always logically equivalent to the statement.
Statement: If I live, then I love.
Contrapositive: If I don't love, then I don't live.

7. **(C)** The negation of $\exists_x : \sim p$ is $\forall_x : p$.
$\exists_x : \sim p =$ Some students do not do their homework
$\forall_x : p =$ All students do their homework

8. **(A)** $\sim (r \wedge \sim t)$ by De Morgan's Law

$$\overline{\sim r \vee t}$$

Solutions to Practice Problems 26.1

1. **(D)** S is a subset of R.

$$S \subset R$$

2. **(B)** The set $\{0\}$ contains the element 0, which is not an element of the original set.

3. **(D)** The elements of P are not elements of R
$\therefore P$ is not a subset of R.

4. **(D)** A is the solution set of

$$|x - 2| = 5$$

$$(x - 2) = -5 \qquad x - 2 = 5$$

$$x - 2 = -5 \qquad\qquad x = 7$$
$$x = -3$$

check	check
$\lvert x - 2\rvert = 5$	$\lvert x - 2\rvert = 5$
$\lvert -3 - 2\rvert = 5$	$\lvert 7 - 2\rvert = 5$
$\lvert -5\rvert = 5$	$5 = 5 \checkmark$
$5 = 5 \checkmark$	

$$A = \{-3, 7\}$$

5. (D)

$M = \{\text{all numbers} > 0\}$

$x^2 + 5x - 24 = 0$

$(x + 8)(x - 3) = 0$

$x = -8 \quad x = 3$

$\therefore N = \{-8, 3\}$

$M \cap N = \{3\}$

6. (E)

$A = \{4 \text{ irrational numbers}\}$

$B = \{\text{all rational numbers}\}$

A and B are disjoint. $A \cap B = \emptyset$ or $\{\ \}$

Practice Problems 26.2

1. (D)

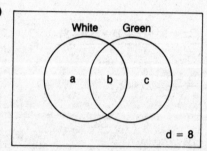

$a + b = 8, \quad b + c = 18, \quad d = 8$

$$\underbrace{\text{Total}}_{28} = \underbrace{a + b}_{8} + c + \underbrace{d}_{8}$$

$$28 = 8 + c + 8$$

$$\therefore c = 12$$

Since $b + c = 18$ and $c = 12$

$$\therefore b = 6$$

2. (C)

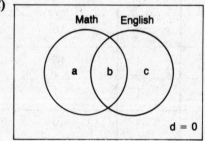

$$d = 0$$

$$a + b = 125$$

$$b + c = 117$$

$$\underbrace{a + b}_{125} + c + \underbrace{d}_{0} = 175$$

$$\therefore c = 50$$

Since $b + c = 117$ and $c = 50$

$$\therefore b = 67$$

3. (A)

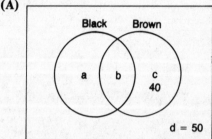

$$a + b + \underbrace{c}_{40} + \underbrace{d}_{50} = 100$$

$$a + b = 10$$

Answer: 10 dogs

4. (E) Let $x =$ number of students taking math

$2x =$ number of students taking English

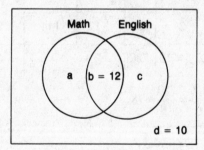

$$b = 12$$

$$a + b = x \Rightarrow a = x - 12$$

$$b + c = 2x \Rightarrow c = 2x - 12$$

$$\underbrace{a + b}_{x} + \underbrace{c}_{2x-12} + \underbrace{d}_{10} = 40$$

$$x + (2x - 12) + 10 = 40$$

$$3x = 42$$

$$x = 14$$

Math $= a + b = x = 14$

Practice Problems 27.1

1. **(B)** The mode is the most frequent value.

$$\text{mode} = 95$$

2. **(D)**

$$\text{mean} = \frac{\text{sum}}{\text{qty of items}}$$

$$85 = \frac{\text{sum}}{4}$$

$$340 = \text{sum}$$

3. **(D)**

$$\text{mean} = \frac{75 + 75 + 85 + 90 + 100}{5}$$

$$= \frac{425}{5} = 85$$

$$\text{median} = \frac{85 + 90}{2} = 87\tfrac{1}{2}$$

$$\text{mode} = 75$$

4. **(B)**

$$\text{mean} = \frac{(2x+1) + (x+1) + (3x-8)}{3}$$

$$= \frac{6x - 6}{3} = 2x - 2$$

Practice Problems 27.2

1. **(D)**

x_i	f_i	$x_i f_i$	$x_i - \bar{x}$	$(x_i - \bar{x})^2$	$f_i(x_i - \bar{x})^2$
60	1	60	−18	324	324
75	4	300	−3	9	36
80	3	240	2	4	12
90	2	180	12	144	288
	$\Sigma f_i = 10$	780			660

$$\bar{x} = \frac{\Sigma x_i f_i}{\Sigma f_i} = \frac{780}{10} = 78$$

$$\sigma = \sqrt{\frac{1}{n}(\Sigma f_i(x_i - \bar{x})^2)}$$

$$= \sqrt{\frac{1}{10}(660)} = \sqrt{66}$$

2. **(E)**

x_i	$x_i - \bar{x}$	$(x_i - \bar{x})^2$
33	1	1
23	−9	81
36	4	16
29	−3	9
36	4	16
36	4	16
33	1	1
29	−3	9
36	4	16
29	−3	9
$\Sigma x_i = 320$		$\Sigma(x_i - \bar{x})^2 = 174$

$$\bar{x} = \frac{\Sigma x_i}{n} = \frac{320}{10} = 32$$

$$\sigma = \sqrt{\frac{1}{10}(174)} = \sqrt{17.4}$$

3. **(D)**

$$Range = x_{max} - x_{min}$$
$$= 99 - 75$$
$$= 24$$

4. **(D)**

x_i	f_i	$x_i f_i$	$x_i - \bar{x}$	$(x_i - \bar{x})^2$	$f_i(x_i - \bar{x})^2$
50	4	200	-10	100	400
58	4	232	-2	4	16
62	3	186	2	4	12
64	6	384	4	16	96
65	2	130	5	25	50
68	1	68	8	64	64
$\Sigma f_i = n$ $= 20$		1200			638

$$\bar{x} = \frac{\Sigma x_i f_i}{\Sigma f_i} = \frac{1200}{20} = 60$$
$$v = \sigma^2 = \frac{1}{n} \Sigma f_i(x_i - \bar{x})^2$$
$$= \frac{1}{20}(638) = 31.9$$

Practice Problems 27.3

1. **(A)**

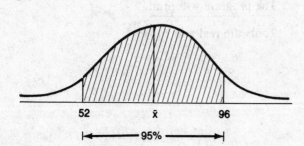

$$\bar{x} = \frac{52 + 96}{2} = 74$$

The 95% region is $\bar{x} \pm 2\sigma$

$$\therefore 96 = 74 + 2\sigma.$$
$$2\sigma = 22$$
$$\sigma = 11$$

2. **(A)**

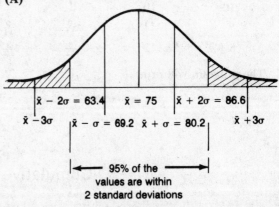

$$\bar{x} \pm 2\sigma$$
$$75 \pm 2(5.8)$$
$$63.4 \text{ to } 86.6$$

The score of 90 is *not* within 2 standard deviations from the mean.

3. **(C)**

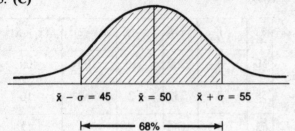

$$\bar{x} - \sigma = 45 \qquad \bar{x} = 50 \qquad \bar{x} + \sigma = 55$$

68%

4. **(C)**

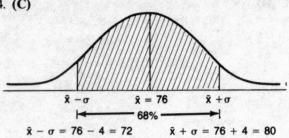

$$\bar{x} - \sigma \qquad \bar{x} = 76 \qquad \bar{x} + \sigma$$

68%

$$\bar{x} - \sigma = 76 - 4 = 72 \qquad \bar{x} + \sigma = 76 + 4 = 80$$

$$\bar{x} - \sigma = 76 - 4 = 72 \qquad \bar{x} + \sigma = 76 + 4 = 80$$

Practice Problems 28.1

1. **(D)**

Values of x	Values of $F = x^2 + 8x - 20$	Compare
0	$F = 0^2 + 8(0) - 20 = -20$	$-20 \neq 0$
1	$F = 1^2 + 8(1) - 20 = -1$	$-11 \neq 0$
2	$F = 2^2 + 8(2) - 20 = 0$	$0 = 0$

$$\frac{\text{The program will print}}{2}$$

This is a root of the equation.

2. **(C)**

Values of A	Values of B	Values of $\dfrac{A}{B}$	Compare
10	10	1	$1 \neq \frac{2}{3}$
9	11	$\frac{9}{11}$	$\frac{9}{11} \neq \frac{2}{3}$
8	12	$\frac{8}{12}$	$\frac{8}{12} = \frac{2}{3}$

$$\frac{\text{The program will print}}{8}$$

3. **(D)**

Values of A	Values of B	Values of C	Values of $A^2 + B^2$
5	12	14	$5^2 + 12^2$

Compare vs. C^2

$$5^2 + 12^2 \neq 14^2$$

The program will print

The values of A, B and C do not represent the lengths of the sides of a right triangle.

4. **(B)**

Values of A	Values of B	Values of C	Values of $B^2 - 4AC$
-2	-4	-2	$16 - 16 = 0$

The program will print

Roots are real and equal.

Cumulative Review Post-Test

1. **(B)**

$$\log r + \log r^2 = \log r \cdot r^2 = \log r^3 = 3 \log r$$

2. **(D)**

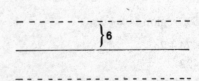

3. **(C)**

line ratio $= \dfrac{\text{little } \Delta}{\text{big } \Delta} = \dfrac{6}{18} = \dfrac{1}{3}$

$\dfrac{6}{18} = \dfrac{4}{x}$

$6x = 4(18)$

$x = 12$

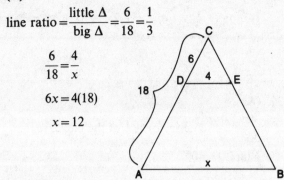

4. **(B)**

$d = \sqrt{(x_2 - x_1)^2 + (y_2 - y_1)^2}$

$= \sqrt{(5-2)^2 + (6-(-1))^2}$

$= \sqrt{9 + 49} = \sqrt{58}$

5. **(C)**

$3^4 = 9^{2x+1}$

$3^4 = (3^2)^{2x+1} = 3^{4x+2}$

$4 = 4x + 2$

$\frac{1}{2} = x$

6. **(B)**

$\dfrac{1}{1+\sqrt{2}}\left(\dfrac{1-\sqrt{2}}{1-\sqrt{2}}\right) = \dfrac{1-\sqrt{2}}{1-2}$

$= \dfrac{1-\sqrt{2}}{-1}$ or $\sqrt{2} - 1$

7. **(A)**

$(n-2)(180) = 1980$

$n - 2 = 11$

$n = 13$

8. **(A)**

$|5 - 2x| = 7$

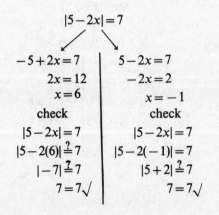

$-5 + 2x = 7$	$5 - 2x = 7$
$2x = 12$	$-2x = 2$
$x = 6$	$x = -1$
check	check
$|5 - 2x| = 7$	$|5 - 2x| = 7$
$|5 - 2(6)| \overset{?}{=} 7$	$|5 - 2(-1)| = 7$
$|-7| \overset{?}{=} 7$	$|5 + 2| \overset{?}{=} 7$
$7 = 7 \checkmark$	$7 = 7 \checkmark$

9. **(D)**

$x^{-1} + y^{-1} = \dfrac{1}{x} + \dfrac{1}{y} = \left(\dfrac{y}{y}\right)\dfrac{1}{x} + \left(\dfrac{x}{x}\right)\dfrac{1}{y}$

$= \dfrac{y + x}{xy}$

10. **(E)**

$x = \sin \theta \quad \dfrac{y}{3} = \csc \theta$

$\sin \theta = \dfrac{1}{\csc \theta} \Rightarrow x = \dfrac{1}{\frac{y}{3}} = \dfrac{3}{y}$

$xy = 3$

hyperbola

11. **(E)** sin and cos are cofunction

$(A + 40) + 10 = 90$

$A = 40$

12. **(B)**

$f(x) = x + \dfrac{1}{x}, \; f\left(\dfrac{1}{x}\right) = \dfrac{1}{x} + \dfrac{1}{\frac{1}{x}}$

$= \dfrac{1}{x} + x = f(x)$

13. **(E)**

If the roots are $x = 3$ and $x = 1$,
the factors are $(x - 3)$ and $(x - 1)$

$(x - 3)(x - 1) = 0$

$x^2 - 4x + 3 = 0$

14. **(C)**

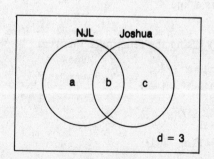

$$a+b+c+\underset{3}{\underbrace{d}}=40$$

$$\therefore a+b+c=37$$

$$a+b=25$$

$$\underline{b+c=30}$$

$$\underset{37}{\underbrace{a+b+c}}+b=55$$

$$\therefore b=18$$

$$b=18,\quad a=7,\quad c=12$$

15. (A)

$$\text{major arc } BC=210°$$

$$\text{minor arc } BC=360°-210°=150°$$

$$m\angle A=\frac{210-150}{2}=30$$

16. (C)

$y=|x-2|$ has critical value at $(x-2)=0$

$\therefore x=2,\quad y=0.$

The legs from left to right have slopes of -1 and 1, respectively.

17. (C) $\tan x+\cot x=1$ is *not* an identity.

18. (B)

$$(1+2i\sqrt{7})(1+2i\sqrt{7})$$

$$=1+2i\sqrt{7}+2i\sqrt{7}+4i^2(7)$$

$$=1+4i\sqrt{7}-28$$

$$=-27+4i\sqrt{7}$$

19. (A) All the terms are linear; the equation represents a line.

20. (A) Using the remainder theorem

$$P(x)=x^5-1$$

$$P(1)=1^5-1=0$$

$$\frac{P(x)}{x-a}=Q(x)+R$$

$$R=P(a)\quad P(1)$$

$$R=0$$

21. (B)

$$(x+2)(x^2-2x+4)=x^3-2x^2+4x+2x^2-4x+8$$

$$=x^3+8$$

22. (A)

$$\frac{4\pi}{36\pi}=\frac{1}{9}\quad\text{The sector is }\tfrac{1}{9}\text{ of the circle.}$$

$$\tfrac{1}{9}(360°)=40°$$

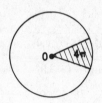

23. (A) By DeMorgan's law

$$\sim(\sim p\vee\sim q)=p\wedge q$$

24. (D)

$$\log 2y=\underset{.3010}{\underline{\log 2}}+\log y$$

$$=.3010+\log y$$

23. (B)

$$x^2-3y^2=13$$

$$x+3y=1\Rightarrow x=1-3y$$

Substitute $1-3y$ for x

$$(1-3y)^2-3y^2=13$$

$$1-6y+9y^2-3y^2=13$$

$$6y^2-6y-12=0$$

$$y^2-y-2=0$$

$$(y-2)(y+1)=0$$

$y=2$	$y=-1$
$x+3y=1$	$x+3y=1$
$x+6=1$	$x-3=1$
$x=-5$	$x=4$
$x=-5$	$x=4$
$y=2$	$y=-1$
$(-5,2)$	$(4,-1)$

26. **(E)**

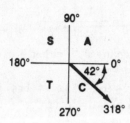

Notes: $\sin(-42°) = \sin 318°$

$$\begin{array}{r} -42 \\ +360 \\ \hline 318 \end{array}$$

$= -\sin 42°$

$= -\cos 58°$

27. **(B)**

$$\cos 30° = \frac{\sqrt{3}}{2}, \ \sin 30° = \frac{1}{2}$$

$$2(\cos 30° + i \sin 30°) = 2\left(\frac{\sqrt{3}}{2} + \frac{i}{2}\right)$$

$$= \sqrt{3} + i$$

28. **(D)** $\frac{1}{4} = .25$ $25\% = .25$

$$\frac{.25 + .25 + .25}{3} = .25$$

29. **(D)** 1. 7 3

$$\begin{array}{r} \overline{3.0000} \\ 1 \end{array}$$

Notes:

$$27\overline{)200}$$
$$189$$
$$343\overline{)1100}$$

345	344	343
×5	×4	×3
1725	1376	1029

1.7 to the nearest tenth.

30. **(C)**

$$\frac{1}{2i} = \frac{1}{2i}\left(\frac{i}{i}\right) = \frac{i}{2i^2} = \frac{i}{-2} \quad \text{or} \quad -\frac{1}{2}i$$

31. **(D)**

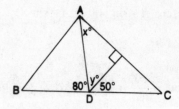

A straight line contains 180°.

$$m\angle y = 180 - (80 + 50) = 50$$

The sum of the measures of the angles of a triangle is 180°.

$$m\angle x + \underbrace{m\angle y}_{50} + 90 = 180$$

$$m\angle x = 40$$

32. **(D)** Both variables are squared and have the same sign, but different coefficients

∴ it is an ellipse.

33. **(C)**
$$[r(\cos\theta + i\sin\theta)]^n$$

$$= r^n(\cos n\theta + i\sin n\theta)$$

$$= \left(\tfrac{1}{2}\right)^3(\cos 3\cdot 10° + i\sin 3\cdot 10°)$$

$$= \tfrac{1}{8}(\cos 30° + i\sin 30°)$$

34. **(C)**

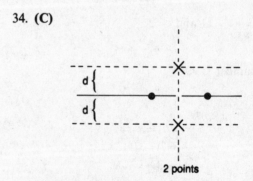

2 points indicated by x_s

35. **(B)**
$$VC = CT = 8$$
$$SD = DT = 10$$
$$CD = CT + TD = 18$$

36. (D)

$$\cos 2A = \cos^2 A - \sin^2 A$$
$$= (\cos A + \sin A)(\cos A - \sin A)$$

37. (D)

$$\text{the period} = \frac{360°}{\text{freq}} = \frac{360°}{\frac{1}{2}} = 720°$$

38. (C)

$$S_\infty = \frac{a}{1-r}$$

$$= \frac{2}{1-\frac{2}{3}} = \frac{2}{\frac{1}{3}} = 6 \quad \text{Note: } r = \frac{\frac{4}{3}}{2} = \frac{4}{6} = \frac{2}{3}$$

39. (C)

$$t + u = 11$$
$$(10u + t) - 34 = 2(10t + u)$$
$$10u + t - 34 = 20t + 2u$$

Combine like terms and rearrange

$$8u - 19t = 34$$

Multiply $t + u = 11$ by -8 and add the two equations

$$8u - 19t = 34$$
$$-8u - 8t = -88$$
$$\overline{ -27t = -54}$$
$$t = 2$$

Since $t + u = 11$ and $t = 2$

$$u = 9$$

The number is 29

40. (B)

$$x_1 t_1 = x_2 t_2$$
$$6(2) = x(4)$$
$$3 = x$$

41. (C)

$$V = \pi r^2 h = 90\pi$$
$$\pi r^2 (10) = 90\pi$$
$$r = 3$$

$$\text{L.A.} = 2\pi r h$$
$$= 2\pi(3)(10)$$
$$= 60\pi$$

42. (C)

$$P(x) = x^4 - 7x^3 + 9x^2 + 13x - 4$$

$$\frac{p}{q} = \frac{\pm 4, \pm 2, \pm 1}{\pm 1}$$

$P(x)$ has 3 sign changes ∴ either 3 or 1 positive root.

$P(-x) = x^4 + 7x^3 + 9x^2 - 13x - 4$ has 1 sign change ∴ only 1 negative root.

$$P(1) = 1 - 7 + 9 + 13 - 4 \neq 0$$
$$P(2) = 16 - 56 + 36 + 26 - 4 \neq 0$$
$$P(3) = 81 - 189 + 81 + 39 - 4 \neq 0$$
$$P(4) = 256 - 448 + 144 + 52 - 4 = 0$$

$x = 4$ is a root, $x - 4$ is a factor

$$
\begin{array}{r|rrrrr}
4 & 1 & -7 & 9 & 13 & -4 \\
  &   & 4 & -12 & -12 & 4 \\
\hline
  & 1 & -3 & -3 & 1 & \boxed{0}
\end{array}
$$

$$P(x) = x^4 - 7x^3 + 9x^2 + 13x - 4$$
$$= (x - 4)(x^3 - 3x^2 - 3x + 1)$$

using the depressed equation

$$D_1(x) = x^3 - 3x^2 - 3x + 1$$

try to locate another root

$$D_1(-1) = -1 - 3 + 3 + 1 = 0$$

∴ $x = -1$ is a root, $x + 1$ is a factor

$$
\begin{array}{r|rrrr}
-1 & 1 & -3 & -3 & 1 \\
   &   & -1 & 4 & -1 \\
\hline
   & 1 & -4 & 1 & \boxed{0}
\end{array}
$$

$$P(x) = (x - 4)(x^3 - 3x^2 - 3x + 1)$$
$$= (x - 4)(x + 1)(\underbrace{x^2 - 4x + 1})$$

$$x = \frac{-b \pm \sqrt{b^2 - 4ac}}{2a}$$

$$= \frac{4 \pm \sqrt{(-4)^2 - 4(1)(1)}}{2(1)}$$

$$= \frac{4 \pm \sqrt{12}}{2} = 2 \pm \sqrt{3}$$

2 rational, 2 irrational roots

43. (B)

$$\text{radians} \times \frac{180}{\pi} = \text{degrees}$$

$$1 \cdot \frac{180}{\pi} = \frac{180}{\pi} \text{ degrees}$$

44. **(E)**

$$\sum_{k=3}^{5} (k^2 - 1) = [3^2 - 1] + [4^2 - 1] + [5^2 - 1]$$
$$= 8 + 15 + 24 = 47$$

45. **(C)**

$$_5C_2(\tfrac{1}{5})^2(\tfrac{4}{5})^3$$

46. **(D)** The negation of $\forall_x: p$ is $\exists_x: \sim p$.

Some squares are not parallelograms.

47. **(D)**

$$\frac{2P + 3Q + x}{3} = A$$
$$x = 3A - 2P - 3Q$$

48. **(E)**

Value of A	Value of B	Value of $F = A^2 - B$
1	1	$1^2 - 1 = 0$

The program will print

2

49. **(A)**

$$\ln x + \ln(x + 2) = \ln 3$$
$$\ln x(x + 2) = \ln 3$$
$$x(x + 2) = 3$$
$$x^2 + 2x - 3 = 0$$
$$(x + 3)(x - 1) = 0$$
$$x = -3 \,\bigg|\, x = 1$$
$$\text{Reject}$$
$$\therefore x = 1$$

50. **(D)** All of the numbers are rational.

Appendix

Tables of Natural Trigonometric Functions
(For use with 9th and 10th Year Mathematics Regents Examinations)

Angle	Sine	Cosine	Tangent	Angle	Sine	Cosine	Tangent
1°	.0175	.9998	.0175	46°	.7193	.6947	1.0355
2°	.0349	.9994	.0349	47°	.7314	.6820	1.0724
3°	.0523	.9986	.0524	48°	.7431	.6691	1.1106
4°	.0698	.9976	.0699	49°	.7547	.6561	1.1504
5°	.0872	.9962	.0875	50°	.7660	.6428	1.1918
6°	.1045	.9945	.1051	51°	.7771	.6293	1.2349
7°	.1219	.9925	.1228	52°	.7880	.6157	1.2799
8°	.1392	.9903	.1405	53°	.7986	.6018	1.3270
9°	.1564	.9877	.1584	54°	.8090	.5878	1.3764
10°	.1736	.9848	.1763	55°	.8192	.5736	1.4281
11°	.1908	.9816	.1944	56°	.8290	.5592	1.4826
12°	.2079	.9781	.2126	57°	.8387	.5446	1.5399
13°	.2250	.9744	.2309	58°	.8480	.5299	1.6003
14°	.2419	.9703	.2493	59°	.8572	.5150	1.6643
15°	.2588	.9659	.2679	60°	.8660	.5000	1.7321
16°	.2756	.9613	.2867	61°	.8746	.4848	1.8040
17°	.2924	.9563	.3057	62°	.8829	.4695	1.8807
18°	.3090	.9511	.3249	63°	.8910	.4540	1.9626
19°	.3256	.9455	.3443	64°	.8988	.4384	2.0503
20°	.3420	.9397	.3640	65°	.9063	.4226	2.1445
21°	.3584	.9336	.3839	66°	.9135	.4067	2.2460
22°	.3746	.9272	.4040	67°	.9205	.3907	2.3559
23°	.3907	.9205	.4245	68°	.9272	.3746	2.4751
24°	.4067	.9135	.4452	69°	.9336	.3584	2.6051
25°	.4226	.9063	.4663	70°	.9397	.3420	2.7475
26°	.4384	.8988	.4877	71°	.9455	.3256	2.9042
27°	.4540	.8910	.5095	72°	.9511	.3090	3.0777
28°	.4695	.8829	.5317	73°	.9563	.2924	3.2709
29°	.4848	.8746	.5543	74°	.9613	.2756	3.4874
30°	.5000	.8660	.5774	75°	.9659	.2588	3.7321
31°	.5150	.8572	.6009	76°	.9703	.2419	4.0108
32°	.5299	.8480	.6249	77°	.9744	.2250	4.3315
33°	.5446	.8387	.6494	78°	.9781	.2079	4.7046
34°	.5592	.8290	.6745	79°	.9816	.1908	5.1446
35°	.5736	.8192	.7002	80°	.9848	.1736	5.6713
36°	.5878	.8090	.7265	81°	.9877	.1564	6.3138
37°	.6018	.7986	.7536	82°	.9903	.1392	7.1154
38°	.6157	.7880	.7813	83°	.9925	.1219	8.1443
39°	.6293	.7771	.8098	84°	.9945	.1045	9.5144
40°	.6428	.7660	.8391	85°	.9962	.0872	11.4301
41°	.6561	.7547	.8693	86°	.9976	.0698	14.3007
42°	.6691	.7431	.9004	87°	.9986	.0523	19.0811
43°	.6820	.7314	.9325	88°	.9994	.0349	28.6363
44°	.6947	.7193	.9657	89°	.9998	.0175	57.2900
45°	.7071	.7071	1.0000	90°	1.0000	.0000	

Table A: Common Logarithms of Numbers*

N	0	1	2	3	4	5	6	7	8	9
10	0000	0043	0086	0128	0170	0212	0253	0294	0334	0374
11	0414	0453	0492	0531	0569	0607	0645	0682	0719	0755
12	0792	0828	0864	0899	0934	0969	1004	1038	1072	1106
13	1139	1173	1206	1239	1271	1303	1335	1367	1399	1430
14	1461	1492	1523	1553	1584	1614	1644	1673	1703	1732
15	1761	1790	1818	1847	1875	1903	1931	1959	1987	2014
16	2041	2068	2095	2122	2148	2175	2201	2227	2253	2279
17	2304	2330	2355	2380	2405	2430	2455	2480	2504	2529
18	2553	2577	2601	2625	2648	2672	2695	2718	2742	2765
19	2788	2810	2833	2856	2878	2900	2923	2945	2967	2989
20	3010	3032	3054	3075	3096	3118	3139	3160	3181	3201
21	3222	3243	3263	3284	3304	3324	3345	3365	3385	3404
22	3424	3444	3464	3483	3502	3522	3541	3560	3579	3598
23	3617	3636	3655	3674	3692	3711	3729	3747	3766	3784
24	3802	3820	3838	3856	3874	3892	3909	3927	3945	3962
25	3979	3997	4014	4031	4048	4065	4082	4099	4116	4133
26	4150	4166	4183	4200	4216	4232	4249	4265	4281	4298
27	4314	4330	4346	4362	4378	4393	4409	4425	4440	4456
28	4472	4487	4502	4518	4533	4548	4564	4579	4594	4609
29	4624	4639	4654	4669	4683	4698	4713	4728	4742	4757
30	4771	4786	4800	4814	4829	4843	4857	4871	4886	4900
31	4914	4928	4942	4955	4969	4983	4997	5011	5024	5038
32	5051	5065	5079	5092	5105	5119	5132	5145	5159	5172
33	5185	5198	5211	5224	5237	5250	5263	5276	5289	5302
34	5315	5328	5340	5353	5366	5378	5391	5403	5416	5428
35	5441	5453	5465	5478	5490	5502	5514	5527	5539	5551
36	5563	5575	5587	5599	5611	5623	5635	5647	5658	5670
37	5682	5694	5705	5717	5729	5740	5752	5763	5775	5786
38	5798	5809	5821	5832	5843	5855	5866	5877	5888	5899
39	5911	5922	5933	5944	5955	5966	5977	5988	5999	6010
40	6021	6031	6042	6053	6064	6075	6085	6096	6107	6117
41	6128	6138	6149	6160	6170	6180	6191	6201	6212	6222
42	6232	6243	6253	6263	6274	6284	6294	6304	6314	6325
43	6335	6345	6355	6365	6375	6385	6395	6405	6415	6425
44	6435	6444	6454	6464	6474	6484	6493	6503	6513	6522
45	6532	6542	6551	6561	6571	6580	6590	6599	6609	6618
46	6628	6637	6646	6656	6665	6675	6684	6693	6702	6712
47	6721	6730	6739	6749	6758	6767	6776	6785	6794	6803
48	6812	6821	6830	6839	6848	6857	6866	6875	6884	6893
49	6902	6911	6920	6928	6937	6946	6955	6964	6972	6981
50	6990	6998	7007	7016	7024	7033	7042	7050	7059	7067
51	7076	7084	7093	7101	7110	7118	7126	7135	7143	7152
52	7160	7168	7177	7185	7193	7202	7210	7218	7226	7235
53	7243	7251	7259	7267	7275	7284	7292	7300	7308	7316
54	7324	7332	7340	7348	7356	7364	7372	7380	7388	7396
N	0	1	2	3	4	5	6	7	8	9

* This table gives the mantissas of numbers with the decimal point omitted in each case. Characteristics are determined from the numbers by inspection.

Table A: Common Logarithms of Numbers*

N	0	1	2	3	4	5	6	7	8	9
55	7404	7412	7419	7427	7435	7443	7451	7459	7466	7474
56	7482	7490	7497	7505	7513	7520	7528	7536	7543	7551
57	7559	7566	7574	7582	7589	7597	7604	7612	7619	7627
58	7634	7642	7649	7657	7664	7672	7679	7686	7694	7701
59	7709	7716	7723	7731	7738	7745	7752	7760	7767	7774
60	7782	7789	7796	7803	7810	7818	7825	7832	7839	7846
61	7853	7860	7868	7875	7882	7889	7896	7903	7910	7917
62	7924	7931	7938	7945	7952	7959	7966	7973	7980	7987
63	7993	8000	8007	8014	8021	8028	8035	8041	8048	8055
64	8062	8069	8075	8082	8089	8096	8102	8109	8116	8122
65	8129	8136	8142	8149	8156	8162	8169	8176	8182	8189
66	8195	8202	8209	8215	8222	8228	8235	8241	8248	8254
67	8261	8267	8274	8280	8287	8293	8299	8306	8312	8319
68	8325	8331	8338	8344	8351	8357	8363	8370	8376	8382
69	8388	8395	8401	8407	8414	8420	8426	8432	8439	8445
70	8451	8457	8463	8470	8476	8482	8488	8494	8500	8506
71	8513	8519	8525	8531	8537	8543	8549	8555	8561	8567
72	8573	8579	8585	8591	8597	8603	8609	8615	8621	8627
73	8633	8639	8645	8651	8657	8663	8669	8675	8681	8686
74	8692	8698	8704	8710	8716	8722	8727	8733	8739	8745
75	8751	8756	8762	8768	8774	8779	8785	8791	8797	8802
76	8808	8814	8820	8825	8831	8837	8842	8848	8854	8859
77	8865	8871	8876	8882	8887	8893	8899	8904	8910	8915
78	8921	8927	8932	8938	8943	8949	8954	8960	8965	8971
79	8976	8982	8987	8993	8998	9004	9009	9015	9020	9025
80	9031	9036	9042	9047	9053	9058	9063	9069	9074	9079
81	9085	9090	9096	9101	9106	9112	9117	9122	9128	9133
82	9138	9143	9149	9154	9159	9165	9170	9175	9180	9186
83	9191	9196	9201	9206	9212	9217	9222	9227	9232	9238
84	9243	9248	9253	9258	9263	9269	9274	9279	9284	9289
85	9294	9299	9304	9309	9315	9320	9325	9330	9335	9340
86	9345	9350	9355	9360	9365	9370	9375	9380	9385	9390
87	9395	9400	9405	9410	9415	9420	9425	9430	9435	9440
88	9445	9450	9455	9460	9465	9469	9474	9479	9484	9489
89	9494	9499	9504	9509	9513	9518	9523	9528	9533	9538
90	9542	9547	9552	9557	9562	9566	9571	9576	9581	9586
91	9590	9595	9600	9605	9609	9614	9619	9624	9628	9633
92	9638	9643	9647	9652	9657	9661	9666	9671	9675	9680
93	9685	9689	9694	9699	9703	9708	9713	9717	9722	9727
94	9731	9736	9741	9745	9750	9754	9759	9763	9768	9773
95	9777	9782	9786	9791	9795	9800	9805	9809	9814	9818
96	9823	9827	9832	9836	9841	9845	9850	9854	9859	9863
97	9868	9872	9877	9881	9886	9890	9894	9899	9903	9908
98	9912	9917	9921	9926	9930	9934	9939	9943	9948	9952
99	9956	9961	9965	9969	9974	9978	9983	9987	9991	9996
N	0	1	2	3	4	5	6	7	8	9

* This table gives the mantissas of numbers with the decimal point omitted in each case. Characteristics are determined from the numbers by inspection.

Table B: Values of Trigonometric Functions

Angle	Sin	Cos	Tan	Cot		
0° 00'	.0000	1.0000	.0000	—	90°	00'
10	.0029	1.0000	.0029	343.77		50
20	.0058	1.0000	.0058	171.89		40
30	.0087	1.0000	.0087	114.59		30
40	.0116	.9999	.0116	85.940		20
50	.0145	.9999	.0145	68.750		10
1° 00'	.0175	.9998	.0175	57.290	89°	00'
10	.0204	.9998	.0204	49.104		50
20	.0233	.9997	.0233	42.964		40
30	.0262	.9997	.0262	38.188		30
40	.0291	.9996	.0291	34.368		20
50	.0320	.9995	.0320	31.242		10
2° 00'	.0349	.9994	.0349	28.636	88°	00'
10	.0378	.9993	.0378	26.432		50
20	.0407	.9992	.0407	24.542		40
30	.0436	.9990	.0437	22.904		30
40	.0465	.9989	.0466	21.470		20
50	.0494	.9988	.0495	20.206		10
3° 00'	.0523	.9986	.0524	19.081	87°	00'
10	.0552	.9985	.0553	18.075		50
20	.0581	.9983	.0582	17.169		40
30	.0610	.9981	.0612	16.350		30
40	.0640	.9980	.0641	15.605		20
50	.0669	.9978	.0670	14.924		10
4° 00'	.0698	.9976	.0699	14.301	86°	00'
10	.0727	.9974	.0729	13.727		50
20	.0756	.9971	.0758	13.197		40
30	.0785	.9969	.0787	12.706		30
40	.0814	.9967	.0816	12.251		20
50	.0843	.9964	.0846	11.826		10
5° 00'	.0872	.9962	.0875	11.430	85°	00'
10	.0901	.9959	.0904	11.059		50
20	.0929	.9957	.0934	10.712		40
30	.0958	.9954	.0963	10.385		30
40	.0987	.9951	.0992	10.078		20
50	.1016	.9948	.1022	9.7882		10
6° 00'	.1045	.9945	.1051	9.5144	84°	00'
10	.1074	.9942	.1080	9.2553		50
20	.1103	.9939	.1110	9.0098		40
30	.1132	.9936	.1139	8.7769		30
40	.1161	.9932	.1169	8.5555		20
50	.1190	.9929	.1198	8.3450		10
7° 00'	.1219	.9925	.1228	8.1443	83°	00'
10	.1248	.9922	.1257	7.9530		50
20	.1276	.9918	.1287	7.7704		40
30	.1305	.9914	.1317	7.5958		30
40	.1334	.9911	.1346	7.4287		20
50	.1363	.9907	.1376	7.2687		10
8° 00'	.1392	.9903	.1405	7.1154	82°	00'
10	.1421	.9899	.1435	6.9682		50
20	.1449	.9894	.1465	6.8269		40
30	.1478	.9890	.1495	6.6912		30
40	.1507	.9886	.1524	6.5606		20
50	.1536	.9881	.1554	6.4348		10
9° 00'	.1564	.9877	.1584	6.3138	81°	00'
10	.1593	.9872	.1614	6.1970		50
20	.1622	.9868	.1644	6.0844		40
30	.1650	.9863	.1673	5.9758		30
40	.1679	.9858	.1703	5.8708		20
50	.1708	.9853	.1733	5.7694		10
10° 00'	.1736	.9848	.1763	5.6713	80°	00'
10	.1765	.9843	.1793	5.5764		50
20	.1794	.9838	.1823	5.4845		40
30	.1822	.9833	.1853	5.3955		30
40	.1851	.9827	.1883	5.3093		20
50	.1880	.9822	.1914	5.2257		10
11° 00'	.1908	.9816	.1944	5.1446	79°	00'
10	.1937	.9811	.1974	5.0658		50
20	.1965	.9805	.2004	4.9894		40
30	.1994	.9799	.2035	4.9152		30
40	.2022	.9793	.2065	4.8430		20
50	.2051	.9787	.2095	4.7729		10
12° 00'	.2079	.9781	.2126	4.7046	78°	00'
	Cos	Sin	Cot	Tan	Angle	

Angle	Sin	Cos	Tan	Cot		
12° 00'	.2079	.9781	.2126	4.7046	78°	00'
10	.2108	.9775	.2156	4.6382		50
20	.2136	.9769	.2186	4.5736		40
30	.2164	.9763	.2217	4.5107		30
40	.2193	.9757	.2247	4.4494		20
50	.2221	.9750	.2278	4.3897		10
13° 00'	.2250	.9744	.2309	4.3315	77°	00'
10	.2278	.9737	.2339	4.2747		50
20	.2306	.9730	.2370	4.2193		40
30	.2334	.9724	.2401	4.1653		30
40	.2363	.9717	.2432	4.1126		20
50	.2391	.9710	.2462	4.0611		10
14° 00'	.2419	.9703	.2493	4.0108	76°	00'
10	.2447	.9696	.2524	3.9617		50
20	.2476	.9689	.2555	3.9136		40
30	.2504	.9681	.2586	3.8667		30
40	.2532	.9674	.2617	3.8208		20
50	.2560	.9667	.2648	3.7760		10
15° 00'	.2588	.9659	.2679	3.7321	75°	00'
10	.2616	.9652	.2711	3.6891		50
20	.2644	.9644	.2742	3.6470		40
30	.2672	.9636	.2773	3.6059		30
40	.2700	.9628	.2805	3.5656		20
50	.2728	.9621	.2836	3.5261		10
16° 00'	.2756	.9613	.2867	3.4874	74°	00'
10	.2784	.9605	.2899	3.4495		50
20	.2812	.9596	.2931	3.4124		40
30	.2840	.9588	.2962	3.3759		30
40	.2868	.9580	.2994	3.3402		20
50	.2896	.9572	.3026	3.3052		10
17° 00'	.2924	.9563	.3057	3.2709	73°	00'
10	.2952	.9555	.3089	3.2371		50
20	.2979	.9546	.3121	3.2041		40
30	.3007	.9537	.3153	3.1716		30
40	.3035	.9528	.3185	3.1397		20
50	.3062	.9520	.3217	3.1084		10
18° 00'	.3090	.9511	.3249	3.0777	72°	00'
10	.3118	.9502	.3281	3.0475		50
20	.3145	.9492	.3314	3.0178		40
30	.3173	.9483	.3346	2.9887		30
40	.3201	.9474	.3378	2.9600		20
50	.3228	.9465	.3411	2.9319		10
19° 00'	.3256	.9455	.3443	2.9042	71°	00'
10	.3283	.9446	.3476	2.8770		50
20	.3311	.9436	.3508	2.8502		40
30	.3338	.9426	.3541	2.8239		30
40	.3365	.9417	.3574	2.7980		20
50	.3393	.9407	.3607	2.7725		10
20° 00'	.3420	.9397	.3640	2.7475	70°	00'
10	.3448	.9387	.3673	2.7228		50
20	.3475	.9377	.3706	2.6985		40
30	.3502	.9367	.3739	2.6746		30
40	.3529	.9356	.3772	2.6511		20
50	.3557	.9346	.3805	2.6279		10
21° 00'	.3584	.9336	.3839	2.6051	69°	00'
10	.3611	.9325	.3872	2.5826		50
20	.3638	.9315	.3906	2.5605		40
30	.3665	.9304	.3939	2.5386		30
40	.3692	.9293	.3973	2.5172		20
50	.3719	.9283	.4006	2.4960		10
22° 00'	.3746	.9272	.4040	2.4751	68°	00'
10	.3773	.9261	.4074	2.4545		50
20	.3800	.9250	.4108	2.4342		40
30	.3827	.9239	.4142	2.4142		30
40	.3854	.9228	.4176	2.3945		20
50	.3881	.9216	.4210	2.3750		10
23° 00'	.3907	.9205	.4245	2.3559	67°	00'
10	.3934	.9194	.4279	2.3369		50
20	.3961	.9182	.4314	2.3183		40
30	.3987	.9171	.4348	2.2998		30
40	.4014	.9159	.4383	2.2817		20
50	.4041	.9147	.4417	2.2637		10
24° 00'	.4067	.9135	.4452	2.2460	66°	00'
	Cos	Sin	Cot	Tan	Angle	

Table B: Values of Trigonometric Functions

Angle		Sin	Cos	Tan	Cot		
24°	00′	.4067	.9135	.4452	2.2460	66°	00′
	10	.4094	.9124	.4487	2.2286		50
	20	.4120	.9112	.4522	2.2113		40
	30	.4147	.9100	.4557	2.1943		30
	40	.4173	.9088	.4592	2.1775		20
	50	.4200	.9075	.4628	2.1609		10
25°	00′	.4226	.9063	.4663	2.1445	65°	00′
	10	.4253	.9051	.4699	2.1283		50
	20	.4279	.9038	.4734	2.1123		40
	30	.4305	.9026	.4770	2.0965		30
	40	.4331	.9013	.4806	2.0809		20
	50	.4358	.9001	.4841	2.0655		10
26°	00′	.4384	.8988	.4877	2.0503	64°	00′
	10	.4410	.8975	.4913	2.0353		50
	20	.4436	.8962	.4950	2.0204		40
	30	.4462	.8949	.4986	2.0057		30
	40	.4488	.8936	.5022	1.9912		20
	50	.4514	.8923	.5059	1.9768		10
27°	00′	.4540	.8910	.5095	1.9626	63°	00′
	10	.4566	.8897	.5132	1.9486		50
	20	.4592	.8884	.5169	1.9347		40
	30	.4617	.8870	.5206	1.9210		30
	40	.4643	.8857	.5243	1.9074		20
	50	.4669	.8843	.5280	1.8940		10
28°	00′	.4695	.8829	.5317	1.8807	62°	00′
	10	.4720	.8816	.5354	1.8676		50
	20	.4746	.8802	.5392	1.8546		40
	30	.4772	.8788	.5430	1.8418		30
	40	.4797	.8774	.5467	1.8291		20
	50	.4823	.8760	.5505	1.8165		10
29°	00′	.4848	.8746	.5543	1.8040	61°	00′
	10	.4874	.8732	.5581	1.7917		50
	20	.4899	.8718	.5619	1.7796		40
	30	.4924	.8704	.5658	1.7675		30
	40	.4950	.8689	.5696	1.7556		20
	50	.4975	.8675	.5735	1.7437		10
30°	00′	.5000	.8660	.5774	1.7321	60°	00′
	10	.5025	.8646	.5812	1.7205		50
	20	.5050	.8631	.5851	1.7090		40
	30	.5075	.8616	.5890	1.6977		30
	40	.5100	.8601	.5930	1.6864		20
	50	.5125	.8587	.5969	1.6753		10
31°	00′	.5150	.8572	.6009	1.6643	59°	00′
	10	.5175	.8557	.6048	1.6534		50
	20	.5200	.8542	.6088	1.6426		40
	30	.5225	.8526	.6128	1.6319		30
	40	.5250	.8511	.6168	1.6212		20
	50	.5275	.8496	.6208	1.6107		10
32°	00′	.5299	.8480	.6249	1.6003	58°	00′
	10	.5324	.8465	.6289	1.5900		50
	20	.5348	.8450	.6330	1.5798		40
	30	.5373	.8434	.6371	1.5697		30
	40	.5398	.8418	.6412	1.5597		20
	50	.5422	.8403	.6453	1.5497		10
33°	00′	.5446	.8387	.6494	1.5399	57°	00′
	10	.5471	.8371	.6536	1.5301		50
	20	.5495	.8355	.6577	1.5204		40
	30	.5519	.8339	.6619	1.5108		30
	40	.5544	.8323	.6661	1.5013		20
	50	.5568	.8307	.6703	1.4919		10
34°	00′	.5592	.8290	.6745	1.4826	56°	00′
	10	.5616	.8274	.6787	1.4733		50
	20	.5640	.8258	.6830	1.4641		40
	30	.5664	.8241	.6873	1.4550		30
	40	.5688	.8225	.6916	1.4460		20
	50	.5712	.8208	.6959	1.4370		10
35°	00′	.5736	.8192	.7002	1.4281	55°	00′
	10	.5760	.8175	.7046	1.4193		50
	20	.5783	.8158	.7089	1.4106		40
	30	.5807	.8141	.7133	1.4019		30
	40	.5831	.8124	.7177	1.3934		20
	50	.5854	.8107	.7221	1.3848		10
36°	00′	.5878	.8090	.7265	1.3764	54°	00′
		Cos	Sin	Cot	Tan	Angle	

Angle		Sin	Cos	Tan	Cot		
36°	00′	.5878	.8090	.7265	1.3764	54°	00′
	10	.5901	.8073	.7310	1.3680		50
	20	.5925	.8056	.7355	1.3597		40
	30	.5948	.8039	.7400	1.3514		30
	40	.5972	.8021	.7445	1.3432		20
	50	.5995	.8004	.7490	1.3351		10
37°	00′	.6018	.7986	.7536	1.3270	53°	00′
	10	.6041	.7969	.7581	1.3190		50
	20	.6065	.7951	.7627	1.3111		40
	30	.6088	.7934	.7673	1.3032		30
	40	.6111	.7916	.7720	1.2954		20
	50	.6134	.7898	.7766	1.2876		10
38°	00′	.6157	.7880	.7813	1.2799	52°	00′
	10	.6180	.7862	.7860	1.2723		50
	20	.6202	.7844	.7907	1.2647		40
	30	.6225	.7826	.7954	1.2572		30
	40	.6248	.7808	.8002	1.2497		20
	50	.6271	.7790	.8050	1.2423		10
39°	00′	.6293	.7771	.8098	1.2349	51°	00′
	10	.6316	.7753	.8146	1.2276		50
	20	.6338	.7735	.8195	1.2203		40
	30	.6361	.7716	.8243	1.2131		30
	40	.6383	.7698	.8292	1.2059		20
	50	.6406	.7679	.8342	1.1988		10
40°	00′	.6428	.7660	.8391	1.1918	50°	00′
	10	.6450	.7642	.8441	1.1847		50
	20	.6472	.7623	.8491	1.1778		40
	30	.6494	.7604	.8541	1.1708		30
	40	.6517	.7585	.8591	1.1640		20
	50	.6539	.7566	.8642	1.1571		10
41°	00′	.6561	.7547	.8693	1.1504	49°	00′
	10	.6583	.7528	.8744	1.1436		50
	20	.6604	.7509	.8796	1.1369		40
	30	.6626	.7490	.8847	1.1303		30
	40	.6648	.7470	.8899	1.1237		20
	50	.6670	.7451	.8952	1.1171		10
42°	00′	.6691	.7431	.9004	1.1106	48°	00′
	10	.6713	.7412	.9057	1.1041		50
	20	.6734	.7392	.9110	1.0977		40
	30	.6756	.7373	.9163	1.0913		30
	40	.6777	.7353	.9217	1.0850		20
	50	.6799	.7333	.9271	1.0786		10
43°	00′	.6820	.7314	.9325	1.0724	47°	00′
	10	.6841	.7294	.9380	1.0661		50
	20	.6862	.7274	.9435	1.0599		40
	30	.6884	.7254	.9490	1.0538		30
	40	.6905	.7234	.9545	1.0477		20
	50	.6926	.7214	.9601	1.0416		10
44°	00′	.6947	.7193	.9657	1.0355	46°	00′
	10	.6967	.7173	.9713	1.0295		50
	20	.6988	.7153	.9770	1.0235		40
	30	.7009	.7133	.9827	1.0176		30
	40	.7030	.7112	.9884	1.0117		20
	50	.7050	.7092	.9942	1.0058		10
45°	00′	.7071	.7071	1.0000	1.0000	45°	00′
		Cos	Sin	Cot	Tan	Angle	

Table C: Logarithms of Trigonometric Functions*

Angle	L Sin	L Cos	L Tan	L Cot	
0° 00'	—	10.0000	—	—	90° 00'
10	7.4637	10.0000	7.4637	12.5363	50
20	7.7648	10.0000	7.7648	12.2352	40
30	7.9408	10.0000	7.9409	12.0591	30
40	8.0658	10.0000	8.0658	11.9342	20
50	8.1627	10.0000	8.1627	11.8373	10
1° 00'	8.2419	9.9999	8.2419	11.7581	89° 00'
10	8.3088	9.9999	8.3089	11.6911	50
20	8.3668	9.9999	8.3669	11.6331	40
30	8.4179	9.9999	8.4181	11.5819	30
40	8.4637	9.9998	8.4638	11.5362	20
50	8.5050	9.9998	8.5053	11.4947	10
2° 00'	8.5428	9.9997	8.5431	11.4569	88° 00'
10	8.5776	9.9997	8.5779	11.4221	50
20	8.6097	9.9996	8.6101	11.3899	40
30	8.6397	9.9996	8.6401	11.3599	30
40	8.6677	9.9995	8.6682	11.3318	20
50	8.6940	9.9995	8.6945	11.3055	10
3° 00'	8.7188	9.9994	8.7194	11.2806	87° 00'
10	8.7423	9.9993	8.7429	11.2571	50
20	8.7645	9.9993	8.7652	11.2348	40
30	8.7857	9.9992	8.7865	11.2135	30
40	8.8059	9.9991	8.8067	11.1933	20
50	8.8251	9.9990	8.8261	11.1739	10
4° 00'	8.8436	9.9989	8.8446	11.1554	86° 00'
10	8.8613	9.9989	8.8624	11.1376	50
20	8.8783	9.9988	8.8795	11.1205	40
30	8.8946	9.9987	8.8960	11.1040	30
40	8.9104	9.9986	8.9118	11.0882	20
50	8.9256	9.9985	8.9272	11.0728	10
5° 00'	8.9403	9.9983	8.9420	11.0580	85° 00'
10	8.9545	9.9982	8.9563	11.0437	50
20	8.9682	9.9981	8.9701	11.0299	40
30	8.9816	9.9980	8.9836	11.0164	30
40	8.9945	9.9979	8.9966	11.0034	20
50	9.0070	9.9977	9.0093	10.9907	10
6° 00'	9.0192	9.9976	9.0216	10.9784	84° 00'
10	9.0311	9.9975	9.0336	10.9664	50
20	9.0426	9.9973	9.0453	10.9547	40
30	9.0539	9.9972	9.0567	10.9433	30
40	9.0648	9.9971	9.0678	10.9322	20
50	9.0755	9.9969	9.0786	10.9214	10
7° 00'	9.0859	9.9968	9.0891	10.9109	83° 00'
10	9.0961	9.9966	9.0995	10.9005	50
20	9.1060	9.9964	9.1096	10.8904	40
30	9.1157	9.9963	9.1194	10.8806	30
40	9.1252	9.9961	9.1291	10.8709	20
50	9.1345	9.9959	9.1385	10.8615	10
8° 00'	9.1436	9.9958	9.1478	10.8522	82° 00'
10	9.1525	9.9956	9.1569	10.8431	50
20	9.1612	9.9954	9.1658	10.8342	40
30	9.1697	9.9952	9.1745	10.8255	30
40	9.1781	9.9950	9.1831	10.8169	20
50	9.1863	9.9948	9.1915	10.8085	10
9° 00'	9.1943	9.9946	9.1997	10.8003	81° 00'
10	9.2022	9.9944	9.2078	10.7922	50
20	9.2100	9.9942	9.2158	10.7842	40
30	9.2176	9.9940	9.2236	10.7764	30
40	9.2251	9.9938	9.2313	10.7687	20
50	9.2324	9.9936	9.2389	10.7611	10
10° 00'	9.2397	9.9934	9.2463	10.7537	80° 00'
10	9.2468	9.9931	9.2536	10.7464	50
20	9.2538	9.9929	9.2609	10.7391	40
30	9.2606	9.9927	9.2680	10.7320	30
40	9.2674	9.9924	9.2750	10.7250	20
50	9.2740	9.9922	9.2819	10.7181	10
11° 00'	9.2806	9.9919	9.2887	10.7113	79° 00'
10	9.2870	9.9917	9.2953	10.7047	50
20	9.2934	9.9914	9.3020	10.6980	40
30	9.2997	9.9912	9.3085	10.6915	30
40	9.3058	9.9909	9.3149	10.6851	20
50	9.3119	9.9907	9.3212	10.6788	10
12° 00'	9.3179	9.9904	9.3275	10.6725	78° 00'
	L Cos	L Sin	L Cot	L Tan	Angle

Angle	L Sin	L Cos	L Tan	L Cot	
12° 00'	9.3179	9.9904	9.3275	10.6725	78° 00'
10	9.3238	9.9901	9.3336	10.6664	50
20	9.3296	9.9899	9.3397	10.6603	40
30	9.3353	9.9896	9.3458	10.6542	30
40	9.3410	9.9893	9.3517	10.6483	20
50	9.3466	9.9890	9.3576	10.6424	10
13° 00'	9.3521	9.9887	9.3634	10.6366	77° 00'
10	9.3575	9.9884	9.3691	10.6309	50
20	9.3629	9.9881	9.3748	10.6252	40
30	9.3682	9.9878	9.3804	10.6196	30
40	9.3734	9.9875	9.3859	10.6141	20
50	9.3786	9.9872	9.3914	10.6086	10
14° 00'	9.3837	9.9869	9.3968	10.6032	76° 00'
10	9.3887	9.9866	9.4021	10.5979	50
20	9.3937	9.9863	9.4074	10.5926	40
30	9.3986	9.9859	9.4127	10.5873	30
40	9.4035	9.9856	9.4178	10.5822	20
50	9.4083	9.9853	9.4230	10.5770	10
15° 00'	9.4130	9.9849	9.4281	10.5719	75° 00'
10	9.4177	9.9846	9.4331	10.5669	50
20	9.4223	9.9843	9.4381	10.5619	40
30	9.4269	9.9839	9.4430	10.5570	30
40	9.4314	9.9836	9.4479	10.5521	20
50	9.4359	9.9832	9.4527	10.5473	10
16° 00'	9.4403	9.9828	9.4575	10.5425	74° 00'
10	9.4447	9.9825	9.4622	10.5378	50
20	9.4491	9.9821	9.4669	10.5331	40
30	9.4533	9.9817	9.4716	10.5284	30
40	9.4576	9.9814	9.4762	10.5238	20
50	9.4618	9.9810	9.4808	10.5192	10
17° 00'	9.4659	9.9806	9.4853	10.5147	73° 00'
10	9.4700	9.9802	9.4898	10.5102	50
20	9.4741	9.9798	9.4943	10.5057	40
30	9.4781	9.9794	9.4987	10.5013	30
40	9.4821	9.9790	9.5031	10.4969	20
50	9.4861	9.9786	9.5075	10.4925	10
18° 00'	9.4900	9.9782	9.5118	10.4882	72° 00'
10	9.4939	9.9778	9.5161	10.4839	50
20	9.4977	9.9774	9.5203	10.4797	40
30	9.5015	9.9770	9.5245	10.4755	30
40	9.5052	9.9765	9.5287	10.4713	20
50	9.5090	9.9761	9.5329	10.4671	10
19° 00'	9.5126	9.9757	9.5370	10.4630	71° 00'
10	9.5163	9.9752	9.5411	10.4589	50
20	9.5199	9.9748	9.5451	10.4549	40
30	9.5235	9.9743	9.5491	10.4509	30
40	9.5270	9.9739	9.5531	10.4469	20
50	9.5306	9.9734	9.5571	10.4429	10
20° 00'	9.5341	9.9730	9.5611	10.4389	70° 00'
10	9.5375	9.9725	9.5650	10.4350	50
20	9.5409	9.9721	9.5689	10.4311	40
30	9.5443	9.9716	9.5727	10.4273	30
40	9.5477	9.9711	9.5766	10.4234	20
50	9.5510	9.9706	9.5804	10.4196	10
21° 00'	9.5543	9.9702	9.5842	10.4158	69° 00'
10	9.5576	9.9697	9.5879	10.4121	50
20	9.5609	9.9692	9.5917	10.4083	40
30	9.5641	9.9687	9.5954	10.4046	30
40	9.5673	9.9682	9.5991	10.4009	20
50	9.5704	9.9677	9.6028	10.3972	10
22° 00'	9.5736	9.9672	9.6064	10.3936	68° 00'
10	9.5767	9.9667	9.6100	10.3900	50
20	9.5798	9.9661	9.6136	10.3864	40
30	9.5828	9.9656	9.6172	10.3828	30
40	9.5859	9.9651	9.6208	10.3792	20
50	9.5889	9.9646	9.6243	10.3757	10
23° 00'	9.5919	9.9640	9.6279	10.3721	67° 00'
10	9.5948	9.9635	9.6314	10.3686	50
20	9.5978	9.9629	9.6348	10.3652	40
30	9.6007	9.9624	9.6383	10.3617	30
40	9.6036	9.9618	9.6417	10.3583	20
50	9.6065	9.9613	9.6452	10.3548	10
24° 00'	9.6093	9.9607	9.6486	10.3514	66° 00'
	L Cos	L Sin	L Cot	L Tan	Angle

* These tables give the logarithms increased by 10. Hence in each case 10 should be subtracted.

Table C: Logarithms of Trigonometric Functions*

Angle	L Sin	L Cos	L Tan	L Cot	
24° 00′	9.6093	9.9607	9.6486	10.3514	66° 00′
10	9.6121	9.9602	9.6520	10.3480	50
20	9.6149	9.9596	9.6553	10.3447	40
30	9.6177	9.9590	9.6587	10.3413	30
40	9.6205	9.9584	9.6620	10.3380	20
50	9.6232	9.9579	9.6654	10.3346	10
25° 00′	9.6259	9.9573	9.6687	10.3313	65° 00′
10	9.6286	9.9567	9.6720	10.3280	50
20	9.6313	9.9561	9.6752	10.3248	40
30	9.6340	9.9555	9.6785	10.3215	30
40	9.6366	9.9549	9.6817	10.3183	20
50	9.6392	9.9543	9.6850	10.3150	10
26° 00′	9.6418	9.9537	9.6882	10.3118	64° 00′
10	9.6444	9.9530	9.6914	10.3086	50
20	9.6470	9.9524	9.6946	10.3054	40
30	9.6495	9.9518	9.6977	10.3023	30
40	9.6521	9.9512	9.7009	10.2991	20
50	9.6546	9.9505	9.7040	10.2960	10
27° 00′	9.6570	9.9499	9.7072	10.2928	63° 00′
10	9.6595	9.9492	9.7103	10.2897	50
20	9.6620	9.9486	9.7134	10.2866	40
30	9.6644	9.9479	9.7165	10.2835	30
40	9.6668	9.9473	9.7196	10.2804	20
50	9.6692	9.9466	9.7226	10.2774	10
28° 00′	9.6716	9.9459	9.7257	10.2743	62° 00′
10	9.6740	9.9453	9.7287	10.2713	50
20	9.6763	9.9446	9.7317	10.2683	40
30	9.6787	9.9439	9.7348	10.2652	30
40	9.6810	9.9432	9.7378	10.2622	20
50	9.6833	9.9425	9.7408	10.2592	10
29° 00′	9.6856	9.9418	9.7438	10.2562	61° 00′
10	9.6878	9.9411	9.7467	10.2533	50
20	9.6901	9.9404	9.7497	10.2503	40
30	9.6923	9.9397	9.7526	10.2474	30
40	9.6946	9.9390	9.7556	10.2444	20
50	9.6968	9.9383	9.7585	10.2415	10
30° 00′	9.6990	9.9375	9.7614	10.2386	60° 00′
10	9.7012	9.9368	9.7644	10.2356	50
20	9.7033	9.9361	9.7673	10.2327	40
30	9.7055	9.9353	9.7701	10.2299	30
40	9.7076	9.9346	9.7730	10.2270	20
50	9.7097	9.9338	9.7759	10.2241	10
31° 00′	9.7118	9.9331	9.7788	10.2212	59° 00′
10	9.7139	9.9323	9.7816	10.2184	50
20	9.7160	9.9315	9.7845	10.2155	40
30	9.7181	9.9308	9.7873	10.2127	30
40	9.7201	9.9300	9.7902	10.2098	20
50	9.7222	9.9292	9.7930	10.2070	10
32° 00′	9.7242	9.9284	9.7958	10.2042	58° 00′
10	9.7262	9.9276	9.7986	10.2014	50
20	9.7282	9.9268	9.8014	10.1986	40
30	9.7302	9.9260	9.8042	10.1958	30
40	9.7322	9.9252	9.8070	10.1930	20
50	9.7342	9.9244	9.8097	10.1903	10
33° 00′	9.7361	9.9236	9.8125	10.1875	57° 00′
10	9.7380	9.9228	9.8153	10.1847	50
20	9.7400	9.9219	9.8180	10.1820	40
30	9.7419	9.9211	9.8208	10.1792	30
40	9.7438	9.9203	9.8235	10.1765	20
50	9.7457	9.9194	9.8263	10.1737	10
34° 00′	9.7476	9.9186	9.8290	10.1710	56° 00′
10	9.7494	9.9177	9.8317	10.1683	50
20	9.7513	9.9169	9.8344	10.1656	40
30	9.7531	9.9160.	9.8371	10.1629	30
40	9.7550	9.9151	9.8398	10.1602	20
50	9.7568	9.9142	9.8425	10.1575	10
35° 00′	9.7586	9.9134	9.8452	10.1548	55° 00′
10	9.7604	9.9125	9.8479	10.1521	50
20	9.7622	9.9116	9.8506	10.1494	40
30	9.7640	9.9107	9.8533	10.1467	30
40	9.7657	9.9098	9.8559	10.1441	20
50	9.7675	9.9089	9.8586	10.1414	10
36° 00′	9.7692	9.9080	9.8613	10.1387	54° 00′
	L Cos	L Sin	L Cot	L Tan	Angle

Angle	L Sin	L Cos	L Tan	L Cot	
36° 00′	9.7692	9.9080	9.8613	10.1387	54° 00′
10	9.7710	9.9070	9.8639	10.1361	50
20	9.7727	9.9061	9.8666	10.1334	40
30	9.7744	9.9052	9.8692	10.1308	30
40	9.7761	9.9042	9.8718	10.1282	20
50	9.7778	9.9033	9.8745	10.1255	10
37° 00′	9.7795	9.9023	9.8771	10.1229	53° 00′
10	9.7811	9.9014	9.8797	10.1203	50
20	9.7828	9.9004	9.8824	10.1176	40
30	9.7844	9.8995	9.8850	10.1150	30
40	9.7861	9.8985	9.8876	10.1124	20
50	9.7877	9.8975	9.8902	10.1098	10
38° 00′	9.7893	9.8965	9.8928	10.1072	52° 00′
10	9.7910	9.8955	9.8954	10.1046	50
20	9.7926	9.8945	9.8980	10.1020	40
30	9.7941	9.8935	9.9006	10.0994	30
40	9.7957	9.8925	9.9032	10.0968	20
50	9.7973	9.8915	9.9058	10.0942	10
39° 00′	9.7989	9.8905	9.9084	10.0916	51° 00′
10	9.8004	9.8895	9.9110	10.0890	50
20	9.8020	9.8884	9.9135	10.0865	40
30	9.8035	9.8874	9.9161	10.0839	30
40	9.8050	9.8864	9.9187	10.0813	20
50	9.8066	9.8853	9.9212	10.0788	10
40° 00′	9.8081	9.8843	9.9238	10.0762	50° 00′
10	9.8096	9.8832	9.9264	10.0736	50
20	9.8111	9.8821	9.9289	10.0711	40
30	9.8125	9.8810	9.9315	10.0685	30
40	9.8140	9.8800.	9.9341	10.0659	20
50	9.8155	9.8789	9.9366	10.0634	10
41° 00′	9.8169	9.8778	9.9392	10.0608	49° 00′
10	9.8184	9.8767	9.9417	10.0583	50
20	9.8198	9.8756	9.9443	10.0557	40
30	9.8213	9.8745	9.9468	10.0532	30
40	9.8227	9.8733	9.9494	10.0506	20
50	9.8241	9.8722	9.9519	10.0481	10
42° 00′	9.8255	9.8711	9.9544	10.0456	48° 00′
10	9.8269	9.8699	9.9570	10.0430	50
20	9.8283	9.8688	9.9595	10.0405	40
30	9.8297	9.8676	9.9621	10.0379	30
40	9.8311	9.8665	9.9646	10.0354	20
50	9.8324	9.8653	9.9671	10.0329	10
43° 00′	9.8338	9.8641	9.9697	10.0303	47° 00′
10	9.8351	9.8629	9.9722	10.0278	50
20	9.8365	9.8618	9.9747	10.0253	40
30	9.8378	9.8606	9.9772	10.0228	30
40	9.8391	9.8594	9.9798	10.0202	20
50	9.8405	9.8582	9.9823	10.0177	10
44° 00′	9.8418	9.8569	9.9848	10.0152	46° 00′
10	9.8431	9.8557	9.9874	10.0126	50
20	9.8444	9.8545	9.9899	10.0101	40
30	9.8457	9.8532	9.9924	10.0076	30
40	9.8469	9.8520	9.9949	10.0051	20
50	9.8482	9.8507	9.9975	10.0025	10
45° 00′	9.8495	9.8495	10.0000	10.0000	45° 00′
	L Cos	L Sin	L Cot	L Tan	Angle

* These tables give the logarithms increased by 10. Hence in each case 10 should be subtracted.